T0217176

Csound

Victor Lazzarini · Steven Yi
John ffitch · Joachim Heintz
Øyvind Brandtsegg · Iain McCurdy

Csound

A Sound and Music Computing System

 Springer

Victor Lazzarini
Department of Music
National University of Ireland
Maynooth, Kildare
Ireland

Steven Yi
Department of Music
National University of Ireland
Maynooth, Kildare
Ireland

John ffitch
Department of Computer Science
University of Bath
Bath
UK

Joachim Heintz
Institut für Neue Musik
Hochschule für Musik, Theater und
    Medien Hannover
Hannover
Germany

Øyvind Brandtsegg
Department of Music
Norwegian University of Science
    and Technology
Trondheim
Norway

Iain McCurdy
Berlin
Germany

ISBN 978-3-319-83277-7   ISBN 978-3-319-45370-5   (eBook)
DOI 10.1007/978-3-319-45370-5

Printed on acid-free paper

This Springer imprint is published by Springer Nature
The registered company is Springer International Publishing AG
The registered company address is: Gewerbestrasse 11, 6330 Cham, Switzerland

This book is dedicated to Richard Boulanger, aka Dr. B, who has been a tireless champion of Csound as a system for sound and music computing. A widely influential educator and composer, he has been instrumental in shaping the development of the software and its community.

Foreword

Csound – A Sound and Music Computing System: this book should be a great asset for all of those who wish to take advantage of the power of computers in the field of sound and music.

I was myself extremely fortunate to be among the first to take advantage of two ancestors of the Csound software, namely MUSIC IV and MUSIC V, the early programs designed by Max Mathews in the 1960s. Thanks to these programs, I could effectively participate in early explorations of sound synthesis for music. This has shaped all my activity in electroacoustic music and mixed music. Csound retains the advantages of MUSIC IV and MUSIC V and adds new possibilities taking advantage of the progress in computer technology and in sonic and musical research.

Today, Csound is in my opinion the most powerful and general program for sound synthesis and processing. Moreover, it is likely to endure, since it is maintained and developed by a team of competent and dedicated persons. The authors of this book are part of this team: they are talented software experts but also composers or sound designers. The book reviews the programs which culminated in the present Csound, and it explains in full detail the recent features. It can thus serve as both an introduction to Csound and a handbook for all its classic and novel resources.

In this foreword, I would like to present some historical recollections of the early days of computer sound synthesis: they may shed light on the developments that resulted in the Csound software and on the *raison d'être* of its specific features.

The first digital recording and the first computer sound synthesis were accomplished in 1957 at Bell Laboratories by Max Mathews and his colleagues. Mathews decided to harness the computer to calculate directly the successive samples of a sound wave, instead of sampling recordings of sounds produced by acoustic vibrations: he wrote the first program for computer music synthesis, called MUSIC I, and his colleague Newman Guttman realised *In The Silver Scale*, a 17-second monophonic computer music piece ... which sounded simplistic and unappealing.

This was a disappointing start: Mathews knew that he had to write more powerful programs. Thus came MUSIC II, with four-voice polyphony. MUSIC II introduced an ingenious new design for a software oscillator that could generate periodic waves by looking up a wavetable with different increments, thus yielding different frequencies. This process saved a lot of computing time, and it was applicable to a huge variety of waveshapes. With this version of the program, Guttman realised a 1-minute piece called *Pitch Study*. Even though the music was still rudimentary, it had some intriguing features. Edgard Varèse was intrigued by the advent of digital synthesis: he hoped it would afford the composer more musical control than the analog devices of electronic music. On April 26, 1959, Varèse decided to present *Pitch Study* to the public in a carte blanche he gave at the Village Gate in New York City, as a sort of manifesto encouraging the continuation of the development of computer sound synthesis at Bell Laboratories.

Clearly, programming permits us to synthesise sounds in many different ways: but Mathews realised that he would have to spend his life writing different programs to implement musical ideas and fulfill the desires of various composers. So he undertook to design a really flexible program, as general as possible – a music *compiler*, that is, a program that could generate a multiplicity of different music programs.

To attain flexibility, Mathews resorted to the concept of *modularity*. By selecting from among a collection of modules and connecting them in various ways, one can implement a large number of possibilities, as in construction sets such as Meccano or Lego.

The modular approach is at work in human languages, in which a small number of basic elements – the phonemes – are articulated into words and phrases, allowing an immense variety of utterances from a limited elementary repertoire. In fact, the idea of a system articulated from a small number of distinct, discontinuous elements – *discrete* in the mathematical sense – had been clearly expressed in the early nineteenth-century by Wilhelm von Humboldt, linguist and brother of the celebrated explorer Alexander von Humboldt. Similarly, chemistry showed that all substances can in principle be synthesised from fewer than one hundred chemical elements, each of them formed of a single type of atoms. Biology also gives rise to an incredible diversity of animals and plants: the common living building blocks of life have only been identified in the last fifty years.

In 1959, Mathews wrote MUSIC III, a compiler implementing a modular approach for the computation of sound waveforms. Users of MUSIC III could perform different kinds of sound synthesis: they had to make their own choice among a repertoire of available modules, called *unit generators*, each of which corresponded to an elementary function of sound production or transformation (oscillator, random number generator, adder, multiplier, filter ...). A user then assembled the chosen modules at will to define *instruments*, as if he or she were patching a modular synthesiser. However, contrary to common belief, Mathews' modular conception did not copy that of synthesisers: on the contrary, it inspired the analog devices built by Moog, Buchla, or Ketoff using voltage control, which only appeared after 1964, while MUSIC III was written in 1959. In fact, Mathews' modular concept has influenced most of the synthesis programs – the next versions MUSIC IV and MUSIC V, but also MUSIC 10, MUSIC 360, MUSIC 11, Cmusic, Csound – and also most analog or digital synthesisers – such as Arp, DX7, 4A, 4B, 4C, 4X, SYTER, compilers for physical modeling such as CORDIS-ANIMA, Genesis, Mimesis, Modalys, and real-time programs such as MaxMSP and Pure Data, much used today and more widely flow-based and object-oriented programming used in languages for electronic circuit simulation or computing software such as MATLAB.

The music compilers introduced by Mathews are software toolboxes. The modules that the user selects and connects are virtual: they correspond to portions of program code. Connections are stipulated by a declarative text that must follow conventions specific to the program. It is more meaningful to represent the connections in terms of a diagram, hence the name of *block diagram compilers*. In MUSIC III and in many later programs, instruments must be activated by *note* statements which specify the parameters left variable, such as starting time, duration, amplitude and pitch. The analogy to traditional instruments is suggestive. It may appear to follow a neo-classical approach: but synthesised notes may last 1 millisecond or 10 minutes, and each one may sound like a chord of several tones or fuse with other notes into a single tone.

In the 1970s, Barry Vercoe wrote MUSIC 11, an efficient synthesis program for the PDP11. Then Csound came around 1986. The advent of the Csound software written by Barry Vercoe was a major step. C, developed at Bell Laboratories, is a high-level language, hence easily readable, but it can also control specific features of a computer processor. In particular, the UNIX operating system – the ancestor of Linux – was written in C. Compositional subroutines can be written in C. Csound is an heir of MUSIC IV rather than MUSIC V: the unit generators transmit their output sample by sample, which makes it easier to combine them.

With MUSIC IV, MUSIC V, and Csound, synthesis can envision building practically any kind of sonic structure, providing if needed additional modules to implement processes not foreseen earlier. These music programs draw on a wide programming potential, and they put at the disposal of the user a variety of tools for virtual sound creation. Compositional control can thus be expanded to the level of the sonic microstructure: beyond composing with ready-made sounds, the musician can compose the sounds themselves.

The issues migrate from hardware to software, from technology to knowledge and know-how. The difficulty is no longer the construction of the tools, but their effective use, which must take into account not only the imperatives of the musical purpose, but also the characteristics of perception. So there is a need to develop the understanding of hearing. One cannot synthesise a sound by just stipulating the desired effect: synthesis requires the description of all its physical parameters, so one should beware of the auditory effect of a physical structure. The musical purpose is the prerogative of the user, but he or she must in any case take perception into account.

Since the beginning of synthesis, the exploration of musical sound with synthesis programs such as MUSIC IV, MUSIC V, and Csound has contributed substantial scientific advances in hearing processes, leading to a better understanding of the perception of musical sound. The physical structure of a synthetic sound is known by construction – the data specifying the synthesis parameters provide a complete structural representation of that sound. Listening to a synthetic sound allows evaluation of the relation between the physical structure and the auditory effect. The experience of computer sound synthesis showed that this so-called *psychoacoustic* relation between cause and effect, between objective structure and auditory sensation, is much more complex that was initially believed.

Exploring synthesis of musical tones showed that prescribed relations between certain physical parameters do not always translate into similar – *isomorphic* – relations between the corresponding perceptual attributes. For instance, the perceived pitch of periodic tones is correlated to the physical frequency: however Roger Shepard could synthesise twelve tones ascending a chromatic scale that seem to go up endlessly when they are repeated. I myself generated glissandi going down the scale but ending higher in pitch than where they started. I also synthesised tones that seem to go down in pitch when one doubles their frequencies. These paradoxes might be called illusory effects, errors of the sense of hearing: but what matters in music is the auditory effect. John Chowning has taken advantage of the idiosyncrasies of spatial hearing to create striking illusory motions of sound sources, a great contribution to kinetic music. He has proposed the expression *sensory esthetics* for a new field of musical inquiry relating to the quest for perceived musicality, including naturalness, exemplified by his simulations of the singing voice. Chowning has analysed by synthesis the capacity of the ear to sort two instrument tones in unison: while the ear fuses coherent vibrations, programming slightly different modulations for such tones makes their vibrations incoherent. This insight can be used to control synthesis so as to make specific sounds emerge from a sonic magma, which Chowning demonstrated in his work *Phoné*.

As Marc Battier wrote in 1992, the use of the software toolboxes such as MUSIC IV, MUSIC V, and Csound has favoured the development of an economy of exchanges regarding sonic know-how. In 1967, Chowning visited Bell Labs to discuss his early experiments on the use of frequency modulation (FM) for sound synthesis, and he communicated his synthesis data. I was impressed by the ease of replicating FM tones, and I took advantage of Chowning's recipes in my 1969 work *Mutations*. That same year 1969, Chowning organised one of the first computer music courses in Stanford, and he invited Mathews to teach the use of MUSIC V. Mathews asked me whether I could pass him some of my synthesis research that he could present. I hastily assembled some examples which I thought could be of interest, and I gave Max a document which I called *An introductory catalogue of computer synthesized sounds* [108]. For each sound example, the catalogue provides the MUSIC V score specifying the parameters of the desired sound (in effect an operational recipe), the audio recording of the sound (initially on an enclosed vinyl disc), and an explanation of the purpose and details. This document was widely diffused by Bell Labs, and it was reprinted without changes in Wergo's *The Historical CD of Digital Sound Synthesis* (1995). In 1973, Chowning published his milestone paper on FM synthesis, including the synthesis recipe for a number of interesting FM tones. In 2000, Richard Boulanger edited *The Csound Book*, comprising tutorials and examples. Together with Victor Lazzarini, in 2011, Boulanger edited *The Audio Programming Book*, with the aim to expand the community of audio developers and stretch the possibilities of the existing programs. Clearly, block diagram compilers such as MUSIC V, Cmusic, and Csound considerably helped this cooperative effort of exploring and increasing the possibilities of computer sound synthesis and processing. The dissemination of know-how provides cues that help users to build their own virtual tools for sound creation with these software toolboxes.

The robustness and precision of structured synthesis programs such as Csound favour musical work survival and reconstitution. In the 1980s, Antonio de Sousa Dias converted several examples from my MUSIC V *Catalogue* into Csound. From the computer score and with some help from the composer, two milestone works by John Chowning have recently been reconstituted, which made it possible to refresh the sound quality by replicating their synthesis with modern conversion hardware: *Stria*, independently by Kevin Dahan and Olivier Baudouin, and *Turenas* by Laurent Pottier, who transcribed it for four players playing digital percussion live.

In the late 1970s, for one of the first composer courses at IRCAM, Denis Lorrain wrote an analysis of my work *Inharmonique* for soprano and computer-synthesised sounds, realized in 1977. Lorrain's IRCAM report (26/80) included the MUSIC V recipes of the important passages, in particular bell-like sounds that were later turned into fluid textures by simply changing the shape of an amplitude envelope. More than 10 years later, with the help of Antonio de Sousa Dias and Daniel Arfib, I could generate such sounds in real-time with the real-time-oriented program MaxMSP discussed below.

MUSIC IV, MUSIC V, Cmusic, and Csound were not initially designed for real-time operation. For many years, real-time synthesis and processing of sounds were too demanding for digital computers. Laptops are now fast enough to make real-time synthesis practical. Real-time performance seems vital to music, but the act of composition requires us to be freed from real-time. One may say that MUSIC IV, MUSIC V, Cmusic, and Csound are composition-oriented rather than performance-oriented.

This might seem a limitation: but real-time operation also has limitations and problems. In the early 1980s, the preoccupation with real-time became dominant for a while. IRCAM claimed that their digital process 4X, then in progress, would make non-real time obsolete, but the 4X soon became technically obsolete, so that most works realised with this processor can only be heard today as recordings. Real-time demands make it hard to ensure the *sustainability* of the music. Boulez's work *Répons*, emblematic of IRCAM, was kept alive after the 4X was no longer operational, but only thanks to man-years of work by dedicated specialists. Nothing is left of Balz Trumpy's *Wellenspiele* real-time version (1978 Donaueschingen), whereas Harvey's *Mortuos Plango* (1980), mostly realised with MUSIC V, remains as a prominent work of that period.

The obsolescence of the technology should not make musical works ephemeral. In the 1980s, the Yamaha DX7 synthesiser was very promising – it was based on timbal know-how developed by Chowning and his collaborators, it was well documented, and it could be programmed to some extent. In 1986 at CCRMA, Stanford, Chowning and David Bristow organised a course to teach composers to program their own sounds on the DX7, but Yamaha soon stopped production to replace the DX7 with completely different models. In contradistinction, Csound remains – it can now emulate the DX7 as well as other synthesisers or Fender Rhodes electric pianos.

A more recent piece of modular software, MaxMSP, designed by Miller Puckette, is a powerful resource for musical performance: indeed it was a milestone for real-time operation. However it is not easy to generate a prescribed score with it. Also, even though it does not require special hardware, real-time oriented software tends to be more fragile and unstable. Csound is precise and efficient, and it can deal with performance in several ways without enslaving the user with the constraints of real-time. It has developed compatibility with the popular and useful MIDI protocol. The Csound library can complement other programs: for instance, Csound can be used as a powerful MaxMSP object. The control and interaction possibilities have been much expanded, and the book explains them in due detail.

It should be made clear that Csound is not limited to synthesis. In the late 1960s, Mathews incorporated in the design of MUSIC V modules that could introduce sampled sounds regardless of their origin: arbitrary sounds could then be digitally processed by block diagram compilers. In GRM, the Paris birthplace of musique concrète, Benedict Mailliard and Yann Geslin wrote in the early 1980s the so-called Studio 123 software to perform sound processing operations, many of which internally used unit generators from MUSIC V. This software was used to produce works such as François Bayle's *Erosphere* and my own piece *Sud*, in which I intimately combined synthetic and acoustic sound material in an effort to marry electronic music and musique concrète; it was a model for the GRM Tools plug-ins. Csound has incorporated and extended sound processing capabilities in the framework of a well-organised and documented logic.

Block diagram compilers therefore supported new ways of thinking for writing music, and they provided means for composing sounds beyond the traditional graphic notation of notes and durations, in ways that can unfurl with the help of specific programs. Certainly, the artistic responsibility for a musical work rests with the composer's own commitment to present his or her sonic result as a piece of music, but the extension of the musical territory has been a collective endeavour of a dedicated community. Csound has grown with this extension, it is now available on several platforms, and it can be augmented by the individual composers and tailored to their needs and their desires.

In my view, Csound is the most powerful software for modular sound synthesis and processing: *Csound – A Sound and Music Computing System* is a timely publication. The book is rightly dedicated to Richard Boulanger, whose indefatigable activity has made Csound the most accomplished development of block diagram compilers. It is also indebted to Max Mathews' generosity and genius of design, and to Barry Vercoe's exigencies for high-level musical software. Last but not least, it owes much to all the contributions of the authors, who are talented as sound designers, developers and musicians.

Marseille, December 2015   *Jean-Claude Risset*

Preface

This book is the culmination of many years of work and development. At the Csound Conference in Hannover, 2011, there was a general agreement among the community that a new book on the many new features of the language was a necessity. For one reason or another, the opportunity of putting together an edited collection of chapters covering different aspects of the software never materialised. As Csound 5 gave way to Csound 6, and the system started expanding into various platforms, mobile, web, and embedded, the need for a publication centred on the system itself became even more evident. This book aims to fill this space, building on the already rich literature in Computer Music, and adding to previous efforts, which covered earlier versions of Csound in good detail.

A major aspect of this book is that it was written by a combination of system developers/maintainers, Lazzarini, Yi and ffitch, and power users/developers, Brandt-segg, Heintz and McCurdy, who together have a deep knowledge of the internal and external aspects of Csound. All the major elements of the system are covered in breadth and detail. This book is intended for both novice and advanced users of the system, as well as developers, composers, researchers, sound designers, and digital artists, who have an interest in computer music software and its applications. In particular, it can be used wholly or in sections, by students and lecturers in music technology programmes.

To ensure its longevity and continued relevance, the book does not cover platform and host-specific issues. In particular, it does not dedicate space to showing how to download and install the software, or how to use a particular feature of a currently existing graphical frontend. As Csound is a programming library that can be embedded in a variety of ways, there are numerous programs employing it, some of which might not be as long-lasting as others. Wherever relevant and appropriate, we will be considering the terminal (command-line interface) frontend, which is the only host that is guaranteed to always be present in a Csound installation. In any case, the operation topics left out of this book are duly covered in many online resources (more details below), which are in fact the most appropriate vehicles for them.

The book is organised in five parts: Introduction; Language; Interaction; Instrument Development; and Composition Case Studies. The two chapters in the Introduction are designed to give some context to the reader. The first one localises Csound in the history of Computer Music, discussing in some detail a number of its predecessors, and introducing some principles that will be used throughout the book. This is followed by a chapter covering key elements of computer music software, as applied to Csound. As it navigates through these, it moves from general considerations to more specific ones that are central to the operation of the system. It concludes with an overview of the Csound application programming interface (API), which is an important aspect of the system for software development. However, the focus of this work is firmly centred on the system itself, rather than embedding or application programming. The next parts of the book delve deeper into the details of using the software as a sound and music computing system.

The second part is dedicated to the Csound language. It aims to be a concise guide to programming, covering basic and advanced elements, in an up-to-date way. It begins with a chapter that covers the ground level of the language, at the end of which the reader should have a good grasp of how to code simple instruments, and use the system to make music. It moves on to discuss advanced data types, which have been introduced in the later versions of the system, and provide new functionality to the language. Core issues of programming, such as branching and loops, as well as scheduling and recursion are covered in the third chapter. This is followed by an overview of instrument graphs and connections. The fifth chapter explores the concept of user-defined opcodes, and how these can be used to extend the language.

The topic of control and interaction is covered in the third part of the book. The first chapter looks at the standard numeric score, and discusses the various types of functionality that it offers to users. The reader is then guided through the MIDI implementation in Csound, which is both simple to use, and very flexible and powerful. The facilities for network control, via the Open Sound Control protocol and other means, is the topic of the third chapter in this part. A chapter covering scripting with a separate general-purpose programming language complements this part. In this, we explore the possibilities of using Python externally, via the Csound API, to control and interact with the system.

The fourth part of this book explores specific topics in instrument development. We look at various types of classic synthesis techniques in its first chapter, from subtractive to distortion and additive methods. The following one examines key time-domain processing elements, studying fixed and variable delay lines and their applications, different types of filtering, and sound localisation. The third chapter introduces sound transformation techniques in the frequency domain, which are a particularly powerful feature of Csound. The more recent areas of granular synthesis and physical models are featured in the two remaining chapters of this part.

The final section of the book is dedicated to composition applications. An interesting aspect of almost all of the developers and contributors to Csound is that they are composers. Although the system has applications that go beyond the usual electronic music composition uses, there is always significant interest from old and new users in its potential as a music composition tool. The six chapters in this section explore the authors' individual approaches to using the software in this context. These case studies allow us to have a glimpse of the wide variety of uses that the system can have.

All the code used as examples in this book is freely available for downloading, pulling, and forking from the Csound GitHub site (http://csound.github.io), where readers will also find the latest versions of the software sources, links to the release packages for various platforms, the Csound Reference Manual, and many other resources. In particular, we would like to mention the Csound FLOSS Manual, which is a community effort led by two of the authors of this book, covering a number of practical and platform/frontend-specific aspects of Csound operation that are beyond the scope of this book. The Csound Journal, a periodic online publication co-edited by another one of the book authors, is also an excellent resource, with articles tackling different elements of the system.

It is also important to mention here that the essential companion to this book, and to the use of Csound, is the Reference Manual. It is maintained to contain a complete documentation of all aspects of the system, in a concise, but also precise, manner. It is important for all users to get acquainted with its layout, and how to find the required information in it. The manual represents a triumph of the collaborative effort of the Csound community, and it contains a wealth of knowledge about the system that is quite remarkable.

The development and maintenance of Csound has been led by a small team of people, including three of the book authors, plus Michael Gogins, who made numerous contributions and has kept the Windows platform versions up to date in a very diligent way, and Andrés Cabrera, who also authored a widely used cross-platform IDE for Csound (CsoundQt). As Free software, Csound is fully open for users to play with it, fork it, copy it and of course, add to it. Over its thirty-odd years of existence, it has benefited from contributions by developers spread all over the world, too many to be listed here (but duly acknowledged in the source code and manual).

At some point, users will realise that Csound can crash. It should not, but it does. The developers are always looking out for flaws, bugs and unprotected areas of the system. We have minimised the occurrence of segmentation faults, but as a programming system that is very flexible and produces 'real', compiled, working programs, it is vulnerable to these. No matter how closely we look at the system, there will always be the chance of some small opportunity for perverse code to be used, which will bring the system down. The development team has introduced a number of safeguards and a very thorough testing program to keep the software well maintained, bug free, and defended from misuse.

An important part of this is the issue tracking system, which is at present handled by the GitHub project. This is a very useful tool for us to keep an eye on problems that have arisen, and the user community is our first line of defence, using and pushing the software to its limits. Reporting bugs, and also asking for features to be implemented, is a good way to help strengthen Csound for all users. The developers work to a reasonably tight schedule, trying to address the issues as they are reported. Once these are fixed, the new code is made available in the develop branch of the source code revision system (git). They become part of the following software release, which happens quite regularly (three or four times yearly).

We hope that the present book is a helpful introduction to the system for new users, and a valuable companion to the ones already acquainted with it.

Maynooth, Rochester, Bath,   *Victor Lazzarini*
Hannover, Trondheim, and Berlin   *Steven Yi*
December 2015   *John ffitch*
                                                                              Joachim Heintz
                                                                          Øyvind Brandtsegg
                                                                                 Iain McCurdy

Acknowledgements

We would like to thank all contributors to the Csound software and its documentation, from developers to users, and educators. Firstly, we should acknowledge the work of Barry Vercoe and his team at the Machine Listening Group, MIT Media Lab, who brought out and shared the early versions of the software. The availability of its source code for FTP download was fundamental to its fantastic growth and development in the three decades that followed.

We are also deeply indebted to Michael Gogins, for many a year a leading figure in the community, one of the main system maintainers, a user and a composer, whose contributions to the system are wide ranging. In fact, he has been single-handedly carrying out the task of keeping the Windows platform version up to date, a sometimes arduous task as the operating system programming support diverges from the more commonly shared tools of the *NIX world. Likewise, we would like to thank Andrés Cabrera for his work on several aspects of the system, and for developing a very good cross-platform IDE, which has allowed many command-line-shy people to approach Csound programming more confidently. We should also acknowledge the many contributions of Istvan Varga and Anthony Kozar to the early development of version 5, and of Maurizio Umberto Puxeddu, Gabriel Maldonado and Matt Ingalls to version 4. Third-party frontend developers Rory Walsh and Stefano Bonetti also deserve recognition for their great work.

A special note of gratitude should also go to a number of power users world-wide, who have kept pushing the boundaries of the system. Particularly, we would like to thank Tarmo Johannes, Menno Knevel, Ben Hackbarth, Anders Gennell, Jan-Jacob Hofmann, Tito Latini, Jim Aikin, Aurelius Prochazka, Dave Seidel, Dave Phillips, Michael Rhoades, Russell Pinkston, Art Hunkins, Jim Hearon, Olivier Baudoin, Richard Van Bemmelen, François Pinot, Jacques Deplat, Stéphane Rollandin, Ed Costello, Brian Carty, Alex Hofmann, Gleb Rogozinsky, Peiman Khosravi, Richard Dobson, Toshihiro Kita, Anton Kholomiov, and Luis Jure. They, and many others through the years, have been instrumental in helping us shape the new versions of the system, and improving its functionality.

We would also like to thank Oscar Pablo de Liscia for reading and commenting on the ATS sections of the text, which has helped us to clarify and explain its mechanisms to the reader. We would like to express our gratitude to Jean-Claude Risset for his inspirational work, and for providing a wonderful preface to this book.

Finally, we should acknowledge the essential part that Richard Boulanger has played as a computer musician and educator. Arguably, this book and the system that it describes would not have been here if it were not for his influential work at the Berklee College of Music, and also world wide, taking his knowledge and experience to a wide audience of composers, sound designers and developers. The authors, and indeed, the Csound community, are very grateful and honoured to have had your help in establishing this system as one of the premier Free computer music software projects in the world.

Contents

Part II The Language

Acronyms

0dbfs	zero decibel full scale
ADC	Analogue-to-Digital Converter
ADSR	Attack-Decay-Sustain-Release
AIFF	Audio Interchange File Format
ALSA	Advanced Linux Sound Architecture
AM	Amplitude Modulation
API	Application Programming Interface
ATS	Analysis Transformation and Synthesis
bpm	beats per minute
CLI	Command-Line Interface
CODEC	Coder-Encoder
cps	cycles per second
DAC	Digital-to-Analogue Converter
DAW	Digital Audio Workstation
dB	Decibel
DFT	Discrete Fourier Transform
EMG	Electromyogram
FDN	Feedback Delay Network
FIR	Finite Impulse Response
FIFO	First In First Out
FLAC	Free Lossless Audio CODEC
FLTK	Fast Light Toolkit
FM	Frequency Modulation
FOF	Fonction D'Onde Formantique, formant wave function
FS	Fourier Series
FT	Fourier Transform
FFT	Fast Fourier Transform
GIL	Global Interpreter Lock
GPU	Graphic Programming Unit
GUI	Graphical User Interface
HRIR	Head-Related Impulse Response

HRTF	Head-Related Transfer Function
Hz	Hertz
IDE	Integrated Development Environment
IDFT	Inverse Discrete Fourier Transform
IFD	Instantaneous Frequency Distribution
IID	Inter-aural Intensity Difference
IIR	Infinite Impulse Response
IO	Input-Output
IP	Internet Protocol
IR	Impulse Response
ITD	Inter-aural Time Delay
JNI	Java Native Interface
LADSPA	Linux Audio Simple Plugin Architecture
LFO	Low Frequency Oscillator
LTI	Linear Time-Invariant
MIDI	Musical Instrument Digital Interface
MIT	Massachusetts Institute of Technology
MTU	Maximum Transmission Unit
OSC	Open Sound Control
PAF	Phase-Aligned Formant
PM	Phase Modulation
PRN	Pseudo-random number
PV	Phase Vocoder
STFT	Short-Time Fourier Transform
TCP	Transport Control Protocol
UDO	User-Defined Opcode
UDP	User Datagram Protocol
UG	Unit Generator
UI	User Interface
VBAP	Vector Base Amplitude Panning
VST	Virtual Studio Technology

Part I
Introduction

Chapter 1
Music Programming Systems

Abstract This chapter introduces music programming systems, beginning with a historical perspective of their development leading to the appearance of Csound. Its direct predecessors, MUSIC I-IV, MUSIC 360 and MUSIC 11, as well as the classic MUSIC V system, are discussed in detail. Following this, we explore the history of Csound and its evolution leading to the current version. Concepts such as unit generators, instruments, compilers, function tables and numeric scores are introduced as part of this survey of music programming systems.

1.1 Introduction

A music programming system is a complete software package for making music with computers [68]. It not only provides the means for defining the sequencing of events that make up a musical performance with great precision, but it also enables us to define the audio signal processing operations involved in generating the sound, to a very fine degree of detail and accuracy. Software such as Csound offers an environment for making music from the ground up, from the very basic elements that make up sound waves and their spectra, to the higher levels of music composition concerns such as sound objects, notes, textures, harmony, gestures, phrases, sections, etc.

   In this chapter, we will provide a historical perspective that will trace the development of key software systems that have provided the foundations for computer music. Csound is based on a long history of technological advances. The current system is an evolution of earlier versions that have been developed over almost 30 years of work. Prior to that, we can trace its origins to MUSIC 11, MUSIC 360 and MUSIC IV, which were seminal pieces of software that shaped the way we make music with computers. Many of the design principles and concepts that we will be studying in this book have been originated or were influenced by these early systems. We will complete this survey with an overview of the most up-to-date developments in Csound.

V. Lazzarini et al.. *Csound*. DOI 10.1007/978-3-319-45370-5_1

1.2 Early Music Programming Languages

The first direct digital synthesis program was MUSIC I, created by Max Mathews in 1957, followed quickly by MUSIC II and MUSIC III, which paved the way to modern music programming systems. Mathews' first program was innovative in that, for the first time, a computer was used to calculate directly the samples of a digital waveform. In his second program, we see the introduction of a key piece of technology, the table-lookup oscillator in MUSIC II, which is one of the most important components of a computer music program. In MUSIC III, however, we find the introduction of three essential elements that are still central to systems such as Csound: the concepts of *unit generator* (UG) and *instrument*, and the *compiler* for music programs [81, 82, 119]. Mathews explains the motivation behind its development:

> I wanted the complexity of the program to vary with the complexity of the musician's desires. If the musician wanted to do something simple, he or she shouldn't have to do very much in order to achieve it. If the musician wanted something very elaborate there was the option of working harder to do the elaborate thing. The only answer I could see was not to make the instruments myself – not to impose my taste and ideas about instruments on the musicians – but rather to make a set of fairly universal building blocks and give the musician both the task and the freedom to put these together into his or her instruments. [110]

The original principle of the UG has been applied almost universally in music systems. A UG is a basic building block for an instrument, which can be connected to other UGs to create a signal path, also sometimes called a synthesis graph. Eventually such a graph is terminated at an output (itself a UG), which is where the generated sound exits the instrument. Conceptually, UGs are black boxes, in that they have a defined behaviour given their parameters and/or inputs, possibly producing certain outputs, but their internals are not exposed to the user. They can represent a process as simple as a mixer (which adds two signals), or as complex as a Fast Fourier Transform (which calculates the spectrum of a waveform). We often speak of UGs as implementing an algorithm: a series of steps to realise some operation, and such an algorithm can be trivial (such as sum), or complicated. UGs were such a good idea that they were used beyond Computer Music, in hardware instruments such as the classic analogue modular systems (Moog, ARP, etc.).

The second concept introduced by MUSIC III, *instruments*, is also very important, in that it provides a structure in which to place UGs. It serves as a way of defining a given model for generating sounds, which is not a black box anymore, but completely configurable by the user. So, here, we can connect UGs in all sorts of ways that suit our intentions. Once that is done, this can 'played', 'performed', by setting it to make sound. In early systems there would have been strict controls on how many copies of such instruments, called *instances*, could be used at the same time. However, in Csound, we can have as many instances as we would like working together (bound only by computing resource requirements).

Finally, the compiler, which Mathews called the *acoustic compiler*, was also a breakthrough, because it allowed the user to make her synthesis programs from these instrument definitions and their UGs into a very efficient binary form. In fact, it enabled an unlimited number of sound synthesis structures to be created in the computer, depending only on the creativity of the designer, making the computer not only a musical instrument, but a musical instrument generator. The principle of the compiler lives on today in Csound, and in a very flexible way, allowing users to create new instruments on the fly, while the system is running, making sound.

1.2.1 MUSIC IV

The first general model of a music programming system was introduced in MUSIC IV, written for the IBM 7094 computer in collaboration with Joan Miller, although the basic ideas had already been present in MUSIC III [100]. MUSIC IV was a complex software package, as demonstrated by its programmer's manual [83], but it was also more attractive to musicians. This can be noted in the introductory tutorial written by James Tenney [122]. The software comprised a number of separate programs that were run in three distinct phases or *passes*, producing at the very end the samples of a digital audio stream stored in a computer tape or disk file. In order to listen to the resulting sounds, users would have to employ a separate program (often in a different computer) to play back these files.

Operation of MUSIC IV was very laborious: the first pass took control data in the form of a numeric computer *score* and associated function-table generation instructions, in an unordered form and stored in temporary tape files. This information, which included the program code itself, was provided in the form of punched cards. The data in this pass were used as parameters to be fed to instruments in subsequent stages. The two elements input at this stage, the score and function-table information, are central to how MUSIC IV operated, and are still employed in music systems today, and in particular, in Csound.

The numeric score is a list of parameters for each instance of instruments, to allow them to generate different types of sounds. For instance, an instrument might ask for its pitch to be defined externally as a parameter in the numeric score. That will allow users to run several copies of the same instrument playing different pitches, each one defined in the numeric score. The start times and durations, plus the requested parameters are given in the score creating an event for a particular instrument. Each one of these is a line in the score, and the list does not need to be entered in a particular time order, as sorting, as well as tempo alterations, can be done later. It is a simple yet effective way to control instruments.

Function tables are an efficient way to handle many types of mathematical operations that are involved in the computing of sound. They are pre-calculated lists of numbers that can be looked up directly, eliminating the need to compute them repeatedly. For instance, if you need to create a sliding pitch, you can generate the numbers that make up all the intermediary pitches in the glissando, place them in a function table, and then just read them. This saves the program from having to calculate them every time it needs to play this sound. Another example is a sound waveform table, also known as a wavetable. If you need a fixed-shape wave in your instrument, you can create it and place it in a table, then your instrument can just read it to produce the signal required.

In MUSIC IV, the first pass was effectively a card-reading stage, with little extra functionality, although external subroutines could be applied to modify the score data before this was saved. Memory for 800 events was made available by the system. The first pass data was then input to the second pass, where the score was sorted in time order and any defined tempo transformations applied, producing another set of temporary files. In the third pass, a synthesis program was loaded, taking the score from the previous stage, and generating the audio samples to be stored in the output file.

The pass 3 program was created by the MUSIC IV acoustic compiler from the synthesis program, called an *orchestra*, made up of instruments and UGs. Some basic conventions governed the MUSIC IV orchestra language, such as the modes of connections allowed, determined by each unit generator and how they were organised syntactically. As a programming language, it featured a basic type system. Data typing is a way of defining specific rules for different types of computation objects (e.g. numbers, text etc.) that a computer language manipulates. In MUSIC IV, there are only a few defined data types in the system: U (unit generator outputs), C (conversion function outputs, from the score), P (note parameters, also from the score), F (function tables) and K (system constants). Unlike in modern systems, only a certain number of parallel instances of each instrument were allowed to be run at the same time, and each score event was required to be scheduled for a specific free instrument instance.

The MUSIC IV compiler understood fifteen unit generators. There were three types of oscillators, which were used to generate basic waveform signals and depended on pre-defined function tables. Two envelope generators were provided to control UG parameters (such as frequency or amplitude). A table-lookup UG was also offered for direct access to tables. Two bandlimited noise units, of sample-hold and interpolating types, complemented the set of generators in the system. Three addition operators, for two, three and four inputs, were provided, as well as a multiplier. Processors included a resonance unit based on ring modulation and a second-order band-pass filter. Finally, an output unit complemented the system:

1. OUT: output unit
2. OSCIL: standard table-lookup oscillator
3. COSCIL: table-lookup oscillator with no phase reset
4. VOSCIL: table-lookup oscillator with variable table number
5. ADD2: add two inputs
6. ADD3: add three inputs
7. ADD4: add four inputs
8. RANDI: interpolating bandlimited noise generator
9. RANDH: sample-and-hold bandlimited noise generator
10. MULT: multiply two inputs
11. VFMULT: table-lookup unit
12. RESON: ring-modulation-based resonant wave generator

13. FILTER: second-order all-pole band-pass filter
14. LINEN: linear envelope generator (trapezoidal)
15. EXPEN: single-scan table-lookup envelope.

   An example from the MUSIC IV manual is shown in listing 1.1, where we can
see the details of the orchestra language. Unit generator output signals are referenced by U names (relating in this case to the order in which they appear), whereas
score parameters and constants are denoted by P and C. This orchestra contains one
simple instrument, WAIL, whose sound is generated by an oscillator (U5) to which
an amplitude envelope (U1) and frequency modulation (U4) are applied. The latter
is a combination of periodic (U2) and random (U3) vibrato signals, mixed together
with the fundamental frequency. The code is finished with the FINE keyword, after
the maximum number of parallel instances for the WAIL instrument is given.

Listing 1.1 MUSIC IV instrument example [83]

```
WAIL      INSTR
          OSCIL  P4,C3,F1
          OSCIL  P6,P7,F1
          RANDI P8,P9
          ADD3   P5,U2,U3
          OSCIL U1,U4,F3
          OUT     U5
          END

WAIL      COUNT  10
          FINE
```

   MUSIC IV was the first fully fledged computer music programming environment. The system allowed a good deal of programmability, which is specially true
in terms of synthesis program design. Following its beginnings at Bell Labs, the
software was ported to the IBM computer installation at Princeton University as
MUSIC IVB [105] and then as MUSIC 4BF [53, 111], written in FORTRAN, one
of the first scientific high-level programming languages. An important feature of
this version was that it also used the FORTRAN language for the programming of
instrument definitions.

1.2.2 MUSIC V

Mathews' work at Bell Labs culminated in MUSIC V [84], written in collaboration
with Joan Miller, Richard Moore and Jean-Claude Risset [52]. This was the final iteration of his MUSIC series, mostly written in the FORTRAN language. This made
it much more portable to other computer installations (including modern operating
systems). It featured the typical three-pass process of MUSIC IV, but with a better integration of these operation stages. The orchestra compilation step was now

combined with pass 3, without the need to generate a separate synthesis program.
FORTRAN conversion subroutines were also integral to the program code. Also
the whole MUSIC V code was written in a single score, which contained both the
note lists and the instruments. Unlike MUSIC IV, there was no maximum instance
count for each instrument. Unit generators could be written either in FORTRAN,
or as separate machine-language subroutines, which could be optimised for specific
computers.

MUSIC V provides simple orchestra data types: P (score parameters), V (scalar
values), B (audio signal buffers) and F (function tables). Audio is generated in a
series of sample blocks (or vectors), which by default hold 512 samples. Vector-
based processing became standard in most modern computer music systems (albeit
with some notable options). MUSIC V has been ported to modern systems using the
Gfortran compiler [16]. This software represents a significant milestone in Computer Music, as it was widely used by composers and researchers. In particular, we
should mention that it provided the means through which Risset developed his *Catalogue of Computer Synthesized Sounds* [108], a fundamental work in digital sound
synthesis. This work stems from his ground-breaking contributions to computer music composition, where we see a perfect marriage of artistic and technical craft.

In listing 1.2, we can observe a simple MUSIC V score, implementing the well-
known Risset-Shepard tones [108]. The instrument employs three interpolating oscillators (IOSs), generating an amplitude envelope (from a bell function), a frequency envelope (a decaying exponential) and a sine wave controlled by these two
signals (in B3 and B4 respectively). Ten parallel oscillators are started, each with a
10% phase offset relative to the preceding one (tables are 512 samples long). Each
NOT in the score defines an oscillator instance, with the first three parameters (P2,
P3, P4) defined as start time (0), instrument (1), and duration (14). Oscillator frequencies are defined by sampling increments (in P6 and P7). The top frequency of
the decaying exponential is $P6 \times f_s/512$, where f_s is the sampling rate. The amplitude and frequency envelopes have a cycle that lasts for $512/(f_s \times P7)$.

Pass I of MUSIC V scans the score (which includes both the instrument definitions and the note list proper) and produces a completely numeric representation of
it. The second pass sorts the note list in time order and applies the CONVT routine,
which can be used to convert frequencies to sampling increments etc. Finally, pass
III schedules the events, calls the unit generators and writes the output.

Listing 1.2 MUSIC V, Risset-Shepard tones, from Risset's catalogue [108]

```
COMMENT -- RISSET CATALOGUE EXAMPLE 513 --
INS 0 1;
IOS P5 P7 B3 F2 P8 ;
IOS P6 P7 B4 F3 P9 ;
IOS B3 B4 B5 F1 P25 ;
OUT B5 B1 ;
END ;
```

```
GEN 0 2 1 512 1 1 ;
GEN 0 7 2 0 ;
GEN 0 7 3 -10;

NOT 0 1 14 100 50 .0001 0 0 ;
NOT 0 1 14 100 50 .0001 51. 51.1 ;
NOT 0 1 14 100 50 .0001 102.2 102.2 ;
NOT 0 1 14 100 50 .0001 153.3 153.3 ;
NOT 0 1 14 100 50 .0001 204.4 204.4 ;
NOT 0 1 14 100 50 .0001 255.5 255.5 ;
NOT 0 1 14 100 50 .0001 306.6 306.6 ;
NOT 0 1 14 100 50 .0001 357.7 357.7 ;
NOT 0 1 14 100 50 .0001 408.8 408.8 ;
NOT 0 1 14 100 50 .0001 459.9 459.9 ;
TER 16 ;
```

1.2.3 MUSIC 360

Another important successor to MUSIC IV was MUSIC 360 [124], written at Princeton University by Barry Vercoe for the large IBM 360 computer [79]. It was directly derived from MUSIC IVB and MUSIC IVBF, and thus related to those systems as a MUSIC IV variant. This program was taken to other IBM 360 and 370 installations, and as it was tied in to those large computer installations, it did not suit smaller institutions, and was not widely available. The development of MUSIC 360 is particularly relevant to Csound, as it is one of its ancestors.

The structure of MUSIC 360 is very similar to its predecessors, utilising the operational principle of three passes discussed above. Here, however, all passes are combined into a single 'load module' (the program) after the orchestra is compiled. In many ways, MUSIC 360 represented a significant advance with regards to MUSIC IV. It allowed any number of parallel instances of instruments to be performed at the same time. An important innovation seen in this system is the clear definition of the initialisation and performance-time stages, with separate processing stages set for each. This is a feature that was successfully embraced by subsequent systems, including Csound. The principle here is that when an instance of an instrument is going to be run, there are a number of operations that do not to be repeated. These then are only run in the initialisation phase. For the actual sound to be computed, the instrument will enter a performance phase, where only the necessary steps to produce the signal are processed.

For instance, let's say we want to notate the pitch of the sound using a system where 60 is middle C and a change of 1 represents a semitone step (61 is C sharp etc.). To do this we would need to perform a mathematical operation to convert this notation into cycles per second (Hz), which is what oscillators expect as a frequency parameter. This operation does not need to be repeated, it only needs to happen once per sound event. So it gets placed in the initialisation stage and its result is then used at the performance time. Many modern systems employ similar principles in their design. It is an obvious optimisation that can save a lot of redundant computation.

Another advanced aspect of the language was that arithmetic expressions of up to 12 terms could be employed in the code, with the basic set of operations augmented by a number of conversion functions. This made the need for different types of addition operators, and separate multipliers, redundant. Data types included 'alpha'-types, which could hold references to unit generator results (both at I-time and P-time), K-types, used to hold values from a KDATA statement (which was used to hold program constants), P-types for note list p-fields, and U-types, which could be used to reference unit generator results (as an alternative to 'alpha' variables). There was scoping control, as symbols could be global or local, which included a facility for accessing variables that were local to a given instrument. The language also supported conditional control of flow, another advanced feature when compared to equivalent systems. Some means of extendability were provided by opcodes that were able to call external FORTRAN subroutines at I- or P-time. In addition, to facilitate programming a macro substitution mechanism was provided. MUSIC 360 was quite a formidable system, well ahead of its competitors, including MUSIC V.

An example of a simple orchestra program featuring an instrument based on an oscillator and trapezoidal envelope combination is shown in listing 1.3. In this example, we can observe the use of U-types, which refer to the output of unit generators previously defined in the code. In this case U1 is a reference to the unit one line above the current. The PSAVE statement is used to indicate which p-fields from score cards will be required in this instrument. The ISIPCH converter is used to convert "octave.pitch_class" notation into a suitable sampling increment, operating at initialisation time only. This pitch notation is very useful because it represents frequencies by an octave number (say 8 is middle-C octave), and a pitch class from 0 to 11 equivalent to note names from C to B in semitones. The term *sampling increment* is a low-level way of defining the frequency of an oscillator, and we will explore its definition later in this book.

OSCIL and LINEN are the truncating oscillator and trapezoidal envelope, respectively, used to generate the audio, the oscillator depending on function table 1, which is defined as a score statement. The syntax is very close to classic Csound code (as we will see later in this book), and unit generators are commonly known here as *opcodes*.

Listing 1.3 MUSIC 360 instrument example [124]

```
        PRINT NOGEN
        ORCH
        DECLARE SR=10000
SIMPL   INSTR   1
        PSAVE  (3,5)
        ISIPCH P5
        OSCIL  P4,U1,1
```

```
LINEN U1,.03,P3,.06
OUT U1
ENDIN
ENDORCH
END
```

1.2.4 MUSIC 11

MUSIC 11 [125], a version of the MUSIC 360 system for the smaller DEC PDP-11 minicomputer [126], was introduced by Barry Vercoe at the MIT Experimental Music Studio. As the PDP 11's popularity grew in the 1970s, and with the introduction of the UNIX operating system, the program was used at various institutions both in the USA and elsewhere for over two decades. Not only were many of the innovative features of MUSIC 360 carried over to the new system, but also important concepts were pioneered here.

One of the main design aspects first introduced in MUSIC 11 was the concept of control (k-) and audio (a-) computation rates. This is a further refinement of the initialisation- and performance-time optimisation principle. This made the system the most computationally efficient software for audio synthesis of its time. It established the main operation principles of the orchestra which divides instrument action times into initialisation and two performance-time rates, which was realised in the three basic data types: i, k (control) and a (audio). Global, local and temporary variables were available (marked as g, l or t).

The control/audio rate mechanism is based on the principle that some signals do not need to be computed as often as others. For instance, envelopes vary much more slowly than actual audio signals, and thus do not require to be updated as often. Sound is computed at a very fast rate (sr, the sampling rate), which often involves periods of a fraction of a millisecond. Control signals in many situations do not require that precision, and can be calculated maybe up to one hundred times more slowly.

Listing 1.4 shows a version of the MUSIC 360 example, now in MUSIC 11 form. Although there are many similarities, some fundamental differences have been introduced in the language. The header declaration now includes the definition of a control rate (kr) and the audio block size (the ratio $\frac{sr}{kr}$, ksmps), as well as the number of output channels to be used (nchnls). The presence of the ksmps parameter indicates explicitly that computation of audio signals is now performed in chunks of this size. We will explore this concept in detail in the next chapter, and also throughout this book.

The type system has been simplified; we observe the presence of the k- and a-type variables, which have now been defined to hold signals (and not just references to unit generator outputs) of control and audio forms, respectively. Also, taking advantage of the introduction of the control rate concept, the oscillator and envelope have had their positions exchanged in the synthesis graph: the envelope is now a control signal generator, rather than an amplitude processor, so the instrument can be run more efficiently.

Listing 1.4 MUSIC 11 instrument example

```
            sr = 10000
            kr = 100
            ksmps = 100
            nchnls = 1

            instr 1
k1          linen p4,.03, p3,.06
a1          oscil  k1, cpspch(p5), 1
            out a1
            endin
```

With the introduction of the concept of control rate two issues arise. Firstly, control signals are liable to produce audio artefacts such as amplitude envelope zipper noise, which are caused by the staircase nature of these signals, introducing discontinuities in the audio output waveform (and a form of aliasing that results in wideband noise). Secondly, score events are quantised at the control rate, which can affect the timing precision in some situations. To mitigate these effects, a balance between efficiency and precision needs to be reached, where the ratio between audio and control rates is small enough to prevent poor results, but high enough to be efficient.

1.3 Csound

Csound[1] is possibly the longest-running heir to these early MUSIC N systems. It was developed in the 1980s, alongside similar systems, such as Cmusic [90] and M4C [7], and Cmix [101]. All of these were developed using the C language, which at the time became the standard for systems implementation. Csound developed beyond its original design into a much larger and multi-functional music programming environment, with the advent of version 5 in 2006, and version 6 in 2013. The full Csound ascendancy is shown in Fig. 1.1 with its approximate dates.

Csound came to light at the MIT Electronic Music Studio in 1986 (MIT-EMS Csound), as a C-language port of MUSIC 11. It inherited many aspects of its parent, but now integrating the orchestra compiler and loader into a single program. The original mit-ems Csound was based on three separate commands, scsort, csound and perf. The first command would sort the score; the second would compile and

---

[1] Despite the different ways in which its name is written down in various places, there is only one correct form: capital 'C' followed by lowercase 'sound'. A fully lowercase variant csound is possible, but only when referring to its command-line frontend (see Chapter 2).

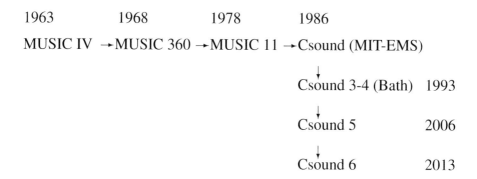

Fig. 1.1 The Csound family tree, from MUSIC IV through to the first release of the MIT-EMS Csound in 1986, and the further versions culminating in its current release

load the orchestra, and run the sorted score on it. The third command was just a convenient tool that called scsort and csound in a sequence.

Csound was originally a very faithful port of MUSIC 11, so much so that even today many programs for that language can still be run on modern versions of the system (the code in listing 1.4 runs perfectly in Csound). Some small differences existed in terms of a collection of new opcodes, and the removal of some others. Also, the separation between temporary and local variables was removed in Csound, and a means of extending the language with new C-code opcodes was provided. However, beyond these small differences, the central concepts were shared between the systems.

In the 1990s, the centre of development of Csound moved from MIT to the University of Bath. Real-time operation had been introduced to the system [129] in the MIT-EMS version. From this, the system developed into an offline composition and real-time synthesis language with widespread applications explored in [15]. The program was also ported to PC-DOS, also with real-time audio via soundblaster soundcards. Separately, at Mills College, a version of the system for the Macintosh platform was developed [54].[2] By the end of the 1990s, the system had been ported to almost all modern general-purpose computing platforms.

In a separate development, Csound was also ported to run on custom DSP hardware, in a closed-source version designated *extended Csound* [127]. This version eventually became part of a commercial project of Analog Devices Inc., to supply simple synthesizers for embedded applications [128]. Meanwhile, a large international developer community was involved in expanding the open-source system, which eventually came under the Lesser GNU Public License (and thus Free software). Many new unit generators were developed for it, culminating in the Csound

---

[2] An earlier port of Csound for the Mac had also been made available from MIT using the original UNIX sources modified for the THINK C compiler.

4.23 version in 2002. The consensus among the community was that the system required a major re-engineering to be able to move forward. A code freeze was established so that the new Csound 5 system could be developed.

1.3.1 Csound 5

Csound 5 was developed with the main goal of providing a clean, re-engineered system that would expose much of its internal operation, moving away from the original monolithic program design that characterised its earlier versions [39]. It was launched in 2006, twenty years after the first MIT-EMS release in 1986. From a language perspective, there were a few significant changes from the original Csound, which brought more flexibility to programming. A plug-in mechanism made it simpler to extend the system with new unit generators, utilities and function table generators. Programmable graphical user interfaces (GUIs) were incorporated into the language. User-defined unit generators were introduced, providing further facilities for structuring the code.

A significant innovation in version 5 is the presence of an Application Programming Interface (API), which allows a lower-level control of system operation. The API made it possible to create Csound-based applications with a variety of programming languages, leading to the development of various third-party host programs. It was now much easier to port it to mobile operating systems [135]), to use it as a plug-in for Digital Audio Workstations (DAWs), and to create custom solutions using Csound as the audio engine.

1.3.2 Csound 6

The next major version of the system, Csound 6, was first released in 2013. The code was again considerably reorganised, and a new language parser was introduced. This is the part of the system that translates the code text into an internal format that can be used by the compiler to create new instruments. The new parser allowed a much simpler means of extending and modifying the language, so a number of new facilities for the programmer were also introduced.

The compiler was significantly modified allowing it to operate in an on-the-fly mode, adding new instruments on demand to a running system. Such changes were also reflected in a newly designed API, which gave developers increased flexibility and new possibilities. Alongside these improvements, ports of the system to new platforms continued, with the addition of support for web-based applications and embedded systems.

As we have seen in this introduction to music programming systems, Csound fits very well the description of a software package for general-purpose music making. However, a more complete picture of what it is today goes beyond this simple definition. In fact, it depends on how a user approaches this software. At the highest level, we can think of it as a synthesiser, or as a sound processor. Computer musicians might look at it as a software package that allows us to use the computer to manipulate sound programmatically.

Software developers, on the other hand, will use Csound as an audio engine, a system component that will provide all the services that are needed to add sound to an application. They will use it by embedding it in their software, which can be done with full control over its operation. So, depending on the perspective, Csound can be a music program, a programming language or a software library.

Figure 1.2 demonstrates these ideas. At the top, we have the level of the music application, where Csound is presented in pre-programmed, pre-packaged forms, with fixed instruments that expose some of their parameters for the user to manipulate. This is the case of the various Csound-based apps that exist for mobile, web and desktop platforms. The next level is the traditional place where music programming systems reside, where musicians, composers and researchers employ the system by developing new instruments and means of controlling them. Finally, at the lowest level, we have Csound as a programming library, the audio engine that developers (and musicians, researchers, etc. working on application programming) use in their projects.

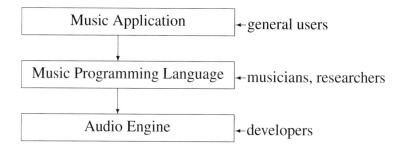

Fig. 1.2 The three system levels in Csound: at the top, we have music applications (Csound-based apps for mobile, web and desktop platforms); the middle level is represented by the music programming system; and the lowest level is that of Csound as a programming library

1.3.3 Compatibility and Preservation

During its thirty-odd years of development, Csound has been gifted a significant number of contributions from a world-wide computer music community. An explicit rule has been followed by developers that new versions of Csound should **always** be backwards-compatible. So this means that any changes to the language or the system can provide new features and facilities, but will not prevent code designed for the MIT-EMS Csound from running in the latest version. There is a significant responsibility on the development team to ensure that the preservation of music made with the software is guaranteed. This approach has enabled even some music made with MUSIC 360 and many MUSIC 11 pieces to be rendered in Csound 6.

The backwards-compatibility principle has a downside. While many of the contributions to the system have had a long-lasting impact, a few were less successful. For instance, some of these were introduced to solve a short-term issue, without a thorough assessment of their implications, introducing less desirable features. Some of these cannot be easily eliminated as part of the evolution of the software, as it is not possible to know whether they have not been used in any work that we want to preserve. Hopefully, this issue is not significant, as through education and documentation we can direct the user to the most up-to-date approaches, and mark as deprecated the less successful legacy elements in the system. Also, since the development of Csound 5, there has been an intense effort to think about the implications of new features and components. This has led to a more cohesive development, which has mostly eliminated problems with unwanted components.

1.4 Conclusions

In this chapter, we have introduced the concept of music programming systems, and examined the history of their development leading to the appearance of Csound. We have seen how the early music languages evolved from the first direct synthesis programs by Max Mathews into MUSIC III and MUSIC IV, which were the model for later systems. Csound itself was the culmination of the developments of MUSIC 360 and MUSIC 11, which were created for specific computers. Other classic systems of note were MUSIC V, written in FORTRAN, which influenced the design of Cmusic, and MUSIC 4C. With the development of computer technology, we also saw the appearance of systems that were directed explicitly at real-time operation, which all of the earlier software could not achieve.

Alongside this survey, we have also introduced some of the key principles that are involved in this technology: computer instruments, numeric scores, function tables, compilers, data types, initialisation and performance times, computation rates etc. These will be followed up and explored in further detail in the following chapters of this book.

Complementing this discussion, the chapter also detailed the main characteristics of the current Csound system, providing an overview of the ways it can be employed in computer applications. We have shown that there are many ways to 'attack' the complexity that it presents to the user, from high to low levels of programming. In this book, we will try to provide a glimpse of all of these levels of operation, while concentrating on the middle-ground approach of music programming through an emphasis on computer music composition.

Chapter 2
Key System Concepts

Abstract This chapter provides an overview of the key principles that underline the operation of Csound: frontends; the sampling theorem; control and audio rates; processing blocks; function table generation; real-time and offline audio; the API. These ideas will be presented as a foundation for the detailed exploration of Csound programming in the subsequent chapters.

2.1 Introduction

In this chapter, we will start to introduce the reader to a number of important principles in computer music. Many of these have been hinted at in the survey of systems, languages and software packages in Chapter 1. Here, we will try to move one step deeper into these ideas. We will also focus on the issues that are central to the operation of Csound, and begin using code examples of the current language to illustrate these, even if not all of the concepts embodied in such examples were fully explained. The discussion will proceed in a mosaic-like fashion, and hopefully the full comprehension of these ideas will emerge as the following chapters tackle them more completely.

A number of key concepts are behind the workings of Csound. These extend from fields such as digital signal processing (e.g. the sampling theorem), to computation structures (sample blocks, buffers etc), and system design (frontends, API). In our discussion of these, they are not going to be organised by the areas they stem from, but by how they figure in the workings of Csound. We assume that the reader has a reasonable understanding of basic acoustics principles (sound waveforms, propagation, frequency, amplitude etc.), but more complex ideas will be introduced from the ground up.

© Springer International Publishing Switzerland 2016
V. Lazzarini et al., *Csound*, DOI 10.1007/978-3-319-45370-5_2

2.2 General Principles of Operation

Csound can be started in a number of different ways. We can provide the code for its instruments, also known as the orchestra, and a numeric score containing the various sound events that will be performed by them. Alternatively, we can supply the instruments, and make Csound wait for instructions on how to play them. These can come from a variety of sources: score lines typed at the terminal; Musical Instrument Digital Interface (MIDI) commands from another program or an external device; Open Sound Control (OSC) commands from a computer network source; etc. We can also submit an orchestra that includes code to instantiate and perform its instruments. Or we can just start Csound with nothing, and send in code whenever we need to make sound.

The numeric score is the traditional way of running Csound instruments, but it is not always the most appropriate. MIDI commands, which include the means of starting (NOTE ON) and stopping (NOTE OFF) instruments might be more suitable in some performance situations. Or if we are using other computers to issue controls, OSC is probably a better solution, as it allows for a simpler way of connecting via a network. Both MIDI and OSC, as well as the numeric score, will be explored in detail later on in this book. Of course, if the user is keen to use code to control the process interactively, she can send instruments and other orchestra code directly to Csound for compilation and performance.

2.2.1 CSD Text Files

Let's examine a simple session using the basic approach of starting Csound by sending it some code. In that case, we often package these two in a single text file using the CSD format. This is made up of a series of tags (much like an XML or HTML file) that identify sections containing the various textual components. The minimum requirement for a CSD file is that it contains a section with one instrument, which can be empty (see listing 2.1).

Listing 2.1 Minimal legal CSD code

```
<CsoundSynthesizer>
<CsInstruments>
instr 1
endin
</CsInstruments>
</CsoundSynthesizer>
```

The relevant tag for instrument (orchestra) code is `<CsInstruments>`, which is closed by `</CsInstruments>`. Everything is enclosed within the `<Csound Synthesizer>` section, and anything outside it is ignored. In order to get sound, we need to give some substance to the instrument, and play it (listing 2.2).

Listing 2.2 Minimal sound-producing CSD code

```
<CsoundSynthesizer>
<CsInstruments>
instr 1
 out rand(1000)
endin
schedule(1,0,1)
</CsInstruments>
</CsoundSynthesizer>
```

In this case, what is happening is this: Csound reads the CSD file, finds the instruments, compiles it, and starts its operation. In the code, there is an instrument defined (with a noise-producing unit generator or opcode), and outside it, an instruction (`schedule`) to run it for a certain amount of time (1 second). Once that is finished, Csound continues to wait for new code, instructions to play the instrument again, etc. If nothing is provided, no sound will be output, but the system will not close.[1]

2.2.2 Using the Numeric Score

Optionally, we could have started Csound with a numeric score in addition to its instruments. This is defined by the `<CsScore>` tag. The example in listing 2.3 is equivalent to the previous one, except for one difference: with the presence of the score, Csound terminates once there are no further events to perform.

Listing 2.3 Minimal sound-producing CSD code with numeric score

```
<CsoundSynthesizer>
<CsInstruments>
instr 1
 out rand(1000)
endin
</CsInstruments>
<CsScore>
i 1 0 1
</CsScore>
</CsoundSynthesizer>
```

You can see from this example that the numeric score has a different syntax to the orchestra code. It is very simple, just a list of space-separated *parameter fields* (p-fields). Ways to use a score and keep Csound open for other external event sources will be discussed later in this book. Both score events and new instruments can be

---

[1] The amplitude 1,000 refers to the default setting of 32,768 for 0 dB full scale. It will produce a white noise of $\frac{1000}{32768} = 0.0305$ or -30 dB. In modern Csound coding practice, the 0 dB value is set to 1 by the statement `0dbfs = 1`, as shown in listing 2.6. More details are given in Section 2.4.2.

supplied to Csound after it has started. How this is done will depend on the way Csound is being used, and how it is being hosted.

2.2.3 Csound Options

Csound's operation is controlled by a series of options. There is a very extensive set of these, and many of them are seldom used. However, it is important to understand how they control the system. There are different ways to set the values for these options, and that also depends on how Csound is hosted. A portable way of making sure the settings are correct for the user's needs is to add them to the CSD file. This can be done under the CsOptions tag as shown in listing 2.4.

Listing 2.4 CSD with options controlling

```
<CsoundSynthesizer>
<CsOptions>
-odac
</CsOptions>
<CsInstruments>
instr 1
 out rand(1000)
endin
</CsInstruments>
<CsScore>
i 1 0 1
</CsScore>
</CsoundSynthesizer>
```

The example shows the setting of an option to run Csound with its output directed to the Digital-to-Analogue Converter (DAC), a generic name for the computer sound card. This is what we need to do if we want to run it in real-time. With no options, by default Csound works in an offline rendering mode, writing the output to a soundfile (named "test" by default).

Options start with a dash (−), for simple single-letter options, or with two dashes (−−) for longer names. A list of all options can be found in the Csound Reference Manual, but in this book we will introduce the most relevant ones, as we come across them.

2.3 Frontends

Users interact with Csound through a program called the *frontend*, which hosts the system. There are many different types of frontends, some designed for different specific tasks, others with a more general-purpose application. The Csound 6 software package provides a few different frontends itself, and there are also a number of third-party ones which are well maintained, and kept up to date with the system.

This is the current list of frontends maintained as part of the Csound project:

- **csound**: a general-purpose command-line interface (CLI) frontend.
- **csound6~**: a Pure Data [103] object that allows Csound to be run inside that system.
- **csound~**: an object for the MaxMSP [136] system.
- **CsLadspa**: a plug-in generator for LADSPA hosts.
- **CsoundVST**: a plug-in for VST hosts.
- **winsound**: a legacy GUI frontend originally for Windows.

Some of these are not distributed in binary form by the project, but are available as source code in the Csound repository. In terms of third-party frontends, there are many options, of which four are of note:

- **CsoundQt**: a general-purpose Integrated Development Environment (IDE) for Csound, which allows the building of user interfaces (UIs) and has facilities for scripting using the Python language.
- **Cabbage**: another IDE designed mostly for the development of plug-ins and stand-alone programs. using Csound, with full support for building UIs.
- **Blue**: a composition system for computer music that uses Csound as its sound engine.
- **WinXsound**: a GUI program with text-editing facilities, built on top of the CLI csound frontend.

2.3.1 The csound *Command*

The csound command , although the most basic of the frontends listed above, provides access to the full functionality of the system. It allows users to run Csound code, and to interact with it in a number of ways. It does not have a graphical interface of its own, although the Csound system includes GUI-building opcodes that can be used for that purpose. An advantage that this frontend has over all the others is that it is present wherever Csound is installed. It is useful to learn how to use it, even if it is not the typical way we interact with the system.

While it is beyond the scope of this book to discuss the operation of different frontends (for which help can be found elsewhere), we would like to provide some basic instructions on the command-line usage. For this, you will need to open up a terminal (also known as a shell, or a command line) where commands can be typed. The basic form of the csound command is:

```
csound [options] [CSD file]
```

The options are the same as discussed before (Sec.2.2). Typically we should supply a CSD file to Csound to get it started, unless we use a specific option to tell Csound to start empty. If that is not the case, a usage message will be printed to the terminal with a list of basic options. We can get a full list of options by using `--help`:

```
csound --help
```

Once started, Csound will behave as discussed in Sec.2.2. It can be stopped at any point by pressing the ctrl and the 'c' keys together (the `ctrl-c` 'kill program' sequence).

2.3.2 Console messages

Csound provides a means of informing the user about its operation through *console* messages. These will be displayed in different ways, depending on the frontend. In the case of graphic frontends, generally there will be a separate window that will hold the message text. In the case of the `csound` command, the console is directed by default to the standard error (stderr), which is in most cases the terminal window. Once the system is started, the console will print basic details about the system, such as version, build date, CSD filename, etc., culminating in the start of performance, when `SECTION 1:` is displayed:

```
time resolution is 1000.000 ns
virtual_keyboard real time MIDI plug-in for Csound
0dBFS level = 32768.0
Csound version 6.07 beta (double samples) Dec 12 2015
libsndfile-1.0.25
UnifiedCSD:  test.csd
...
SECTION 1:
```

Following this, as instruments start playing, we will get some information about them and their output. It is possible to write code to print custom messages to the console. We can suppress most of the console displays by adjusting the messaging level with the option `-m N`, where `N` controls how much is printed (0 means reducing messages to the minimum).

2.4 Audio Computation, the Sampling Theorem, and Quantisation

Earlier in this chapter, we introduced the idea of *signals*, in particular audio signals, without actually explaining what these are. We expect that the reader would identify this with the sound that the program is generating, but it is important to define it more precisely. Computers, through programs like Csound, process sound as a digital signal. This means that this signal has two important characteristics: it is sampled in time, and quantised in value.

Let's examine each one of these features separately. The sound that we hear can be thought of as a signal, and it has a particular characteristic: it is continuous, as far as time is concerned. This means, for instance, that it can be measured from one infinitesimal instant to another, it 'exists' continuously in time. Digital computers cannot deal with this type of signal, because that would required an infinite amount of memory. Instead, this signal is encoded in a form that can be handled, by taking samples, or measurements, regularly in time, making a discrete (i.e. discontinuous) representation of a sound recording.

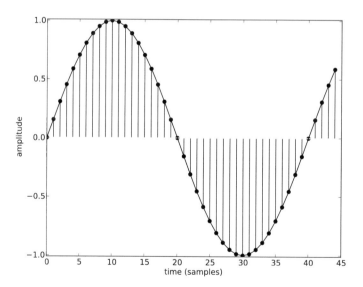

Fig. 2.1 A sampled waveform and its underlying continuous-time form. The vertical lines represent the times at which a measurement is made, and the dots represent the actual samples of the waveform

In Fig. 2.1, a sample sine wave and its underlying continuous-time form is shown. The vertical lines represent the times at which a measurement is made, and the dots represent the actual samples of the waveform. Note that, as far as the discrete representation is concerned, the signal is not defined in the times between each sample. However, it can be reconstructed perfectly, as denoted by the continuous plot.

Equally, the sound waveform in the air can vary by infinitesimal differences as time progresses, and computers cannot, for the same reasons, deal with that. So when a measurement is made to produce a sampled representation, it needs to place that value in a finite grid of values. We call this *quantisation*. Two sampled numbers that are very close together might be represented by a single quantised output, and that will depend on how fine the grid used is. We can think of sampling and quantisation as slicing the sound waveform in two dimensions, in time, and in amplitude.

2.4.1 Aliasing

There are some key implications that arise as a result of these two characteristics of digital audio. First and foremost, we need to understand that a discrete representation is not the same thing as the continuous signal it encodes. However, we can define the conditions in which the two can be considered equivalent.

With regards to sampling in time, the first thing to recognise is that we are introducing a new quantity (or parameter) into the process, which determines how often we will be taking measurements. This is called the sampling rate (sr) or sampling frequency, and is measured in samples per second, or in Hertz (Hz, which is another way of saying 'something per second'). Sound itself is also composed of time-varying quantities, which can also be measured in Hz. The sampling process introduces a complex interaction between the frequency of the components of a sound wave and the sr. This relationship is captured by the sampling theorem, also know as the Nyquist(-Shannon) theorem, which tells us that

> In order to encode a signal containing a component with (absolute) frequency X, it is required to use a sampling rate that is at least equivalent to 2X. [95, 116]

This places some limits in terms of the sampling rate used and the types of signals we want to use. The main implication of this is that if any component in a digital audio signal exceeds the sampling rate, it will be folded over, *aliased*, into the range of possible frequencies, which extends from 0 Hz to $\pm \frac{sr}{2}$. These aliased components can appear as a form of noise in the digital signal (if they are numerous), or unwanted/unrelated inharmonic components of an audio waveform. Within the stated limits, we can, for all practical purposes, accept that the digitally encoded signals are the same as their original form (in terms of their frequency content). In Csound, the default sampling rate is set at 44,100 Hz, but this can be modified by setting the `sr` constant at the top of the orchestra code (listing 2.5), or by using the relevant option.

Listing 2.5 Setting the sr to 48,000 Hz

```
<CsoundSynthesizer>
```

```
<CsInstruments>
sr = 48000
instr 1
 out oscili(1000,440)
endin
schedule(1,0,1)
</CsInstruments>
</CsoundSynthesizer>
```

How do we avoid aliasing? There are two main cases where this can be present: when we convert an original ('analogue') signal into its digital form (through an Analogue-to-Digital Converter, the ADC, a general name for the computer input sound card), and when we generate the sound directly in digital form. In the first case, we assume that ADC hardware will deal with the unwanted high-frequency components by eliminating them through a low-pass filter (something that can cut signals above a certain frequency), and so there is no need for any further action. In the second case, we need to make sure that the process we use to generate sounds does not create components with frequencies beyond $\frac{sr}{2}$ (which is also known as the Nyquist frequency). For instance, the code in listing 2.5 observes this principle: the sr is 48,000 Hz, and the instrument generates a sine wave at 440 Hz, well below the limit (we will study the details of instruments such as this one later in the book). If we observe this, our digital signal will be converted correctly by a Digital-to-Analogue Converter (DAC, represented by the computer sound card) (within its operation limits) to an analogue form. It is important to pay attention to this issue, especially in more complex synthesis algorithms, as aliasing can cause significant signal degradation.

2.4.2 Quantisation Precision

From the perspective of quantisation, we will also be modifying the original signal in the encoding process. Here, what is at stake is how precisely we will reproduce the waveform shape. If we have a coarse quantisation grid, with very few steps, the waveform will be badly represented, and we will introduce a good amount of modification into the signal. These are called quantisation errors, and they are responsible for adding noise to the signal. The finer the grid, the less noise we will introduce. However, this is also dependent on the amount of memory we have available, as finer grids will require more space per sample.

The size of each measurement in bytes determines the level of quantisation. If we have more bits available, we will have better precision, and less noise. The current standard for audio quantisation varies between 16 and 64 bits, with 24-bit encoding being very common. Internally, most Csound implementations use 64-bit floating-point (i.e. decimal-point) numbers to represent each sample (double precision), although in some platforms, 32 bits are used (single precision). Externally, the encoding will depend on the sound card (in case the of real-time audio) or the soundfile format used.

Generally speaking quantisation size is linked to the maximum allowed absolute amplitude in a signal, but that is only relevant if the generated audio is using an integer number format. As Csound uses floating-point numbers, that is not very significant. When outputting the sound, to the sound card or to a soundfile, the correct conversions will be applied, resolving the issue.

However, for historical reasons, Csound has set its default maximum amplitude (also known as '0dB full scale') to the 16-bit limit, 32768. This can be redefined by setting the 0dbfs constant in the Csound code (or the relevant option). In listing 2.6, we see the maximum amplitude set to 1, and the instrument generating a signal whose amplitude is half scale.

Listing 2.6 Setting the 0dbfs to 1

```
<CsoundSynthesizer>
<CsInstruments>
0dbfs=1
instr 1
 out rand(0.5)
endin
schedule(1,0,1)
</CsInstruments>
</CsoundSynthesizer>
```

In summary, when performing audio computation, we will be dealing with a stream of numbers that encodes an audio waveform. Each one of these numbers is also called a *sample*, and it represents one discrete measurement of an continuous (analogue) signal. Within certain well-defined limits, we can assume safely that the encoded signal is equal to its intended ('real-world') form, so that when it is played back, it is indistinguishable from it.

2.4.3 Audio Channels

A digital audio stream can accommodate any number of channels. In the case of multiple channels, the samples for these are generally arranged in *interleaved* form. At each sample (measurement) point, the signal will contain one value for each channel, making up a *frame*. For example, a two-channel stream will have twice the number of samples as a mono signal, but the same number of sample frames. Channels in a frame are organised in ascending order. Csound has no upper limit on the number of channels it can use, but this will be limited by the hardware in case of real-time audio. By default, Csound works in mono, but this can be changed by setting a system parameter, nchnls, which sets both the input and output number of channels. If a different setting is needed for input, nchnls_i can be used. Listing 2.7 shows how to use the nchnls parameter for stereo output.

Listing 2.7 Setting the nchnls to 2

```
<CsoundSynthesizer>
<CsInstruments>
nchnls=2
instr 1
 out rand(1000), oscili(1000,440)
endin
schedule(1,0,1)
</CsInstruments>
</CsoundSynthesizer>
```

2.5 Control Rate, ksmps and Vectors

As discussed before in Sec. 1.2, the idea of having distinct audio and control signals is an enduring one. It is fundamental to the operation of Csound, and it has some important implications that can inform our decisions as we design our instruments. The first one of these is very straightforward: control rate signals are digital signals just like the ones carrying audio. They have the same quantisation level, but a lower sampling rate. This means that we should not use them for signals whose frequencies are bound to exceed the stated limit. In other words, they are suited to slow-varying quantities.

In Fig. 2.2, we demonstrate this by showing a 100 Hz envelope-shaped sine wave. The one-second envelope (top panel) has three stages, the shortest of which lasts for 100 ms. The waveform, on the other hand has cycles that last 10 ms each. The output signal is shown in the bottom panel. This illustrates the fact that the rates of change of audio and control signals have different timescales, and it is possible to compute the latter at a lower frequency

Internally, the control rate (kr) is also what drives the computation performed by all unit generators. It defines a fundamental cycle in Csound, called the k-cycle, whose duration is equivalent to one control period. The audio cycle determined by the sampling rate becomes a subdivision of this. In other words, at every control period, a control signal will contain one sample, and an audio signal will contain one or more samples, the number of which will be equivalent to the ratio $\frac{kr}{sr}$. This ratio has to be integral (because we cannot have half a sample), and is called ksmps, the number of audio samples in a control period.

Another way of looking at control and audio signals is this: in the course of a computation cycle, the former is made up of a single sample, while the latter contains a block of samples. A common name given to this block is a *vector*, while the single value is often called a *scalar*. It is very important to bear this fact in mind when we start looking at how Csound is programmed. The default control rate in Csound is 4,410 Hz, but this can also be determined by the code (listing 2.8), or by a command-line option.

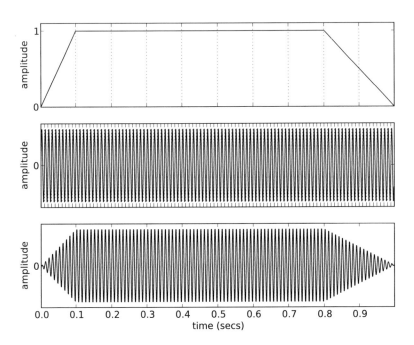

Fig. 2.2 A comparison of control and audio signal timescales. On the top plot, we have a control signal, the envelope; in the middle, an audio waveform; and at the bottom, the envelope-shaped signal. The gridlines show the difference in timescales between an audio waveform cycle and an envelope stage

Listing 2.8 Setting the kr to 441Hz

```
<CsoundSynthesizer>
<CsInstruments>
kr = 441
instr 1
 out oscili(1000,440)
endin
schedule(1,0,1)
</CsInstruments>
</CsoundSynthesizer>
```

The kr cannot be arbitrary: only values that ensure an integral number of ksmps are allowed. So often we want to set the ksmps directly instead, which is shown in listing 2.9, where $ksmps = 64$ makes $kr = 689.0625$ (at $sr = 44,100$). The control rate can have a fractional part, as implied by this example.

Listing 2.9 Setting ksmps to 64

```
<CsoundSynthesizer>
<CsInstruments>
ksmps = 64
instr 1
 out oscili(1000,440)
endin
schedule(1,0,1)
</CsInstruments>
</CsoundSynthesizer>
```

One important aspect is that when two control and audio signals are mixed in some operation, the former will be constant for a whole computation (ksmps) block, while the latter varies sample by sample. Depending on the $\frac{sr}{kr}$ ratio, this can lead to artefacts known as zipper noise. This is a type of aliasing in the audio signal caused by the stepping of the control signals, which are staircase-like. Zipper noise will occur, for instance, in control-rate envelopes, when ksmps is large.

In Fig. 2.3, we see an illustration of this. Two envelopes are shown with two different control rates, applied to a waveform sampled at 44,100 Hz. The topmost plot shows an envelope whose control rate is 441 Hz (ksmps=100), above its resulting waveform. Below these, we see a control signal at 44.1 Hz (ksmps=1000) and its application in the lower panel. This demonstrates the result of using a control rate that is not high enough, which is shown to introduce a visible stepping of the amplitude. This will cause an audible zipper noise in the output signal. It is important to make sure the control rate is high enough to deal with the envelope transitions properly, i.e. a short attack might require some careful consideration. Of course, envelope generators can also be run at the sampling rate if necessary.

The final implication is that, as the fundamental computation cycle is determined by the kr, event starting times and durations will be rounded up to an even number of these k-periods. If the control rate is too slow, the timing accuracy of these events can be affected. In the examples shown in listings 2.8 and 2.9, events will be rounded up to a 2.27 and 1.45 ms time grid.

2.6 Instruments, Instances, and Events

We have already outlined that one of the main structuring pieces in Csound is the *instrument*. This is a model, or recipe, for how a sound is to be processed. It contains a graph of interconnected opcodes (unit generators), which themselves embody their own model of sound generation. In order for anything to happen, we need an *instance* of an instrument. This is when the computation structures and operations defined in an instrument get loaded into the audio engine so that they can produce something. Instances are created in response to *events*, which can originate from

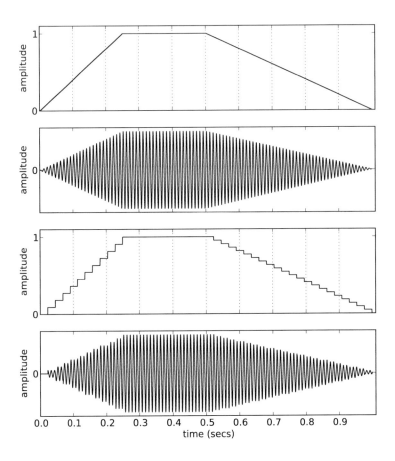

Fig. 2.3 A comparison of two different $\frac{sr}{kr}$ ratios and the resulting zipper artefacts. The illustration shows two envelopes with the resulting application of these to an audio waveform. The top example uses a $\frac{sr}{kr}$ ratio (ksmps) of 100, whereas the other control signal has its ksmps set to 1000. The zipping effect is clearly seen on the lower plot, which results into an audible broadband noise

various sources: the numeric score, MIDI commands, real-time inputs and orchestra opcodes.

2.6.1 The Life Cycle of an Instrument

An instrument passes through a number of stages in its complete life cycle: it is compiled from a text representation into a binary form, and then loaded by the engine to perform audio processing. Finally, it is de-instantiated when it stops making sound.

Compilation

Instruments start life as plain text, that contains Csound orchestra code. The next stage is the compilation. Any number of compilations can be made as Csound runs, although the first one is slightly special in that it allows for system constants to be set (e.g. `sr`, `kr` etc.), if necessary. Compilation is broken down into two steps: parsing and compilation proper. The first takes the code text, identifies every single element of it, and builds a tree containing the instrument graph and its components. Next, the compilation translates this tree into a series of data structures that represent the instrument in a binary form that can be instantiated by the engine. All compiled instruments are placed in an ordered list, sorted by instrument number.

Performance

When an event for a given instrument number is received, the audio engine searches for it in the list of compiled instruments, and if it finds the requested one, instantiates it. This stage allocates memory if no free space for the instance exists, issuing the console message

```
new alloc for instr ...,
```

with the name of the instrument allocated. This only happens if no free instrument 'slots' exist. If a previously run instrument has finished, it leaves its allocated space for future instances. The engine then runs the initialisation routines for all opcodes in the instrument. This is called the init-pass. Once this is done, the instance is added to a list of running instruments, making it perform audio computation.

Csound performance is bound to the fundamental k-cycle described in Section 2.5. The engine runs an internal loop (the *performance loop*) that will go into each instrument instance and call the perform routine of each opcode contained in it. The sequence in which this occurs is: ascending instrument number; for each instrument, instantiation time order (oldest first); and opcode order inside an instrument (line order). For this reason, opcode position inside an instrument and instrument number can both be significant when designing Csound code. The audio engine loop repeats every k-cycle, but instruments can also subdivide this by setting a local ksmps, which will trigger an inner loop operating over a shorter period for a given

instance. The life cycle of an instrument, from text to performance, is depicted in
Fig. 2.4.

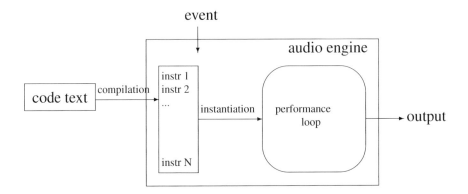

Fig. 2.4 The life cycle of an instrument: on the left, it begins as code in text form; then it is
compiled and added to the list of instruments in the audio engine; once an event is received for that
instrument, it gets instantiated and produces an output, running in the performance loop

De-instantiation

An instrument instance will either run until the event duration elapses, or if this duration is undefined, until a turnoff command is sent to it. In any case, the instrument
can also contain code that determines a *release* or extra time period, in which case
it will carry on for a little longer. Undefined duration instances can originate from
MIDI NOTEON commands, in which case a corresponding NOTEOFF will trigger
its turnoff, or via an event duration set to -1. In this case the instance can be turned
off from an event whose instrument number is negative and corresponds to a currently running instance. On turnoff, an instance is deallocated, and for any opcodes
that have them, deallocation routines are run. The memory for a given instrument is
not recovered immediately, so it can be used for a future instance. Instruments exist
in the engine until they are replaced by a new compilation of code using the same
number or explicitly removed.[2]

---

[2] See opcode `remove`.

2.6.2 Global Code

Code that exists outside instruments is global. Only init-time operations are allowed here, and the code is executed only once, immediately after compilation. It can be used very effectively for one-off computation needs that are relevant to all instrument instances, and to schedule events. Data computed here can be accessed by instrument instances via global variables. In addition, system constants (sr, kr, ksmps, nchnls, 0dbfs) can be set here, but are only used in the first compilation (and ignored thereafter). This is because such attributes cannot be changed during performance.

2.7 Function Tables

Another key concept in Csound is the principle of *function tables*. These have existed in one form or another since the early days of computer music. They were introduced to provide support for a fundamental unit generator, the table-lookup oscillator (which will be discussed in detail in a subsequent chapter), but their use became more general as other applications were found. In a nutshell, a function table is a block of computer memory that will contain the results of a pre-calculated mathematical expression. There are no impositions on what this should be: it may be as simple as a straight-line function, or as involved as a polynomial. Tables generally hold the results of a one-dimensional operation, although in some cases multiple dimensions can also be stored.

The actual contents of a table are very much like a data array: a series of contiguous values stored in memory (arrays as data structures will be discussed in the next chapter). Each one of these values is a floating-point number, usually using double precision (64-bit), but that is dependent on the platform and version of Csound. Tables can be accessed via a variety of means, of which the simplest is direct lookup: an index indicating the position to be read is used to read a given value from the table. Fig. 2.5 illustrates this: a function table with 18 points, whose position 9 is being looked up, yielding 0.55 as a result. Note that indexing is zero-based, i.e. the first position is index 0, and the last is the table size - 1. Csound provides unit generators for direct lookup, which can be used for a variety of applications. Similarly, there are several other opcodes that use function tables as an efficient way of handling pre-defined calculations, and will employ a lookup operation internally.

2.7.1 GEN Routines

The data stored in tables is computed at the time they are created. This is done by invoking a GEN routine, which implements the mathematical operations needed for that. The contents of a table will depend on the routine used. For instance, the GEN function might be asked to construct one cycle of a waveform and place it in

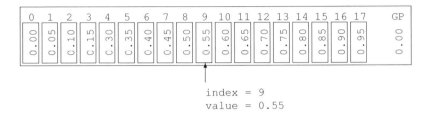

```
index = 9
value = 0.55
```

Fig. 2.5 An example of a function table with 18 points, whose position 9 is being looked up, yielding 0.55 as a result

the table. Or we might want to define an envelope shape with a certain number of segments. We could store a sequence of pitch values to use in a loop, or to be read in random order.

2.7.2 Normalisation

Once the data is computed, we can choose to store it in raw form, i.e. the exact results, or we can normalise these prior to keeping them. Normalisation scales the function to a given range (e.g. 0 to 1, -1 to 1), which is useful for certain applications. In the Csound Reference Manual, this is called re-scaling. For instance, when storing an audio signal, such as a waveform, it is often useful to keep it normalised so that when we read it, we can apply any amplitude to it. If, however, we want to store some frequency values in Hz to use in an instrument, then it will be best to turn off re-scaling. The default for Csound, as we will see, is to normalise.

2.7.3 Precision

A table is defined by one very important parameter: its size. This can be set to anything, but as we will see later, some opcodes are designed to work with tables of specific sizes, and thus we need to be careful about this. For instance, some unit generators require tables to be set with a power-of-two size. From the perspective of signal quality, longer tables are more desirable. They will hold better precisely calculated results. On the other hand, more memory is consumed, which in most modern systems is not a significant issue.

2.7.4 Guard Point

Regardless of its size, every table created in Csound will include an extra point (or position) at the end, called a *guard point*, which is also illustrated in Fig. 2.5. This is to facilitate the implementation of an operation called interpolation, which provides the means for finding an in-between value when the position sought is not a whole number. This extra point can be a copy of the first position on the table (as in Fig. 2.5), or it can extend the contour calculated for the table (called an *extended guard point*). If the table is to be read iteratively (wrapping around at the ends), then we should create it with an ordinary guard point (the default). In applications using a one-shot lookup, extended guard points are needed.

2.7.5 Table types

There are various uses for function tables, and several types of data can be held in them. A few examples of these are:

- **Wavetables**: tables holding one or more cycles of a waveform. Typical applications are as sources for oscillators. For instance, a table might contain one cycle of a wave constructed using a Fourier Series, with a certain number of harmonics. Another example is to load into a table the samples of a soundfile, which are then available to instruments for playback.
- **Envelopes**: these are segments of curves, which can be of any shape (e.g. linear, exponential), and which can be looked up to provide control signals.
- **Polynomials**: polynomial functions are evaluated over a given interval (e.g. -1 to 1), and can be used in non-linear mapping operations.
- **Sequences**: tables can be used to store a sequence of discrete values for parameter control. For instance you might to store a pattern of pitches, durations, intensities, etc, to be used in a composition.
- **Audio storage**: we can also write to tables directly from instruments, and use them to store portions of audio that we want to play back later, or to modify. This can be used, as an example, for audio 'scratching' instruments, or to create bespoke audio delays.

Beyond these examples, there are many other applications. Csound has over 40 different GEN routines, each one with many different uses. These will be discussed in more detail in the relevant sections of this book.

2.8 Audio Input and Output

Audio input and output (IO) is the key element in a music programming system. In Csound, it is responsible for delivering audio to instruments and collecting their output, connecting to whichever sources and destinations are being used. These can be, for instance, a computer sound card or an audio file, depending on the options used when Csound is started. These will control whether the software is run in a real-time mode, or will work offline with soundfiles. It is also possible to combine real-time output with soundfile input, and vice versa.

In addition to the main IO, discussed in this section, there are other forms of IO that are implemented by specific opcodes. These can be used, for example, to write to a given network address or to open soundfiles for reading or writing.

2.8.1 Audio Buffers

Audio IO employs a software device called a *buffer*. This is a block of memory that is used to store computation results before they can be sent to their destination. The reason these are used is that it is more efficient to read and write chunks of data rather than single numbers. The norm is to iterate over a block of samples to produce an output, and place that result in a buffer, then write data (also in blocks) from that buffer to its destination. Similarly, for input, we accumulate a certain number of samples in a buffer, then read blocks out of it when needed. This mechanism is used regardless of the actual source or destination, but it has slightly different details depending on these.

2.8.2 The Audio IO Layers

Csound's main IO consists of a couple of layers that work together to provide access to the inner sound computation components (the audio engine). The outer level is the one used to access the actual IO device, whatever form it might take. It is composed of buffers that are used to accumulate the audio data produced by the software before reading or writing can be performed. The size of these buffers can be defined by options that are passed to Csound. Depending on the type of IO device, one or two buffers are used, a *software* buffer and a *hardware* buffer. We will discuss the particular details of these two buffers when exploring real-time and offline audio IO operations below. The relevant options to set the size of these buffers are

```
-b N
-B N
```

or in long option form

```
--iobufsamps=N
--hardwarebufsamps=N
```

In the case of the software buffer (-b or --iobufsamps), N refers to the number of sample *frames* (see Section 2.4.3), whereas for the hardware buffer (-B or --hardwarebufsamps), the size N is defined in samples.

The innermost layer is where Csound's fundamental k-cycle loop (see Sec. 2.5) operates, consuming and producing its audio samples. Here we have one buffer for each direction, input or output, called *spin* and *spout*, respectively. These buffers are ksmps sample frames long. For input, they hold the most current audio input block, and can be accessed by the Csound input unit generators. The output buffer adds the output blocks from all active instruments. The input buffer is updated each k-cycle period, by reading ksmps sample frames from the input software buffer. Once all instruments are processed, the spout buffer is written to the output software buffer. If used, a hardware buffer provides the final connection to the IO device used. Figure 2.6 shows the inner and outer layers of the Csound main IO system.

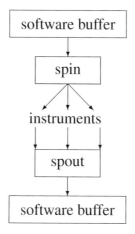

Fig. 2.6 Csound software and inner engine buffers: the audio samples are stored in the input buffer; passed on to the spin buffer at every k-cycle and accessed by instruments, which write to spout; this is then copied into the output buffer

The outer buffers can read/write to soundfiles, or to the audio device (ADC/-DAC). As already explained in Sect. 2.2, the relevant option for output is $-o$. More precisely

```
-i fnam sound input filename
-o fnam sound output filename
```

or in long format

```
--input=FNAME          Sound input filename
--output=FNAME         Sound output filename
```

Here, *filename* is taken in the UNIX sense of a logical device, which can be either a proper disk file, or in the case of real-time audio, adc for input or dac for output. Depending on the platform, it should be possible to pipe the output to a different program as well. For file input, the data will be streamed into the input buffer until the end of file (EOF) is reached. Beyond this, no more data will be read, as there is no mechanism to loop back to the beginning of the file (to do this, there are specialised opcodes, which we will study later in this book). The $-i$ option can be useful for offline batch processing of soundfiles.

2.8.3 Real-Time Audio

Real-time audio in Csound is implemented externally to the main system via a series of backend plug-ins (Fig. 2.7). These are interfaces to the various systems provided by the platforms that support real-time IO. Across the major desktop platforms, Csound provides a standard plug-in based on the cross-platform third-party library *portaudio*. In addition to this, there is also a plug-in that interfaces with the Jack IO kit, which exists in the Linus and OSX platforms. There are also OS-specific plug-ins: AuHAL and AlSA, under OSX and Linux respectively. Finally, some platforms implement exclusive audio IO solutions, which are the only option provided by the system. This the case for iOS, Android and Web-based implementations.

Fig. 2.7 The Csound real-time audio system based on plug-ins that provide backends across various platforms

The main consideration for real-time audio is latency. This is the time it takes for a sound to appear at the physical audio output. It can refer to the full round-trip (bidirectional) period from the sound going into the software and coming out of the speakers, or it can relate to the time from a control input (say a key pressed) until the corresponding effect happens at the output (single direction). Musicians are very sensitive to latencies, and to improve performance, we attempt to minimise it. The time lag will depend primarily on the buffer size, but it will also depend on the hardware and software platform. Smaller buffers will imply shorter latencies.

We can estimate the latency a buffer adds to the signal by dividing the number of frames in it by the sr. For instance, a buffer of 128 sample frames will have an inherent latency of 2.9 milliseconds. Only the outer buffers add latency, the spin/spout do not contribute to it (but their size will limit how small the IO buffers can be, as these cannot be less than 1 ksmps). Very low-latency buffers might cause samples to be dropped out, causing clicks in the output audio (known as 'drop-outs'). This is because the system cannot cope with producing the audio in time to fill the output buffer. In general, this is what limits how smaller a buffer can be, and it will depend on the platform used. Experimentation is often required to achieve low-latency audio.

In addition, drop-outs will also occur when the processing requested exceeds the computing capacity of the platform. In that case, increasing the amount of buffering will not eliminate the issues. Measures will need to be taken to reducing the computation load. These might include reducing the sr, increasing ksmps, simplifying the instrument code, reducing the number of allowed parallel instrument instances (limiting polyphony in MIDI-based performances) or a combination of these.

If present, the portaudio backend is used by default. On Linux, the alsa plug-in is loaded if portaudio is not found, otherwise a dummy IO module is employed. This does not output any audio, but runs the synthesis system under a timer, mimicking the behaviour of a sound card. For all of Csound's IO modules, the `-odac` and `-iadc` options will open the default playback and record devices. In the following sections, we will discuss the details of each one of the main backends provided by Csound.

Portaudio

The portaudio plug-in comes in two different forms, blocking and non-blocking (callback-based). The former employs a simple method to place the audio out of Csound: a subroutine is called, which writes the audio to the soundcard, blocking execution until the operation is finished. It works in a similar fashion with input audio. Both operations operate on the software buffer, which is read/written as a block to/from the system. A performance price is paid for all this simplicity: in general it is not possible to employ small buffers for low latency (drop outs resulting) with blocking operation.

The non-blocking mode of the portaudio plug-in performs much better with regards to latency. It works in asynchronous mode: the audio is written to the soundcard inside a callback routine. This is invoked by the system whenever new data needs to be sent to the output or copied from the input. In parallel to this, Csound does its processing, consuming/filling a buffer that will be filled/consumed by the callback. This buffer has the same size as Csound's software buffer, from which it gets its data, and to which it sends its samples.

Latency in both modes is dependent on the size of the software buffer (`-b` option), which is set in sample frames. The hardware buffer size (`-B`) is only used to suggest a latency (in samples) to the portaudio library. The relevant options for loading this module in Csound are

```
-+rtaudio=pa_cb
-+rtaudio=pa_bl
```

for callback and blocking modes, respectively. Specific audio devices in this backend can be accessed as `dac0`, `dac1` etc. (for output), and `adc0`, `adc1` etc. (for input).

Jack

The Jack IO kit is an audio system that allows program inputs and outputs to be interconnected. It can also provide a low-latency route to the soundcard. Jack works as a server with any number of clients. Through the use of a dedicated plug-in, Csound can be used as a client to the system. This uses a callback system with circular buffering, whose size is set to $-B$ samples. The software buffer writes $-b$ samples to this. The circular buffer needs to be at least twice the size of this buffer, e.g. -b 128 -B 256. The software buffer also cannot be smaller than the Jack server buffer.

This module can be loaded with

```
-+rtaudio=jack
```

Different audio destinations and sources can be selected with `dac:<destination>` or `adc:<source>`, which will depend on the names of the desired Jack system devices.

ALSA

ALSA is the low-level audio system on the Linux platform. It can provide good performance with regards to latency, but it is not as flexible as Jack. It is generally single-client, which means that only one application can access a given device at a time. Csound will read/write $-b$ samples from/to the soundcard at a time, and it will set the ALSA hardware buffer size to $-B$ samples. The rule of thumb is to set the latter to twice the value of the former (as above, with Jack).

The alsa module is selected with the option:

```
-+rtaudio=alsa
```

The destinations and sources are selected by name with the same convention as in the jack module: `dac:<destination>` or `adc:<source>`, which will refer to the specific ALSA device names.

AuHAL

The AuHAL module provides a direct connection to OSX's CoreAudio system. It is also based on a callback system, using a circular buffer containing $-B$ sample frames, with a software buffer size set to $-b$ sample frames. As with Jack, the optimal configuration is to set the former to twice the value of the latter. The AuHAL backend allows very small buffer sizes, which can be used for very low latencies.

The module is loaded with

```
-+rtaudio=auhal
```

As with portaudio, specific audio devices can be accessed as `dac0`, `dac1`, etc (for output), and `adc0`, `adc1` etc. (for input).

2.8.4 Offline Audio

Audio can be computed offline by Csound. In this case, we do not have the constraints imposed by real-time operation, such as latency and computing capacity. In addition, the software will use all the processing available to generate its output as quickly as possible. This can be very useful for post-processing of recorded audio, or for composing fixed-media pieces. As noted in Chapter 1, originally all music programming systems operated purely offline, and Csound was originally designed for this type of use. As it evolved into the modern system, it kept this working model intact as its default option.

While buffering is still employed in this mode, it is less critical. The -b and -B options still refer to software and hardware (disk) buffer sizes, but its default values (1,024 and 4,096, respectively) are valid across all platforms and rarely need to be modified. The soundfile interface used by Csound is based on the third-party *libsndfile* library, which is the standard across open-source systems. Thus, Csound will be able to deal with all file formats supported by that library, which extend from the major uncompressed types (RIFF-Wave, AIFF etc.) to the ones employing compressed data (ogg, FLAC etc.).

As mentioned above, in offline mode, Csound's main software buffers will read or write to disk files. The input will draw audio from a file until it reaches the end-of-file, and will be silent thereafter. This reading begins at the same time as Csound starts processing audio and is independent from any instrument instance, because it feeds the main buffer, not a particular instance. So if an instrument uses the main inputs, it will start processing the input file at the time it is instantiated, and not necessarily at the beginning of the file (unless the instance is running from time zero). Similarly, the output is directed to a file and this is only closed when Csound finishes performance (or we run out of disk space). The relevant options for offline audio are listed in Table 2.1. These include the various possibilities for output soundfile format, and their different encoding precision settings.

2.9 Csound Utilities

The Csound software distribution also includes various utility routines that implement spectral analysis, noise reduction, soundfile amplitude scaling and mixing, and other operations. These can be accessed via the -U option:

```
-U <utility name> <arguments>
```

where the arguments will depend on the particular utility being used. Frontend hosts have full access to these via the API, with some of them providing graphical interfaces to these routines.

Option	Description
`-i FILE, --input=FILE`	input soundfile name
`-o FILE, --output=FILE`	output soundfile name
`-8, --format=uchar`	precision is set to 8-bit (unsigned)
`-c, --format=schar`	precision is set to 8-bit (signed)
`-a, --format=alaw`	a-law compressed audio format
`-u, --format=ulaw`	u-law compressed audio format
`-s, --format=short`	precision is set to 16-bit (integer)
`-3, --format=24bit`	precision is set to 24-bit, with formats that support it
`-l, --format=long`	precision is set to 32-bit (integer), with formats that support it
`-f, --format=float`	precision is set to single-precision (32-bit) floating point, with formats that support it
`--format=double`	precision is set to double-precision (64-bit) floating point, with formats that support it
`-n, --nosound`	no sound, bypasses writing of sound to disk
`-R, --rewrite`	continually rewrite the header while writing the soundfile (WAV/AIFF formats)
`-K, --nopeaks`	do not generate any PEAK chunks, in formats that support this
`-Z, --dither--triangular, --dither--uniform`	switch on dithering of audio conversion UDO from internal floating point to 32-, 16- and 8-bit formats. In the case of -Z the next digit should be a 1 (for triangular) or a 2 (for uniform)
`-h, --noheader`	no header in soundfile, just audio samples
`-W, --wave, --format=wave`	use a RIFF-WAV format soundfile
`-A, --aiff, --format=aiff`	use an AIFF format soundfile
`-J, --ircam, --format=ircam`	use an IRCAM format soundfile
`--ogg`	use the ogg compressed file format
`--vbr-quality=X`	set variable bit-rate quality for ogg files
`--format=type`	use one of the libsndfile-supported formats. Possibilities are: `aiff, au, avr, caf, flac, htk, ircam, mat4, mat5, nis, paf, pvf, raw, sd2, sds, svx, voc, w64, wav, wavex, xi`

Table 2.1 Offline audio options. These include the various possibilities for output soundfile format, and their different sample precision settings

2.10 Environment Variables

Csound takes notice of a number of environment variables. These are used to configure a system, and can be set by the user to define certain default file locations, etc. Environment variables are system-dependent, and the methods to set them will depend on the operating system. In UNIX-like systems, these can be set in shell configuration files such as .profile, .bashrc (on the bash shell) or .cshrc (C shell) using commands such as export or setenv. Generally these will only affect any programs run from the shell. Some OSs allow the user to set these environment vars for the whole system. GUI frontends, such as CsoundQt, will also allow these to be set inside the program as part of their configuration options.

The environment variables used by Csound are:

* OPCODE6DIR64 and OPCODE6DIR: these indicate the place where Csound will look for plug-ins, in double-precision or single-precision versions of Csound. In fully installed systems, they do not need to be set, as Csound will look for plug-ins in their default installed places (system-dependent).
* SFDIR: soundfile directory, where Csound will look for soundfiles by default. If not set, Csound will look for soundfiles in the current directory (which can be the directory where the CSD file is located). Note that soundfiles (and other files) can also be passed to Csound with their full path, in which case SFDIR is not used.
* SSDIR: sound sample directory, where Csound will look for sound samples by default. As above, if not set, Csound will look for sound samples in the current directory (which can be the directory where the CSD file is located).
* SADIR: sound analysis directory, similar to the above, but for sound analysis files.
* INCDIR: include directory, where Csound will look for text files included in the code (with the #include preprocessor directive).

2.10.1 Configuration File

A file named .csoundrc can be used by Csound to hold default options for your system. It should reside in the user home directory (the topmost user directory), and contain any options that the user wants to keep as default, in a single line, such as:

```
-o dac -i adc -b 128 -B 512
```

These options will be used whenever Csound is run without them. If the same options are already passed to a frontend, either in the CSD or as command-line parameters, the configuration file will be ignored (the order of precedence is: parameters to program; CSD options; configuration file options).

2.11 The Csound API

The API underpins the operation of all Csound-based applications. While the main details of its operation are beyond the scope of this book, we will introduce it here in order to give the reader an insight into the lower levels of operation inside the software. An API is the public face of a programming library, which is a component of an application that provides support for specific tasks. Developers would like to use software libraries for almost all of their work, as they save the need to reinvent every step of each fundamental operation they want to use. So these components are pre-packaged bundles of functionality that can be reused whenever they are required.

Csound was originally a monolithic program, whose only interfaces were those provided by its language, and a few extra controls (MIDI and events typed at the terminal or piped from another program). It evolved into a software library with the exposure of some of its internal operations in an early form of the API. From this stage, it was then more formally developed as a library, and the functionality exposed by the API was enhanced. One key aspect of these changes is that the internals were modified to make the library *reentrant*. This allowed the Csound engine to be treated like an object, which could be instantiated multiple times. So a program using the library, such as MaxMSP or PD, or a DAW, can load up several copies of a Csound-based plug-in without them interfering with each other.

The Csound API is written in the C language, but is also available in C++ through a very thin layer. From these two basic forms, various language wrappers have been created for Java (via its Java Native Interface, JNI), Python, Lua, Closure (via JNI), and others. So programmers can access the functionality via their preferred language. In particular, scripting languages interface very well with Csound through the API, allowing composers to make use of extended possibilities for algorithmic approaches. This will be explored in later sections of this book.

2.11.1 A Simple Example

The API can be used to demonstrate the operation stages of Csound, exposing some of the internals to the reader. The presentation should be clear enough even for non-programmers. We show here a simple example in C++ that will guide us through the steps from start to completion. This code implements a command-line program that is very similar to the `csound` frontend discussed in Section 2.3. It takes input from the terminal, runs the audio engine, and closes when the performance is finished. The program code is shown in listing 2.10.

Listing 2.10 Simple Csound API example in C++

```
1  #include <csound.hpp>
2
3  int main(int argc, char** argv){
```

```
4    Csound  csound; // csound object
5    int  error;      // error code
6
7    // compile CSD and start the engine
8    error = csound.Compile(argc, argv);
9
10   // performance loop
11   while(!error)
12       error = csound.PerformKsmps();
13
14   return 0;
15 }
```

This example uses the bare minimum functionality, but demonstrates some important points. First, in a C++ program, Csound is a class, which can be instantiated many times. In line 4, we see one such object, csound, which represents the audio engine. This can be customised through different options etc, but in this simple program, we move straight to the next step (line 8). This takes in the command-line arguments and passes them to the engine. The parameters argv and argc contain these arguments, and how many of them there are, respectively. This allows Csound to compile a CSD code (or not, depending on the options), and once this is done, start. Note that these are other forms of the Compile() method, as well as other methods to send orchestras for compilation, and other ways of starting the engine. This is the most straightforward of them.

If there were no errors, the program enters the *performance loop*, which has been discussed in some detail in Section 2.6. This does all the necessary processing to produce one ksmps of audio, and places that into the output buffer. The method returns an error code, which is used to check whether the loop needs to continue to the next iteration. The end of performance can be triggered in a number of ways, for instance, if the end of the numeric score is reached; via a 'close csound' event; or with a keyboard interrupt signalling KILL (ctrl-c). In this case, PerformKsmps() returns a non-zero code, and the program closes. The C++ language takes care of all the tidying up that is required at the end, when the csound object is destroyed.

To complete this overview, we show how the C++ code can be translated into a scripting language, in this case, Python (listing 2.11). We can very easily recognise that it is the same program, but with a few small changes. These are mainly there to accommodate the fact that C/C++ pointers generally do not exist outside these languages. So we need an auxiliary object to hold the command-line argument list and pass it to Compile() in a form that can be understood. These small variations are inevitable when a C/C++ API is wrapped for other languages. Otherwise, the code corresponds almost on a line-by-line basis.

Listing 2.11 Simple Csound API example in Python

```
1  import csnd6
2  import sys
3
```

```
 4  csound = csnd6.Csound()
 5  args   = csnd6.CsoundArgVList()
 6  for arg in sys.argv: args.Append(arg)
 7
 8  error = csound.Compile(args.argc(), args.argv())
 9
10  while (not error):
11      error = csound.PerformKsmps()
```

2.11.2 Levels of Functionality

The example above shows a very high level use of the API. Depending on the application, much lower-level access can be used. The whole system is fully configurable. The API allows new opcodes and GEN routines to be added, new backend plug-ins to be provided for audio, MIDI and utilities. We can access each sample that is computed in the spout and output buffers. It is also possible to fill the input and spin buffers with data via the API. Each aspect of Csound's functionality is exposed. The following sections provide an introduction to each one of the areas covered by the interface.

Instantiation

The fundamental operations of the API are to do with setting up and creating instances of Csound that can be used by the host. It includes functions for the initialisation of the library, creation and destruction of objects, and ancillary operations to retrieve the system version.

Attributes

It is possible to obtain information on all of the system attributes, such as sr, kr, ksmps, nchnls and 0dbfs. There are functions to set options programmatically, so the system can be fully configured. In addition, it is possible to get the current time during performance (so that, for instance, a progress bar can be updated).

Compilation and performance

The API has several options for compiling code. The simplest takes in command-line parameters and compiles a CSD ready for performance as seen in listings 2.10 and 2.11. Other functions will take in orchestra code directly, as well as a full CSD file. At lower levels it is possible to pass to Csound a parsed tree structure, to parse code into a tree, and to evaluate a text.

Performance can be controlled at the k-cycle level (as in the examples in listings 2.10 and 2.11), at the buffer level or at the performance loop level. This means that we can process as little as one ksmps of audio, a whole buffer, or the whole performance from beginning to end, in one function call. There are also functions to start and stop the engine, to reset it and to do a clean up.

General IO

There are specific functions to set the main Csound inputs and outputs (file or device names), file format, MIDI IO devices and filenames.

Real-time audio and MIDI

The API provides support for setting up a real-time audio IO backend, for hosts that need an alternative to the plug-ins provided by the system. It is also possible to access directly the spin, spout, input and output audio buffers. Similarly, there is support for interfacing with Csound's MIDI handling system, so applications can provide their own alternatives to the MIDI device plug-ins.

Score handling

Functions for reading, sorting, and extracting numeric scores are provided. There are transport controls for offsetting and rewinding playback, as well as for checking current position.

Messages and text

It is possible to redirect any messages (warnings, performance information etc.) to other destinations. By default these go to the terminal, but the host might want to print them to a window, or to suppress them. The API also allows users to send their own text into Csound's messaging system.

Control and events

Csound has a complete software bus system that allows hosts to interface with engine objects. It is possible to send and receive control and audio data to/from specifically named channels. This is the main means of interaction between Csound and applications that embed it. In addition, it is possible to send events directly to instantiate instruments in the engine. The API allows programs to kill specific instances of instruments. It can register function callbacks to be invoked on specific key presses, as well as a callback to listen for and dispatch events to Csound.

Tables

Full control of tables is provided. It is possible to set and get table data, and to copy in/out the full contents from/to arrays. The size of a function table is also accessible.

Opcodes

The API allows hosts to access a list of existing opcodes, so they can, for instance, check for the existence of a given unit generator, or printout a list of these. New opcodes can also be registered with the system, allowing it to be easily extended.

Threading and concurrency

Csound provides auxiliary functionality to support concurrent processing: thread creation, spinlocks, mutexes, barriers, circular buffers, etc. These can be used by hosts to run processing and control in parallel.

Debugger

An experimental debugger mode has been developed as part of the Csound library. The API gives access to its operation, so that breakpoints can be set and performance can be stepped and paused for variable inspection.

Miscellaneous

A number of miscellaneous functions exist in the API, mostly to provide auxiliary operations in a platform-independent way. Examples of these are library loading, timers, command running and environmental variable access. Csound also allows global variables to be created in an engine object, so that data can be passed to/from functions that have access to it. Finally, a suite of utilities that come with the system can also be manipulated via the API.

2.12 Conclusions

In this chapter, we introduced a number of key concepts that are relevant to Csound. They were organised in a mosaic form, and presented at a high level, in order to give a general overview of the system. We began by looking at how the system operates, and discussed CSD files, numeric scores and options. The principle of Csound as a library, hosted by a variety of frontends, was detailed. The `csound` command was shown to be a basic but fully functional frontend.

Following this system-wide introduction, we looked at fundamental aspects of digital audio, and how they figured in Csound. Sampling rate and quantisation were discussed, and the conditions for high-quality sound were laid out. We also discussed the concept of control rate, k-cycles and ksmps blocks, which are essential for the operation of the software.

The text also described in some detail concepts relating to instruments, instances and events. We outlined the life-cycle of an instrument from its text code form to compilation, instantiation, performance and destruction. The basic principle of function tables was explored, with a description of the typical GEN routines that Csound uses to create these.

Completing the overview of fundamental system operation, we looked at audio input and output. The two main modes of operation, real-time and offline, were introduced. The principle of audio IO plug-ins that work as the backend of the system was outlined. This was followed by the description of the common audio modules included in Csound. Offline audio and soundfile options were detailed.

The chapter closed with an overview of the Csound API. Although the details of its operation are beyond the scope of this book, we have explored the key aspects that affect the use of the system. A programming example in C++ and Python was presented to show how the high-level operation of Csound can be controlled. To conclude, the different levels of functionality of the API were discussed briefly. This chapter concludes the introductory part of this book. In the next sections, we will explore Csound more deeply, beginning with a study of its programming language.

Part II
The Language

Chapter 3
Fundamentals

Abstract This chapter will introduce the reader to the fundamental aspects of
Csound programming. We will be examining the syntax and operation of instruments, including how statements are executed and parameters passed. The principle
of variables and their associated data types will be introduced with a discussion
of update rates. The building block of instruments, *opcodes*, are explored, with a
discussion of traditional and function-like syntax. Selected key opcodes that implement fundamental operations are detailed to provide an introduction to synthesis
and processing. The Csound orchestra preprocessor is introduced, complementing
this exploration of the language fundamentals.

3.1 Introduction

The Csound language is domain specific. In distinction to a general-purpose language, it attempts to target a focused set of operations to make sound and music,
even though it has capabilities that could be harnessed for any type of computation,
and it is a *Turing-complete* language [29]. Because of this, it has some unique features, and its own way of operating. This chapter and the subsequent ones in Part II
will introduce all aspects of programming in the Csound language, from first principles. We do not assume any prior programming experience, and we will guide the
reader with dedicated examples for each aspect of the language. In this chapter, we
will start by looking at the fundamental elements of the system, and on completion,
we will have covered sufficient material to enable the creation of simple synthesis
and processing programs.

The code examples discussed here can be typed using any plain text (ASCII)
editor, in the `<CsInstruments>` section of a CSD file (see Section 2.2). Csound
code is case sensitive, i.e. `hello` is not the same as `Hello`. It is important that only
ASCII text, with no extra formatting data, is used, so users should avoid using word
processors that might introduce these. Alternatively, users can employ their Csound

© Springer International Publishing Switzerland 2016   53
V. Lazzarini et al., *Csound*, DOI 10.1007/978-3-319-45370-5_3

IDE/frontend of choice. The examples presented in this chapter are not dependent on any particular implementation of the system.

3.2 Instruments

The basic programming unit in Csound is the *instrument*. In Part I, we have already explored the history of the concept, and looked at some operational aspects. Now we can look closely at its syntax. Instruments are defined by the keywords `instr` and `endin`. The general form is shown in listing 3.1.

Listing 3.1 Instrument syntax

```
instr <id>[,<id2>, ...]

endin
```

An instrument has an identification name, `<id>` in listing 3.1, which can be either numeric or textual. Optionally, as indicated in square braces in the example, it can have alternative names. Csound allows an unspecified number of instrument definitions, but each instrument in a single compilation has to have unique names. In subsequent compilations, if an existing name is used, it replaces the old definition. The keyword `endin` closes an instrument definition. They cannot be nested, i.e. one instrument cannot be defined inside another. Both `instr` and `endin` need to be placed on their own separate lines. In order to perform computation, an event needs to be scheduled for it. The opcode `schedule` can be used for this purpose, and its syntax is

```
schedule(id,istart,idur, ...)
```

where `id` is the instrument name, `istart` the start time and `idur` the duration of the event. Times are defined in seconds by default. Extra parameters to instruments may be used.

3.2.1 Statements

Csound code is made of a series of sequentially executed statements. These will involve either a mathematical expression, or an opcode, or both. Statements are terminated by a newline, but can extend over multiple lines in some cases, where multiple opcode parameters are used. In this case, lines can be broken after a comma, a parenthesis or an operator. Execution follows line order, from the topmost statement to `endin`. Any number of blank lines are allowed, and statements can start at any column. For code readability,we recommend to indent by one or more columns any code inside instruments (and keep the statements aligned to this). Listing 3.2 shows an example of three legal statements in a instrument, which is then scheduled to run.

Listing 3.2 Statements in an instrument

```
instr 1
 print 1
 print 1+1
 print 3/2
endin
schedule(1,0,1)
```

This instrument will print some numbers to the console, in the order the statements are executed:

```
SECTION 1:
new alloc for instr 1:
instr 1:   1 = 1.000
instr 1:   #i0 = 2.000
instr 1:   #i1 = 1.500
```

The `print` opcode prints the name of the argument (which can be a *variable*, see Sec. 3.3) and its value. The first print statement has 1 as argument, and the subsequent ones are expressions, which get calculated and put into synthetic (and hidden) variables that are passed as arguments (`#i0`, `#i1`). It is possible to see that the statements are executed in line order.

3.2.2 Expressions

Csound accepts expressions of arbitrary size combining constants, variables (see Sec. 3.3) and a number of arithmetic operators. In addition to the ordinary addition, subtraction, multiplication and division (`+`, `−`, `*` and `/`), there is also exponentiation (`^`) and modulus (or remainder, `%`). Normal precedence applies, from highest to lowest exponentiation: multiplication, division and modulus; addition and subtraction. Operators at the same level will bind left to right. Operations can be grouped with parentheses, as usual, to control precedence. The following operators are also available for short hand expressions:

```
a += b   => a = a + b
a -= b   => a = a - b
a *= b   => a = a * b
a /= b   => a = a / b
```

3.2.3 Comments

Csound allows two forms of comments, single and multiple line. The former, which is the traditional way of commenting code, is demarcated by a semicolon (;) and runs until the end of the line. A multiple-line comment is indicated by the /* and */ characters, and includes all text inside these. These comments cannot be nested. Listing 3.3 shows an example of these two types of comment. Csound can also use C++ style single-line comments introduced by //.

Listing 3.3 Comments in an instrument

```
instr 1
 /*
    the line below shows a
    statement and a single-line comment.
 */
 print 1 ; this is a print statement
 ; print 2  ; this never gets executed
endin
schedule(1,0,1)
```

When this code is run, as expected, only the first statement is executed:

```
SECTION 1:
new alloc for instr 1:
instr 1:  1 = 1.000
```

3.2.4 Initialisation Pass

The statements in an instrument are actually divided into two groups, which get executed at different times. The first of these is composed of the *initialisation* statements. They are run first, but only once, or they can be repeated explicitly via a reinitialisation call. These statements are made up of expressions, init-time opcodes, or both. In the example shown in listings 3.2 and 3.3, the code consists solely of init-time statements, which, as shown by their console messages, are only executed once.

3.2.5 Performance Time

The second group of statements are executed during *performance* time. The main difference is that these are going to be iterated, i.e. repeated, for the duration of the event. We call a repetitive sequence of statements a *loop*. In Csound, there is a fundamental operating loop that is implicit to performance-time statements in an instrument, called the k-cycle. It repeats at $\frac{1}{kr}$ seconds of output audio (*kr* is the control rate, see Chapter 2), which is called a *k-period*. Performance-time statements are executed after init-pass, regardless of where they are in an instrument. Listing 3.4 shows an example.

Listing 3.4 Performance-time and init-pass statements

```
instr 1
 printk 0, 3
 print 1
 print 2
 printk 0, 4
endin
schedule(1,0,1)
```

If we run this code, we will see the messages on the console show the two init-time statements first, even though they come after the first `printk` line. This op-code runs at performance time, printing its second argument value at regular times given by the first. If this is 0, it prints every k-period. It reports the times at which it was executed:

```
SECTION 1:
new alloc for instr 1:
instr 1:   1 = 1.000
instr 1:   2 = 2.000
 i   1 time       0.00023:     3.00000
 i   1 time       0.00023:     4.00000
 i   1 time       0.00045:     3.00000
 i   1 time       0.00045:     4.00000
 ...
```

For clarity of reading, sometimes it might be useful to group all the init-pass statements at the top of an instrument, and follow that with the performance-time ones. This will give a better idea of the order of execution (see listing 3.4)

Listing 3.5 Performance-time and init-pass statements, ordered by execution time

```
instr 1
 print 1
 print 2
 printk 0, 3
 printk 0, 4
endin
```

Note that many opcodes might produce output at performance time only, but they are actually run at initialisation time as well. This is the case with unit generators that need to reset some internal data, and run single-pass computation, whose results are used at performance time. This is the case of the majority of opcodes, with only a small set being purely perf-time. If, for some reason, the init-time pass is bypassed, then the unit generators might not be initialised properly and will issue a performance error.

3.2.6 Parameters

An instrument can be passed an arbitrary number of parameters or arguments. Minimally, it uses three of these, which are pre-defined as instrument number (name), start time and duration of performance. These are parameters 1, 2 and 3 respectively. Any additional parameters can be used by instruments, so that different values can be set for each instance. They can be retrieved using the p(x) opcode, where x is the parameter number. An example of this is shown in listing 3.6.

Listing 3.6 Instrument parameters

```
instr 1
 print p(4)
 print p(5)
endin
schedule(1,0,1,22,33)
schedule(1,0,1,44,55)
```

This example runs two instances of instrument 1 at the same time, passing different parameters 4 and 5. The console messages show how the parameters are set per instance:

```
SECTION 1:
new alloc for instr 1:
instr 1:   #i0 = 22.000
instr 1:   #i1 = 33.000
new alloc for instr 1:
instr 1:   #i0 = 44.000
instr 1:   #i1 = 55.000
```

3.2.7 Global Space Code

Code that exists outside instruments is deemed to be in *global space*. It gets executed only once straight after every compilation. For this reason, performance-time code is not allowed at this level, but any i-pass code is. In the examples above, we have shown how the schedule opcode is used outside an instrument to start events. This is a typical use of global space code, and it is perfectly legal, because the operation of schedule is strictly i-time. It is also possible to place the system constants (sr, kr, ksmps, 0dbfs, nchnls, and nchnls_i) in global space, if we need to override the defaults, but these are only effective in the first compilation (and ignored thereafter). Global space has been called *instr 0* in the past, but that term does not define its current uses appropriately.

3.3 Data Types and Variables

Data types are used to distinguish between the different objects that are manipulated by a program. In many general-purpose languages, types are used to select different numeric representations and sizes, e.g. characters, short and long integers, single- and double-precision floating-point numbers. In Csound all numbers are floating point (either single- or double-precision, depending on the platform). The fundamental distinction for simple numeric types is a different one, based on *update rates*. As we have learned above, code is executed in two separate stages, at initialisation and performance times. This demands at least two data types, so that the program can distinguish what gets run at each of these stages. Furthermore, we have also seen in Chapter 2 that Csound operates with two basic signal rates, for control and audio. The three fundamental data types for Csound are designed to match these principles: one init-pass and two perf-time types.

We give objects of these types the name *variables*, or sometimes, when referring to perf-time data, *signals*. In computing terms, they are memory locations that we can create to store information we computed or provided to the program. They are given unique names (within an instrument). Names can be arbitrary, but with one constraint: the starting letter of a variable name will determine its type. In this case, we will have i for init time, k for control rate, and a for audio rate (all starting letters are lower case). Some examples of variable names:

```
i1, ivar, indx, icnt
k3, kvar, kontrol, kSig
a2, aout, aSig, audio
```

Before variables are used as input to an expression or opcode, they need to be declared by initialising them or assigning them a value. We will examine these ideas by exploring the three fundamental types in detail.

3.3.1 Init-Time Variables

Csound uses the i-var type to store the results of init time computation. This is done using the assignment operator (=), which indicates that the left-hand side of the operator will receive the contents of the right-hand side. The computation can take the form of opcodes or expressions that work solely at that stage. In the case of expressions, only the ones containing constants and/or i-time variables will be executed at that stage. The code in listing 3.7 shows an example of i-time expressions, which are stored in the variable ires and printed. Note that variables can be reused liberally.

Listing 3.7 Using i-time variables with expressions

```
instr 1
 ires = p(4) + p(5)
 print ires
 ires = p(4)*p(5)
 print ires
endin
schedule(1,0,1,2,3)
```

In the case of opcodes, some will operate at i-time only, and so can store their output in an i-variable (the Reference Manual can be consulted for this). For example, the opcode date, which returns the time in seconds since the Epoch (Jan, 1 1970) is an example of this.

Listing 3.8 Using i-time variables with opcodes

```
instr 1
 iNow date
 print iNow
endin
schedule(1,0,0)
```

This example will print the following to the console:

```
SECTION 1:
new alloc for instr 1:
instr 1:   iNow = 1441142199.000
```

Note that the event duration (p3) can be set to 0 for i-time-only code, as it will always execute, even if performance time is null.

A special type of i-time variable, which is distinct to i-vars, is the p-type. These are used to access instrument parameter values directly: p1 is the instrument name, p2 its start time, p3 the duration and so on. Further pN variables can be used to access other parameters used. This is an alternative to using the p() opcode, but it also allows assignment at i-time. So an instrument can modify its parameters on the fly. In the case of p1 and p2, this is meaningless, but with p3, for instance, it is possible to modify the event's duration. For instance, the line

```
p3 = 10
```

makes the instrument run for 10 seconds, independently of how long it was scheduled for originally. The p-type variables can be used liberally in an instrument:

Listing 3.9 Using p-variables

```
instr 1
 print p4 + p5
endin
schedule(1,0,1,2,3)
```

3.3.2 Control-Rate Variables

Control-rate variables are only updated at performance time. They are not touched at init-time, unless they are explicitly initialised. This can be done with the `init` operator, which operates only at i-time, but it is not always necessary:

```
kval init 0
```

At performance time, they are repeatedly updated, at every k-cycle. Only expressions containing control variables will be calculated at this rate. For instance,

```
kval = 10
kres = kval*2
```

will execute at every k-cycle, whereas the code

```
ival = 10
kres = ival*2
```

will be executed mostly at i-time. The first line is an assignment that will be run at the i-pass. The second is an expression involving an i-var and a constant, which is also calculated at i-time, and the result stored in a synthetic i-var. The assignment happens at every k-cycle. Synthetic variables are created by the compiler to store the results of calculations, and are hidden from the user.

The i-time value of a k-var can be obtained using the `i()` operator:

```
ivar = i(kvar)
```

This only makes sense if the variable has a value at that stage. There are two situations when this can happen with instrument variables:

1. the k-var has been initialised:

   ```
   kvar init 10
   ivar = i(kvar)
   ```

2. the current instance is reusing a slot of an older instance. In that case, the values of all variables at the end of this previous event are kept in memory. As the new instance takes this memory space up, it also inherits its contents.

There is no restriction in assigning i-var to k-var, as the former is available (as a constant value) throughout performance. However, if an expression is calculated only at i-time, it might trigger an opcode to work only at i-time. Sometimes we need it to be run at the control rate even if we are supplying an unchanging value. This is the case, for instance, for random number generators. In this case, we can use the `k()` converter to force k-time behaviour.

Listing 3.10 Producing random numbers at i-time

```
instr 1
 imax = 10
 printk 0.1, rnd(imax)
```

```
endin
schedule(1,0,1)
```

For instance, the code on listing 3.10, using the `rnd()` function to produce random numbers, produces the following console output:

```
i    1 time      0.00023:      9.73500
i    1 time      0.25011:      9.73500
i    1 time      0.50000:      9.73500
i    1 time      0.75011:      9.73500
i    1 time      1.00000:      9.73500
```

This is a fixed random number calculated at i-time.

Listing 3.11 Producing random numbers at k-rate

```
instr 1
 imax = 10
 printk 0.1, rnd(k(imax))
endin
schedule(1,0,1)
```

If we want it to produce a series of different values at control rate, we need the code in listing 3.11, which will print:

```
i    1 time      0.00023:      9.73500
i    1 time      0.25011:      9.10928
i    1 time      0.50000:      3.58307
i    1 time      0.75011:      8.77020
i    1 time      1.00000:      7.51334
```

Finally, it is important to reiterate that k-time variables are signals, of the control type, sampled at `kr` samples per second. They can also be called *scalars*, as they contain only a single value in each k-period.

3.3.3 Audio-Rate Variables

Audio-rate variables are also only updated at performance time, except for initialisation (using `init`, as shown above). The main difference to k-vars is that these are *vectors*, i.e. they contain a block of values at each k-period. This block is `ksmps` samples long. These variables hold audio signals sampled at `sr` samples per second.

Expressions involving a-rate variables will be iterated over the whole vector every k-period. For instance, the code

```
a2 = a1*2
```

will loop over the contents of `a1`, multiply them by 2, and place the results in a2, `ksmps` operations every k-period. Similarly, expressions involving k- and a-vars will be calculated on a sample-by-sample basis. In order to go smoothly from a scalar to a vector, k-rate to a-rate, we can interpolate using the `interp` opcode, or the `a()` converter. These interpolate the k-rate values creating a smooth line between them, placing the result in a vector. This *upsamples* the signal from *kr* to *sr* samples per second. It is legal to assign a k-rate variable or expression directly to an a-rate, in which case the whole vector will be set to a single scalar value. The `upsamp` opcode does this in a slightly more efficient way.

With audio-rate variables we can finally design an instrument to make sound. This example will create a very noisy waveform, but it will demonstrate some of the concepts discussed above. The idea is to create a ramp that goes from $-A$ to A, repeating at a certain rate to give a continuous tone. To do this, we make the audio signal increment every k-cycle by *incr*, defined as:

$$incr = \frac{1}{kr} \times f_0, \qquad (3.1)$$

recalling that $\frac{1}{kr}$ is one k-period. If $f_0 = 1$, starting from 0 we will reach 1 after `kr` cycles, or 1 second. The ramp carries on growing after that, but if we apply a modulo operation (`%1`), it gets reset to 0 when it reaches 1. This makes the ramp repeat at 1 cycle per second (Hz). Now we can set f_0 to any frequency we want, to change the pitch of the sound.

The ramp goes from 0 to 1, but we want to make it go from $-A$ to A, so we need to modify it. First, we make it go from -1 to 1, which is twice the original range, starting at -1. Then we can scale this by our target A value. The full expression is:

$$out = (2 \times ramp - 1) \times A \qquad (3.2)$$

The resulting instrument is shown in listing 3.12. We use the `out` opcode to place the audio in the instrument output. Two parameters, 4 and 5 are used for amplitude (A) and frequency, respectively. Note the use of the `a()` converter to smooth out the ramp. Without it, the fixed value from the i-time expression `(1/kr)*p5` would be used. The converter creates a vector containing a ramp from 0 to its argument, which is scaled by p5 and added to the `aramp` vector, creating a smoothed, rather than a stepped output (see Fig. 3.1). The other statements translate the other elements discussed above. The pitch is set to A 440 Hz, and the amplitude is half-scale (the constant `0dbfs` defines the full scale value).

Listing 3.12 A simple sound synthesis instrument

```
instr 1
 aramp init 0
 out((2*aramp-1)*p4)
 aramp += a(1/kr)*p5
 aramp = aramp%1
endin
schedule(1,0,10,0dbfs/2,440)
```

This example shows how we can apply the principles of audio and control rate, and some simple mathematical expressions to create sounds from scratch. It is noisy because the ramp waveform is not bandlimited and causes aliasing. There are more sophisticated means of creating similar sounds, which we will examine later in this book.

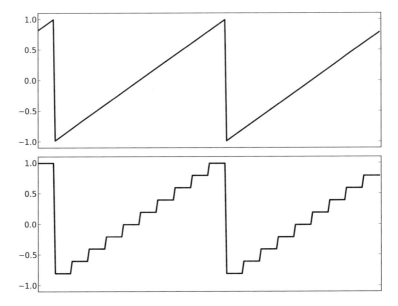

Fig. 3.1 Two plots comparing ramp waves generated with the control to audio-rate converter a () (top), and without it (bottom). The typical control-rate stepping is very clear in the second example

Finally, audio variables and expressions cannot be assigned directly to control variables. This is because we are going from many values (the vector), to a single one (a scalar). For this, we can apply the k () converter, which *downsamples* the signal from *sr* to *kr* samples per second.[1]

3.3.4 Global Variables

All variables looked at so far were *local* to an instrument. That means that their *scope* is limited to the instrument where they appear, and they cannot be seen outside it. Csound allows for *global* variables, which, once declared, exist outside instruments and can be seen by the whole program. These can be first declared in an instrument or in global space, remembering that only init-pass statements can be used outside instruments. In this case, the declaration uses the init opcode. To define a variable as global, we add a g to its name:

---

[1] Other converters also exist, please see the Reference Manual for these.

```
gi1, givar, gindx, gicnt
gk3, gkvar, gkontrol, gkSig
ga2, gaout, gaSig, gaudio
```

Only one copy of a global variable with a given name will exist in the engine, and care should be taken to ensure it is handled properly. For instance, if an instrument that writes to a global audio signal stops playing, the last samples written to it will be preserved in memory until this is explicitly cleared. This can lead to garbage being played repeatedly if another instrument is using this variable in its audio output.

Global variables are useful as a means of connecting signals from one instrument to another, as patch-cords or busses. We will examine this functionality later on in Chapter 6.

3.4 Opcodes

The fundamental building block of an instrument is the *opcode*. They implement the unit generators that are used for the various functional aspects of the system. It is possible to write code that is largely based on mathematical expressions, using very few opcodes (as shown in listing 3.12), but that can become complex very quickly. In any case, some opcodes always have to be present, because they handle input/output (IO) and events, and although we have not introduced the idea formally, we have used `print`, `printk` and `out` for data output, and `schedule` for events.

3.4.1 Structure

An opcode has a structure that is composed of two elements:

1. **state**: internal data that the opcode maintains throughout its life cycle
2. **subroutines**: the code that is run at initialisation and/or performance time. Generally, an opcode will have distinct subroutines for these execution stages.

We can think of opcodes as black box-type (i.e. opaque, you can only see input and output arguments) instruments: recipes for performing some processing, which will only execute when instantiated. When an opcode is placed in an instrument, which is then scheduled, the engine will

1. Allocate memory for its state, if necessary.
2. Run the init-time subroutine, if it exists. This will generally modify the opcode state (e.g. initialising it).
3. Run the perf-time subroutine, if it exists, in the performance loop, repeating it at every k-period. Again, the opcode state will be updated as required every time this subroutine is run.

Each opcode that is placed in an instrument will have its own separate state, and will use the subroutines that have been assigned to it.

3.4.2 Syntax

There is a straightforward syntax for opcodes, which is made up of output variable(s) on the left-hand side of the opcode name, and input arguments (parameters) on the right-hand side. Multiple output variables and arguments are separated by commas

```
[var, ...]    opname    [arg, ...]
```

The number and types of arguments and outputs will depend on the opcode. Some opcodes might have optional inputs, and allow for a variable number of outputs. The Reference Manual provides this information for all opcodes in the system. In any case, expressions that can be evaluated to a given argument type are accepted as inputs to opcodes.

An alternative syntax allows some opcodes to be used inline inside expressions:

```
[var =]   opname[:r]([arg, ...])
```

In this case, we see that the opcode arguments, if any, are supplied inside parentheses, and its output needs to be assigned to an output variable (if needed). The whole opcode expression can now be used as one of the arguments to another opcode, or as part of an expression. If it is used on its own and it produces an output, we need to use the assignment operator (=) to store it in a variable. If used as an input to an opcode, we might need to explicitly determine the output rate (r) with :i, :k or :a after the opcode name (see Sec. 3.4.4). It is also important to note that this syntactical form is only allowed for opcodes that have either one or no outputs. Table 3.1 shows some examples of equivalence between the original and alternative syntax forms.

Table 3.1 Comparison between original and alternative syntaxes for opcodes

original	alternative
print 1	print(1)
printk 0,kvar	printk(0,kvar)
schedule 1,0,1	schedule(1,0,1)
out asig	out(asig)
a1 oscili 0dbfs, 440	a1 = oscili(0dbfs, 440)

Both forms can be intermixed in a program, provided that their basic syntax rules are observed correctly. In this book, we will use these liberally, so that the reader can get used to the different ways of writing Csound code.

3.4.3 Functions

Csound also includes a set of `functions`. Functions do not have *state* like op-codes. A function is strictly designed to produce an output in response to given in-puts, and nothing else. Opcodes, on the other hand, will store internal data to manage their operation from one execution time to another. Function syntax is similar to the second opcode syntax shown above:

```
var = function(arg[, ...])
```

The main difference is that functions will have at least one argument and produce one output. They can look very similar to opcodes in the alternative ('function-like') usage, but the fact that they operate slightly differently will be important in some situations.

A typical example is given by trigonometric functions, such as `sin()` and `cos()`. We can use these in an instrument to generate sinusoidal waves, which are a fundamental type of waveform. This is easily done by adapting the audio ex-ample in listing 3.12, where we created a ramp waveform. We can use this ramp as an input to the `sin()` function, and make it generate a sine wave. For this, we only need to keep the ramp ranging from 0 to 1, and scale it to a full cycle in radians (2π).

Listing 3.13 A sine wave synthesis instrument

```
instr 1
 i2pi = 6.28318530
 aramp init 0
 out(sin(aramp*i2pi)*p4)
 aramp += a(1/kr)*p5
 aramp = aramp%1
endin
schedule(1,0,10,0dbfs/2,440)
```

As the input to `sin()` rises from 0 to 2π, a sine waveform is produced at the output. This will be a clean sinusoidal signal, even though we are using very basic means to generate it. A plot of the output of instr 1 in listing 3.13 is shown in Fig.3.2.

3.4.4 Initialisation and Performance

Opcodes can be active at the init-pass and/or performance time. As noted before, some will be exclusively working at i-time. Usually an opcode that is used at perf-time will also include an initialisation stage that is used to set or reset its internal state to make it ready for operation. Some of them have options to skip this initiali-sation, if the state from a previously running instance is to be preserved as the new one takes over its space. There are some cases, however, where there is no activity at the init-pass. In general, we should not expect that a k-var output of an opcode

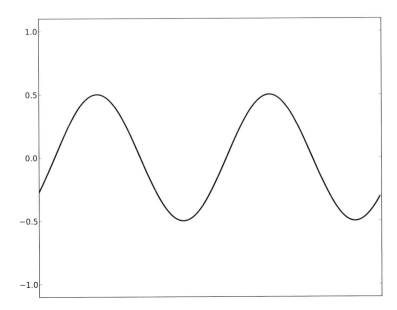

Fig. 3.2 A sinusoidal wave, the output of the instr 1 in listing 3.13

has any meaningful value at i-time, even if an opcode has performed some sort of initialisation procedure.

There are many opcodes that are *polymorphic* [40], meaning that they can have different forms depending on their output or input types. For instance, an opcode such as `oscili` will have four different implementations:

1. `k oscili k,k`: k-rate operation
2. `a oscili k,k`: a-rate operation, both parameters k-rate
3. `a oscili k,a`: first parameter k-rate, second a-rate
4. `a oscili a,k`: first parameter a-rate, second k-rate
5. `a oscili a,a`: both parameters a-rate.

These forms will be selected depending on the input and output parameters. In normal opcode syntax, this is done transparently. If we are using the alternative function-like syntax, then we might need to be explicit about the output we want to produce by adding a rate hint (`:r`, where r is the letter indicating the required form). Many functions will also be polymorphic, and specifying the correct input type is key to make them produce output of a given rate, as demonstrated for `rnd()` in listings 3.10 and 3.11.

3.5 Fundamental Opcodes

The core of the Csound language is its opcode collection, which is significantly large. A small number of these are used ubiquitously in instruments, as they perform fundamental signal processing functions that are widely used. We will discuss here the syntax and operation of these key opcodes, with various examples of how they can be employed.

3.5.1 Input and Output

A number of opcodes are designed to get audio into and out of an instrument. Depending on the number of channels, or how we want to place the audio, we can use different ones. However, it is possible to just use a few general purpose opcodes, which are

```
asig in   ; single-channel input
ain1[, ...] inch kchan1[,...]   ; any in channels
out asig[,...]   ;   any number of out channels
outch kchan1, asig1 [,...,...] ; specify out channel
```

For input, if we are using channel 1 only, we can just use `in`. If we need more input channels, and/or switch between these at k-time (using the `kchan` parameter), we can use `inch`. For audio output, `out` will accept any number of signals up to the value of `nchnls`. All the examples in this chapter used the default number of channels (one), but if we want to increase thia, we can set the system constant `nchnls`, and add more arguments to `out`. A specific output channel can be assigned with `outch`, for easy software routing. Refer to Section 2.8 for more details on the audio IO subsystem.

3.5.2 Oscillators

Oscillators have been described as the workhorse of sound synthesis, and this is particularly the case in computer music systems. It is possible to create a huge variety of instruments that are solely based on these opcodes. The oscillator is a simple unit generator designed to produce periodic waveforms. Given that many useful signals used in sound and music applications are of this type, it is possible to see how widespread the use of these opcodes can be. The oscillator works by reading a function that is stored in a table (see Section 2.7) repeatedly at a given frequency. This allows it to produce any signal based on fixed periodic shapes.

There are four operational stages in an oscillator [67]:

1. **table lookup**: read a sample of a function table at a position given by an index n.

2. **scaling**: multiply this sample by a suitable amplitude and produce an output.
3. **increment**: increment the position of index n for a given frequency f_0 depending on the sampling rate (sr) and table size (l). This means adding $incr$ to n for each sample:

$$incr = \frac{l}{sr} \times f_0 \qquad (3.3)$$

This step is called *phase* or *sampling increment* (phase is another name for index position).
4. **wrap-around**: when n exceeds the table size it needs to be wrapped back into its correct range $[0, l)$. This a modulus operation, which needs to be general enough to work with positive and negative index values.

It is possible to see that we have effectively implemented an oscillator in our first two audio examples (listings 3.12 and 3.13), without a table lookup but using all the other steps, 2, 3 and 4. In these instruments, the increment is calculated at the k-rate, so we use `kr` as the sampling rate, and its maximum value is 1 ($l = 1$ and $sr = kr$ makes eq. 3.3 the same as eq. 3.1). In the first example, the process is so crude that we are replacing step 1 by just outputting the index value, which is one of the reasons it sounds very noisy. In the second case, we do a direct `sin()` function evaluation. This involves asking the program to compute the waveform values on the spot. Table lookup replaces this by reading a pre-calculate sine function that is stored on a table.

There are three types of original oscillator opcodes in Csound. They differ in the way the table-lookup step is performed. Since table positions are integral, we need to decide what to do when an index is not a whole number, and falls between two table positions. Each oscillator type treats this in a different way:

1. `oscil`: this opcode truncates its index position to an integer, e.g index 2.3 becomes 2.
2. `oscili`: performs a linear interpolation between adjacent positions to find the output value. Index 2.3 then makes the table lookup step take 70% (0.7) of the value of position 2 and 30% (0.3) of the next position, 3.
3. `oscil3`: performs a cubic interpolation where the four positions around the index value are used. The expression is a fourth-order polynomial involving the values of these four positions. So, for index 2.3, we will read positions 1,2,3 and 4 and combine them, with 2 and 3 contributing more to the final lookup value than 1 and 4.

There are two consequences to using these different types: (a) the lookup precision and quality of the output audio increase with interpolation [89]; (b) more computation cycles are used when interpolating. It is a consensus among users that in modern platforms the demands of low-order interpolation are not significant, and that we should avoid using truncating oscillators like `oscil`, because of their reduced audio quality.[2]

---

[2] In addition to these oscillators, Csound offers `poscil` and `poscil3`, discussed later in the book.

The full syntax description of an oscillator is (using `oscili` as an example):

```
xsig   oscili   xamp, xfreq [,ifn, iph]
```

or, using the alternative form:

```
[xsig  =]   oscili[:r](xamp, xfreq [,ifn, iph])
```

Oscillators are polymorphic. They can work at audio or control rates (indicated by the `xsig` in the description), and can take amplitude (`xamp`) and frequency (`xfreq`) as i-time, k-rate or a-rate values (however the output cannot be a rate lower than its input). Amplitudes, used in step 2 above, will range between 0 and `0dbfs`. Frequencies are given in Hz (cycles per second). There are two optional arguments: `ifn`, which is an optional table number, and `iph` the oscillator starting phase, set between 0 and 1. The default values for these are -1 for `ifn`, which is the number of a table containing a sine wave created internally by Csound, and 0 for phase (`iph`). With an oscillator, we can create a sine wave instrument that is equivalent to the previous example in listing 3.13, but more compact (and efficient).

Listing 3.14 A sine wave synthesis instrument, using `oscili`

```
instr 1
    out(oscili(p4, p5))
endin
schedule(1,0,10,0dbfs/2,440)
```

Connecting oscillators: mixing, modulating

Opcodes can be connected together in a variety of ways. We can add the output of a set of oscillators to mix them together.

Listing 3.15 Mixing three oscillators

```
instr 1
    out((oscili(p4, p5) +
    oscili(p4/3,p5*3)  +
    oscili(p4/5,p5*5))/3)
endin
schedule(1,0,10,0dbfs/2,440)
```

The oscillators are set to have different amplitudes and frequencies, which will blend together at the output. This example shows that we can mix any number of audio signals by summing them together. As we do this, it is also important to pay attention to the signal levels so that they do not exceed 0dbfs. If that happens, Csound will report *samples out of range* at the console, and the output will be distorted (clipped at the maximum amplitude). In listing 3.15, we multiply the mix by 1/3, so that it is scaled down to avoid distortion.

We can also make one oscillator periodically modify one of the parameters of another. This is called amplitude or frequency *modulation*, depending on which pa-

rameter the signal is applied to. Listing 3.16 shows two instruments, whose amplitude (1) and frequency (2) are modulated. The resulting effects are called *tremolo*
and *vibrato*, respectively.

Listing 3.16 Two modulation instruments

```
instr 1
   out(oscili(p4/2 + oscili:k(p4/2, p6), p5))
endin
schedule(1,0,5,0dbfs/2,440,3.5)

instr 2
   out(oscili(p4, p5 + oscili:k(p5/100,p6)))
endin
schedule(2,5,5,0dbfs/2,440,3.5)
```

Note that we are explicit about the rate used for the modulation opcodes. Here
we are using the control rate since we have low-frequency oscillators (LFOs), whose
output can be classed as control signals. LFOs are widely used to modulate parameters of sound generators[3]. When modulating parameters with audio-range frequencies (> 20 Hz), we are required to use a-rate signals. This is the case of frequency
modulation synthesis, studied in Chapter 12.

Phasors

Steps 3 and 4 in the oscillator operation, the phase increment and modulus, are
so fundamental that they have been combined in a specific unit generator, called
phasor:

```
xsig   phasor   xfreq [,iph]
```

The output of this opcode is a ramp from 0 to 1, repeated at xfreq Hz (similar to
the code in listing 3.12). Phasors are important because they provide a normalised
phase value that can be used for a variety of applications, including constructing
an oscillator from its constituent pieces. For example, yet another version of a sine
wave oscillator (cf. listing 3.14) can be created with the following code:

```
instr 1
  i2pi = 6.28318530
  out(p4*sin(i2pi*phasor(p5)))
endin
```

---

[3] However, you should be careful with zipper noise as discussed in a previous chapter. If ksmps is
large (e.g. above 64 samples), then it is advisable to use audio signals as modulators. For a clean
result that is independent of the control rate, always use audio-rate modulators and envelopes.

3.5.3 Table Generators

Oscillators can read arbitrary functions stored in tables. By default, they read table -1, which contains a sine wave. However, Csound is capable of generating a variety of different function types, which can be placed in tables for oscillators to use. This can be done with the opcode `ftgen`:

```
ifn ftgen inum, itime, isize, igen, ipar1 [, ipar2, ...]
```

This generator executes at init-time to compute a function and store it in the requested table number. It can take several arguments:

inum: table number
itime: generation time
isize: table size
igen: function generator routine (GEN) code
ipar1, ...: GEN parameters.

It returns the number of the table created (stored in ifn). If the inum argument is 0, Csound will find a free table number and use it, otherwise the table is placed in the requested number, replacing an existing one if there is one. The generation time can be 0, in which case the table is created immediately, or it can be sometime in the future (itime > 0). Most use cases will have this parameter set to 0.

The table size determines how many positions (points, numbers) the table will hold. The larger the table, the more finely defined (and precise) it will be. Tables can be constructed with arbitrary sizes, but some opcodes will only work with table lengths that are set to power-of-two or power-of-two plus one. The oscil, oscili and oscil3 opcodes are examples of these.

The size will also determine how the table guard point (see Section 2.7) is set. For all sizes except power-of-two plus one, this is a copy of the first point of the table ($table[N] = table[0]$, where N is the size of the table), otherwise the guard point is *extended*, a continuation of the function contour. This distinction is important for opcodes that use interpolation and read the table in a once-off fashion, i.e. not wrapping around. In these cases, interpolation of a position beyond the length of the table requires that the guard point is a continuation of the function, not its first position. In listing 3.17, we see such an example. The table is read in a once-off way by an oscillator (note the frequency set to $\frac{1}{p3}$, i.e. the period is equal to the duration of the sound), and is used to scale the frequency of the second oscillator. The table size is 16,385, indicating an extended guard point, which will be a continuation of the table contour.

Listing 3.17 Two modulation instruments

```
ifn ftgen 1,0,16385,-5,1,16384,2
instr 1
 k1 oscili 1,1/p3,1
 a1 oscili p4,p5*k1
 out a1
```

```
endin
schedule(1,0,10,0dbfs/2,440)
```

The type of function is determined by the GEN used. The `igen` parameter is a code that controls two aspects of the table generation: (a) the function generator; (b) whether the table is re-scaled (normalised) or not. The absolute value of `igen` (i.e. disregarding its sign) selects the GEN routine number. Csound has over thirty types of this, which are described in the Reference Manual. The sign of `igen` controls re-scaling: positive turns it on, negative suppresses it. Normalisation scales the created values to the 0 to 1 range for non-negative functions, and -1 to 1 for bipolar ones. The table in listing 3.17 uses GEN 5, which creates exponential curves, and it is not re-scaled (negative GEN number).

The following arguments (`ipar`, . . .) are parameters to the function generator, and so they will be dependent on the GEN routine used. For instance, GEN 7 creates straight lines between points, and its arguments are (start value, length in points, end value) for each segment required. Here is an example of a trapezoid shape using three segments:

```
isize = 16384
ifn ftgen 1, 0,  isize,  7,
     0, isize*0.1, 1,isize*0.8, 1, isize*0.1, 0
```

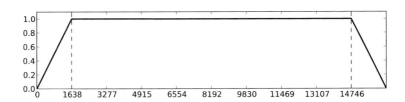

Fig. 3.3 Plot of a function table created with GEN 7 in listing 3.18

This function table goes from 0 to 1 in 10% (0.1) of its length, stays at 1 for a further 80%, and goes back down to 0 in the final 10% portion. Since tables can be used freely by various instrument instances, we should create them in global space. Listing 3.18 shows an example of an instrument that uses this function table. We will use an oscillator to read it and use its output to control the amplitude of a sine wave.

Listing 3.18 An instrument using a function table to control amplitude

```
isize = 16384
ifn ftgen 1, 0,  isize,  7,
     0, isize*0.1, 1,isize*0.8, 1, isize*0.1, 0
instr 1
 kenv  oscili  p4, 1/p3, 1
```

```
asig   oscili   kenv, p5
 out   asig
endin
schedule(1,0,10,0dbfs/2,440)
```

This example shows how oscillators can be versatile units. By setting the frequency of the first oscillator to $\frac{1}{p3}$, we make it read the table only once over the event duration. This creates an amplitude *envelope*, shaping the sound. It will eliminate any clicks at the start and end of the sound. The various different GEN types offered by Csound will be introduced in later sections and chapters, as they become relevant to the various techniques explored in this book.

3.5.4 Table Access

Another fundamental type of unit generator, which also uses function tables, is the table reader. These are very basic opcodes that will produce an output given an index and a table number. Like oscillators, they come in three types, depending on how they look up a position: `table` (non-interpolating), `tablei` (linear interpolation) and `table3` (cubic interpolation). Table readers are more general-purpose than oscillators, and they can produce outputs at i-time, k-rate and a-rate, depending on what they are needed for. Their full syntax summary is (using `table` as an example).

```
xvar   table   xndx, ifn[, imode, ioff, iwrap]
```

The first two arguments are required: the index position (`xndx`) and function table number (`ifn`) to be read. The optional arguments define how `xndx` will be treated:

`imode` sets the indexing mode: 0 for raw table positions (0 to size-1), 1 for normalised (0-1).
`ioff` adds an offset to `xndx`, making it start from a different value.
`iwrap` switches wrap-around (modulus) on. If 0, limiting is used, where index values are pegged to the ends of the table if they exceed them.

All optional values are set to 0 by default.

Table readers have multiple uses. To illustrate one simple use, we will use it to create an arpeggiator for our sine wave instrument. The idea is straightforward: we will create a table containing various pitch values, and we will read these using a table reader to control the frequency of an oscillator.

The pitch table can be created with GEN 2, which simply copies its arguments to the respective table positions. To make it simple, we will place four interval ratios desired for the arpeggio: 1 (unison), 1.25 (major third), 1.5 (perfect fifth) and 2 (octave). The table will be set not to be re-scaled, so the above values can be preserved. This function table is illustrated by Fig 3.4.

```
ifn ftgen   2,0,4,-2,1,1.25,1.5,2
```

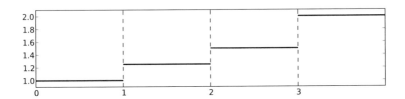

Fig. 3.4 Function table created with GEN2 in listing 3.19

We can read this function table using a `phasor` to loop over the index at a given rate (set by `p6`), and use a `table` opcode. Since this truncates the index, it will only read the integral positions along the table, stepping from one pitch to another. If we had used an interpolating table reader, the result would be a continuous slide from one table value to another. Also because `phasor` produces an output between 0 to 1, we set `imode` to 1, to tell `table` to accept indexes in that range:

```
kpitch table phasor(p6), 2, 1
```

Now we can insert this code into the previous instrument (listing 3.18), and connect `kpitch` to the oscillator frequency.

Listing 3.19 An arpeggiator instrument

```
isize = 16384
ifn ftgen 1, 0,   isize,   7,
   0, isize*0.1, 1,isize*0.8, 1, isize*0.1, 0
ifn ftgen   2,0,4,-2,1,1.25,1.5,2

instr 1
   kenv     oscili  p4, 1/p3, 1
   kpitch   table   phasor:k(p6), 2, 1
   asig     oscili  kenv, p5*kpitch
   out   asig
endin
schedule(1,0,10,0dbfs/2,440,1)
```

If we replace the means we are using to read the table, we can change the arpeggiation pattern. For instance, if we use an oscillator, we can speed up/slow down the reading, according to a sine shape:

```
kpitch   table   oscil:k(1,p6/2), 2, 1, 0, 1
```

Note that because the oscillator produces a bipolar signal, we need to set `table` to wrap-around (`iwrap=1`), otherwise we will be stuck at 0 for half the waveform period.

Table writing

In addition to being read, any existing table can be written to. This can be done at perf-time (k-rate or a-rate) by the `tablew` opcode:

```
tablew xvar, xndx, ifn [, ixmode] [, ixoff] [, iwgmode]
```

> `xvar` holds the value(s) to be written into the table. It can be an i-time variable, k-rate or a-rate signal.
> `xndx` is the index position(s) to write to, its type needs to match the first argument.
> `imode` sets the indexing mode: 0 for raw table positions (0 to size -1), 1 for normalised (0-1).
> `ioff` adds an offset to `xndx`.
> `iwgmode` controls the writing. If 0, it limits the writing to between 0 and table size (inclusive); 1 uses wrap-around (modulus); 2 is guard-point mode where writing is limited to 0 and table size -1, and the guard point is written at the same time as position 0, with the same value.

Note that `tablew` only runs at perf-time, so it cannot be used to write a value at i-time to a table. For this, we need to use `tableiw`, which only runs at initialisation.

3.5.5 Reading Soundfiles

Another set of basic signal generators is the soundfile readers. These are opcodes that are given the name of an audio file in the formats accepted by Csound (see Section 2.8.4), and source their audio signal from it. The simplest of these is `soundin`, but a more complete one is provided by `diskin` (and `diskin2`, which uses the same code internally):

```
ar1[,ar2,ar3, ... arN] diskin Sname[,kpitch, iskipt,
                               iwrap,ifmt,iskipinit]
```

where `Sname` is the path/name of the requested soundfile in double quotes. This opcode can read files from multiple channels (using multiple outputs), and will resample if the file sr differs from the one used by Csound (which `soundin` cannot do). It can also change the playback pitch (if `kpitch != 1`) like a varispeed tape player, negative values allow for reverse readout. The `iwrap` parameter can be used to wrap-around the ends of the file (if set to 1). The number of channels and duration of any valid soundfile can also be obtained with

```
ich  filenchnls Sname
ilen filen Sname
```

An example of these opcodes is shown below in listing 3.20.

Listing 3.20 A soundfile playback instrument

```
nchnls=2
instr 1
 p3 = filelen(p4)
 ich = filenchnls(p4)
 if ich == 1 then
  asig1 diskin   p4
  asig2 = asig1
 else
  asig1, asig2 diskin p4
 endif
 out asig1,asig2
endin
schedule(1,0,1,"fox.wav")
```

Note that we have to select the number of outputs depending on the file type using a control-of-flow construct (see Chapter 5). The duration (p3) is also taken from the file.

3.5.6 Pitch and Amplitude Converters

Csound provides functions to convert different types of pitch representation to frequency in cycles per second (Hz), and vice-versa. There are three main ways in which pitch can be notated:

1. **pch**: octave point pitch class
2. **oct**: octave point decimal
3. **midinn**: MIDI note note number.

The first two forms consist of a whole number, representing octave height, followed by a specially interpreted fractional part. Middle-C octave is represented as 8; each whole-number step above or below represents an upward or downward change in octave, respectively. For **pch**, the fraction is read as two decimal digits representing the 12 equal-tempered pitch classes from .00 for C to .11 for B.

With **oct**, however, this is interpreted as a true decimal fractional part of an octave, and each equal-tempered step is equivalent to $\frac{1}{12}$, from .00 for C to $\frac{11}{12}$ for B. The relationship between the two representations is then equivalent to the factor $\frac{100}{12}$.

The concert pitch A 440 Hz can be represented as 8.09 in pch notation and 8.75 in oct notation. Microtonal divisions of the pch semitone can be encoded by using more than two decimal places. We can also increment the semitones encoded in the fractional part from .00 to .99 (e.g. 7.12 => 8.00; 8.24 => 10.00). The third notation derives from the MIDI note number convention, which encodes pitches between 0 and 127, with middle C set to 60.

These are the functions that can be used to convert between these three notations:

```
octpch (pch): pch to oct.
cpspch (pch): pch to cps (Hz).
pchoct (oct): oct to pch.
cpsoct (oct): oct to cps (Hz).
octcps (cps): cps (Hz) to oct.
cpsmidinn (nn): midinn to cps (Hz).
pchmidinn (nn): midinn to pch.
octmidinn (nn): midinn to oct.
```

The following code example demonstrate the equivalence of the three pitch representations:

```
instr 1
  print cpspch (8.06)
  print cpsoct (8.5)
  print cpsmidinn (66)
endin
schedule (1,0,1)
```

When we run this code, the console prints out the same values in Hz:

```
new alloc for instr 1:
instr 1:   #i0 = 369.994
instr 1:   #i1 = 369.994
instr 1:   #i2 = 369.994
```

It is worth noting that while the majority of the pitch converters are based on a twelve-note equal temperament scale, Csound is not so restricted. The opcode cps2pch convert from any octave-based equal temperament scale to Hertz, while cpsxpch and cpstun provide for non-octave scales as well as table-provided scales. Details can be found in the reference manual.

For amplitudes, the following converters are provided for transforming to/from decibel (dB) scale values:

ampdb (x): dB scale to amplitude.

$$\mathrm{ampdb}(x) = 10^{\frac{x}{20}} \tag{3.4}$$

dbamp (x): amplitude to dB

$$\mathrm{dbamp}(x) = 20\log_{10} x \tag{3.5}$$

ampdbfs (x): dB scale to amplitude, scaled by 0dbfs

$$\mathrm{ampdbfs}(x) = \mathrm{ampdb}(x) \times 0\mathrm{dbfs} \tag{3.6}$$

These converters allow the convenience of setting amplitudes in the dB scale, and frequencies in one of the three pitch scales, as shown in the example below:

```
instr 1
 out(oscili(ampdbfs(p4),cpspch(p5)))
endin
schedule(1,0,1,-6,8.09)
```

3.5.7 Envelope Generators

The final set of fundamental Csound opcodes are the envelope generators. As we have seen earlier, function tables can be used to hold shapes for parameter control. However, these suffer from a drawback: they will expand and contract, depending on event duration. If we create an envelope function that rises for 10% of its length, this will generally get translated in performance to 10% of the sound duration, unless some more complicated coding is involved. This rise time will not be invariant, which might cause problems for certain applications.

Envelope generators can create shapes whose segments are fixed to a given duration (or relative, if we wish, the flexibility is there). These can be applied at control or audio rate to opcode parameters. Csound has a suite of these unit generators, and we will examine a few of these in detail. The reader is then encouraged to look at the Reference Manual to explore the others, whose operation principles are very similar.

The basic opcode in this group is the trapezoid generator `linen`, which has three segments: rise, sustain and decay. Its syntax is summarised here:

```
xsig linen   xamp,irise,idur,idec
```

The four parameters are self-describing: `irise` is the rise time, `idur` is the total duration and `idec` the decay time, all of these in seconds. The argument `xamp` can be used as a fixed maximum amplitude value, or as an audio or control signal input. In the first case, `linen` works as a signal generator, whose output can be sent to control any time-varying parameter of another opcode. If, however, we use a signal as input, then the opcode becomes a processor, shaping the amplitude of its input.

This envelope makes a signal or a value rise from 0 to maximum in `irise` seconds, after which it will maintain the output steady until $dur - idec$ seconds. Then it will start to decay to 0 again. Listing 3.21 shows an example of its usage, as a signal generator, controlling the amplitude of an oscillator. Its arguments are taken from instrument parameters, so they can be modified on a per-instance basis.

Listing 3.21 Linen envelope generator example

```
instr 1
   kenv   linen   ampdbfs(p4),p6,p3,p7
   asig   oscili   kenv, cpspch(p5)
   out   asig
endin
schedule(1,0,10,-6,8.09,0.01,0.1)
```

Alternatively, we could employ it as a signal processor (listing 3.22). One important difference is that in this arrangement `linen` works at the audio rate, whereas in listing 3.21, it produces a k-rate signal. Depending on how large `ksmps` is, there might be an audible difference, with this version being smoother sounding.

Listing 3.22 Linen envelope processor example

```
instr 1
    aosc    oscili ampdbfs(p4),cpspch(p5)
    asig    linen aosc,p6,p3,p7
    out     asig
endin
schedule(1,0,10,-6,8.09,0.01,0.1)
```

Another form of envelope generators are the line segment opcodes. There are two types of these: linear and exponential. The first creates curves based on constant differences of values, whereas the other uses constant ratios. They are available as a simple single-segment opcodes, or with multiple stages:

```
xsig line   ipos1,idur,ipos2
xsig expon  ipos1,idur,ipos2
xsig linseg ipos1,idur,ipos2,idur2,ipos3[,...]
xsig expseg ipos1,idur,ipos2,idur2,ipos3[,...]
```

The exponential opcodes cannot have 0 as one of its `ipos` values, as this would imply a division by 0, which is not allowed. These unit generators are very useful for creating natural-sounding envelope decays, because exponential curves tend to match well the way acoustic instruments work. They are also very good for creating even glissandos between pitches. A comparison of linear and exponential decaying envelopes is shown in Fig. 3.5. Listing 3.23 shows two exponential envelopes controlling amplitude and frequency.

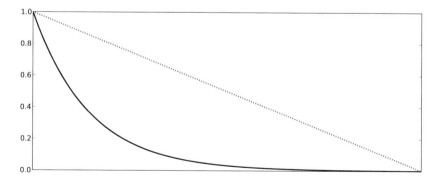

Fig. 3.5 Linear (dots) and exponential (solid) decaying curves.

Listing 3.23 Exponential envelope example

```
instr 1
   kenv   expon 1, p3, 0.001
   kpit   expon 2,p3,1
   asig   oscili kenv*ampdbfs(p4),kpit*cpspch(p5)
   out    asig
endin
schedule(1,0,1,-6,8.09)
```

In addition to these two types of curves, the `transeg` opcode can produce linear and exponential (concave or convex) curves, defined by an extra parameter for each segment.

3.5.8 Randomness

Another class of opcodes provided random values from a variety of distributions. They have many uses but a major one is in providing control signals, especially for humanising a piece. For example, in an envelope definition it might be desirable for the durations of the segments to vary a little to reduce the mechanical effect of the notes. This can be achieved by adding a suitable random value to the average value desired. All the opcodes described here are available at initialisation, control or audio rate, but the commonest uses are for control.

The simplest of these opcodes is `random`, which uses a pseudo-random number generator to deliver a value in a range supplied with a uniform distribution; that is all values within the range are equally likely to occur. A simple example of its use might be as shown in listing 3.24, which will give a pitch between 432 Hz and 452 Hz every time an instance of instrument 1 is started.

Listing 3.24 Random pitch example

```
instr 1
   aout oscili 0.8, 440+random(-8, 12)
   outs aout
endin
```

More realistic would be to be to think about a performer who tries to hit the correct pitch but misses, and then attempts to correct it when heard. The pitch attempts are not too inaccurate from the target, and it is unlikely that the performer will stray far away. A simple model would be that the error is taken from a normal distribution (also called a Gaussian distribution or bell curve), centred on the target pitch. A normal distribution looks like Fig. 3.6.

The scenario can be coded as in listing 3.25. A similar method can thicken a simple oscillation by adding normally distributed additional oscillations as shown in instrument 2. The `gauss` opcode takes an argument that governs the spread of the curve, all effective values being within ± that value from zero.

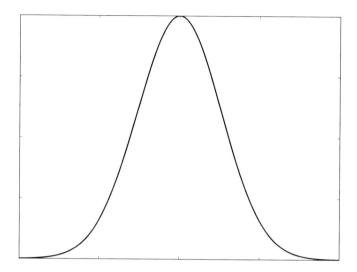

Fig. 3.6 Normal distribution.

Listing 3.25 Normal distribution of error pitch

```
instr 1
   kerr  gauss  10
   aout  oscili 0.8, 440+kerr
   outs aout
endin
instr 2
   kerr1 gauss 10
   kerr2 gauss 10
   a0    oscili 0.6, 440
   a1    oscili 0.6, 440+kerr1
   a2    oscili 0.6, 440+kerr2
   outs    a0+a1+a2
endin
```

Different distribution curves can be used for different effects. In addition to the uniform distribution of `random` and the normal distribution of `gauss` Csound provides many others [80]:

- **Linear**: The distribution is a straight line from 2 at zero to zero at 1. The `linrand` opcode delivers this scaled so values come from zero to `krange`, and small values are more likely. It could for example be used for adding a tail to a note so durations were non-regular as in instrument 1 in listing 3.26.
- **Triangular**: Centred on zero the distribution goes to zero at ±`krange`. A simple example is shown as instrument 2 in listing 3.26.

- **Exponential**: The straight lines of the uniform, linear and triangular distribution are often unnatural, and a curved distribution is better. The simplest of these is the exponential distribution $\lambda e^{-\lambda x}$ which describes the time between events which occur independently at a fixed average rate. Random values from this distribution (Fig. 3.7) are provided by the `exprand` opcode, with an argument for λ. The larger the value of λ the more likely that the provided value will be small. The average value is $\frac{1}{\lambda}$ and while theoretically the value could be infinite in practice it is not. The value is always positive. A version of this distribution symmetric about zero is also available as `bexprand`.
- **Poisson**: A related distribution gives the probability that a fixed number of events occur within a fixed period. The opcode `poisson` returns an integer value of how many events occurred in the time-window, with an argument that encodes the probability of the event. Mathematically the probability of getting the integer value j is $e^{-\lambda}\left(\frac{\lambda^j}{j!}\right)$.
- **Cauchy**: Similarly shaped to the normal distribution, the `cauchy` distribution make values away from zero more likely; that is the distribution curve is flatter.
- **Weibull**: A more variable distribution is found in the `weibull` opcode, which takes two parameters. The first, σ, controls the spread of the distribution while the second, τ, controls the shape. The graphs are shown in Fig. 3.8, for three values of τ. When the second parameter is one this is identical to the exponential distribution; smaller values favour small values and larger delivers more large values.
- **Beta**: Another distribution with a controllable shape is the beta distribution (`betarand`). It has two parameters, one governing the behaviour at zero and the other at one, $x^{a-1}(1-x)^{b-1}/B$ where B is just a scale to ensure the distribution gives a probability in the full range (area under the curve is one). This gives bell-curve-like shapes when a and b are both greater than 1 (but with a finite range), when both are 1 this is a uniform distribution, and with a and b both less than 1 it favours 0 or 1 over intermediate values.

Listing 3.26 Example of use of linrand

```
instr 1
   itail linrand 0.1
   p3 += itail
   a1     oscili  0.8, 440
   out    a1
endin
instr 2
   itail trirand 0.1
   p3 += itail
   a1     oscili  0.8, 440
   out    a1
endin
```

All these opcodes use a predictable algorithmic process to generate pseudo-random values which is sufficient for most purposes, and has the feature of being

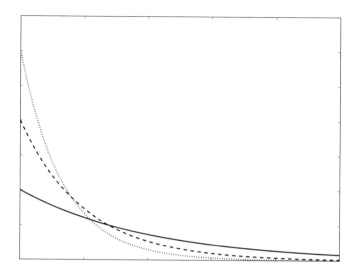

Fig. 3.7 Exponential distribution, with $\lambda = 0.5$ (solid), 1.0 (dash) and 1.5 (dots)

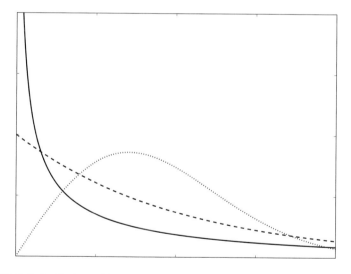

Fig. 3.8 Weibull distribution, with $\tau = 0.5$ (solid), 1.0 (dash) and 2.0 (dots)

repeatable. If a better chance of being random is required the uniformly distributed
opcode `urandom` can be used; it uses other activity on the host computer to inject
more randomness.

When considering any process that derives from random values there is the question of reproducibility. That is, if one synthesises it again does one want or expect
the results to be identical or different. Both are reasonable expectations, which is
why Csound provides a `seed` opcode. This allows control over the sequence of
pseudo-random numbers, by selecting a starting value, or seeding from the computer clock which is at least difficult to predict. It is also possible to find the current
state of the PRN generator with the `getseed` opcode.

3.6 The Orchestra Preprocessor

Csound, like other programming languages, includes support for preprocessing of
its orchestra code. This allows the user to make a number of modifications to text,
which are made before compilation. There are seven preprocessor statements in
Csound, all starting with #:

```
#include
#define
#undef
#ifdef
#ifndef
#else
#end
```

Include

The include statement copies the contents of an external text file into the code, at
the exact line where it is found. For instance, the following code

```
#include "instr1.inc"
schedule(1,1,2,-12,8.00)
```

where the file *instr1.inc* contains listing 3.23, will play two sounds with instr 1 (the
first one corresponding to the `schedule` line in the included file). The file name
should be in double quotes. If the included file is in the same directory as the code,
or the same working directory as a running Csound program, then its name will be
sufficient. If not, the user will need to pass the full path to it (in the usual way used
by the OS platform in which Csound is running).

Define

The Csound preprocessor has a simple macro system that allows text to be replaced throughout a program. We can use it in two ways, with or without arguments. The first form, which is the simplest, is

```
#define   NAME   #replacement text#
```

where $NAME will be substituted by the replacement text, wherever it is found. This can be used, for instance to set constants that will be used throughout the code.

Listing 3.27 Macro example

```
#define   DURATION #p3#
instr 1
   kenv   expon 1, $DURATION, 0.001
   kpit   expon 2, $DURATION, 1
   asig   oscili kenv*ampdbfs(p4),kpit*cpspch(p5)
   out    asig
endin
schedule(1,0,1,-6,8.09)
```

Macros replace the text where it is found, so we can concatenate it with other text characters by terminating its call with a full stop, for instance

```
#define   OCTAVE   #8.#
schedule(1,0,1,-6,$OCTAVE.09)
```

Macros can be used for any text replacement. For instance, we could use it to define what type of envelopes we will use:

```
#define   ENV #expon#
instr 1
   kenv   $ENV 1, p3, 0.001
   kpit    $ENV 2, p3, 1
   asig   oscili kenv*ampdbfs(p4),kpit*cpspch(p5)
   out    asig
endin
```

Using full capitals for macro names is only a convention, but it is a useful one to avoid name clashes. The second form of #define allows for arguments:

```
#define   NAME(a' b')   #replacement text#
```

where the parameters a, b etc. can be referred to in the macro as $a, $b etc. For instance, listing 3.28 shows how a macro can take arguments to replace a whole opcode line.

Listing 3.28 Macro example, with arguments

```
#define   OSC(a'b) #oscili $a, $b#
instr 1
```

```
    kenv   expon 1, p3, 0.001
    kpit    expon 2, p3, 1
    asig   OSC(kenv*ampdbfs(p4) ' kpit*cpspch(p5))
    out    asig
endin
schedule(1,0,1,-6,8.09)
```

Csound also includes a number of internally defined numeric macros that are useful for signal processing applications, e.g. the value of π, e etc. (see Table 3.2).

Table 3.2 Numeric macros in Csound

macro	value	expression
$M_E	2.7182818284590452354	e
$M_LOG2E	1.4426950408889634074	$log_2(e)$
$M_LOG10E	0.43429448190325182765	$log_{10}(e)$
$M_LN2	0.69314718055994530942	$log_e(2)$
$M_LN10	2.30258509299404568402	$log_e(10)$
$M_PI	3.14159265358979323846	π
$M_PI_2	1.57079632679489661923	$\pi/2$
$M_PI_4	0.78539816339744830962	$\pi/4$
$M_1_PI	0.31830988618379067154	$1/\pi$
$M_2_PI	0.63661977236758134308	$2/\pi$
$M_2_SQRTPI	1.12837916709551257390	$2/\sqrt{\pi}$
$M_SQRT2	1.41421356237309504880	$\sqrt{2}$
$M_SQRT1_2	0.70710678118654752440	$1/\sqrt{\pi}$

Finally, a user defined macro can be undefined:

```
#undefine   NAME
```

Conditionals

To complement the preprocessor functionality, Csound allows for *conditionals*. These control the reading of code by the compiler. They use a token definition via #define, a check for it, and an alternative branch:

```
#define   NAME
#ifdef NAME
#else
#end
```

A negative check is also available through #ifndef. We can use these conditionals to switch off code that we do not need at certain times. By using #define PRINTAMP in listing 3.29, we can make it print the amplitudes rather than the frequencies.

Listing 3.29 Preprocessor condititionals

```
#define PRINTAMP #1#
instr 1
  iamp = ampdbfs(p4)
  ifreq = cpspch(p5)
#ifdef PRINTAMP
  print iamp
#else
  print ifreq
#end
 out(oscili(iamp,ifreq)
endin
schedule(1,0,1,-6,8.09)
```

3.7 Conclusions

This chapter explored the essential aspects of Csound programming, from instrument definitions to basic opcodes. We started the text without any means of making sound with the software, but at the end we were to create some simple music-generating instruments. As the scope of this discussion is not wide enough to include synthesis techniques (that is reserved for later), the types of sounds we were able to produce were very simple. However, the instrument designs showed the principles that underlie programming with Csound in a very precise way, and these will be scaled up as we start to cover the various sound processing applications of the system.

With a little experimentation and imagination, the elements presented by this chapter can be turned into very nice music-making tools. Readers are encouraged to explore the Csound Reference Manual for more opcodes to experiment with and add to their instruments. All of the principles shown here are transferable to other unit generators in the software. As we progress to more advanced concepts, it is worth coming back to some of the sections in this chapter, which explore the foundations on which the whole system is constructed.

Chapter 4
Advanced Data Types

Abstract This chapter will explore the more advanced data types in Csound. We will begin by discussing Strings and manipulating texts. Next, we will explore Csound's spectral data types for signal analysis and transformations. Finally, we will discuss arrays, a data type that acts as a container for other data types.

4.1 Introduction

In Section 3.1, we discussed Csound's basic data types: i-, k- and a-types. These allow us to express numeric processing of signals that cover a large area of sound computing and are the foundation signals used in Csound.

In this chapter, we will cover more advanced data types. These include strings, spectral-domain signals and arrays. These allow us to extend what we can represent and work with in Csound, as well as to express new kinds of ideas.

4.2 Strings

Strings are ordered sets of characters and are often thought of as "text". They may be constant strings–text surrounded by quotes–or may be values held in S-type variables. They are useful for specifying paths to files, names of channels, and printing out messages to the user. They can also be used to create ad hoc data formats.

© Springer International Publishing Switzerland 2016
V. Lazzarini et al., *Csound*, DOI 10.1007/978-3-319-45370-5_4

4.2.1 Usage

String constants are defined using a starting quote, then the text to use for the string, followed by a closing quote. The quotes are used to mark where the string starts and ends, and are not part of the string's value.

```
prints   "Hello World"
```

In the above example, "Hello World" defines a string constant. It is used with the `prints` opcode to print out "Hello World" to the console. Besides the value, strings also have a *length* property. This may be retrieved using the `strlen` opcode, and may be useful for certain computations.

```
print   strlen("Hello World")
```

Executing the code in the above example will print out 11.000, which is the number of characters in the text "Hello World". String constants are the most common use of strings in Csound.

```
print   strlen("\t\"Hello World\"\n")
```

Certain characters require *escaping*, which means they are processed in a special way. Escaped characters start with a backslash and are followed by the character to escape. The valid escape characters are given in Table 4.1.

Table 4.1 Escape sequences

Escape Sequence	Description
\a	alert bell
\b	backspace
\n	newline
\r	carriage return
\t	tab
\\	a single backslash
\"	double quote
\{	open brace
\}	close brace
nnn	ASCII character code in octal number format

In addition to single-line strings delimited by quotes, one can use double braces to define multi-line strings. In the following example, a multi-line string is used to print four lines of text. The text is delimited by opening double braces ({{) and closing double braces (}}). All text found within those delimiters is treated as a string text values and is not processed by Csound. For example, on line 2, the semicolon and the text following it are read in as part of the text and are not processed as a comment.

```
prints {{Hello
World ; This is not a comment
Hello
World}}
```

Defining string constants may be enough for many use cases. However, if one wants to define a string value and refer to it from multiple locations, one can store the string into an S-type variable.

```
Svalue = "Hello World"
prints Svalue
```

In the above example, we have defined a string variable called `Svalue` and assigned it the value "Hello World". We then pass that variable as an argument to the `prints` opcode and we get the same "Hello World" message printed to the console as we did when using a string constant. The difference here from the original example is that we are now storing the string value into a variable which we can now refer to in our code by name.

Listing 4.1 String processing

```
Sformat = "Hello \%s\n"
Smessage sprintf Sformat, "World"
Smessage2 sprintf Sformat, "Csounder"
prints Smessage
prints Smessage2                              .
```

Here we are using the `Sformat` as a template string for use with the `sprintf` opcode. The `Sformat` is used twice with two different values, "World" and "Csounder". The return values from the calls to `sprintf` are stored in two variables, `Smessage` and `Smessage2`. When the `prints` opcode is used with the two variables, two messages are printed, "Hello World" and "Hello Csounder".

Csound includes a whole suite of opcodes for string data processing. These can be used for concatenation, substitution, conversion from numeric data, etc.:

- `strlen` - obtains the length of a string
- `strcat` - concatenates two strings
- `strcpy` - string copy
- `strcmp` - string comparison.
- `sprintf` - formatting/conversion of strings
- `puts`, `prints`, `printf` - printing to console
- `strindex`, `strrindex` - searching for substrings
- `strsub` - copying substrings
- `strtod` - string to floating point conversion
- `strtol` - string to integer conversion
- `strlower` - lower case conversion
- `strupper` - upper case conversion

The following example demonstrates string processing at i-time.

Listing 4.2 String processing example

```
instr 1
 S1 = p4
 S2 sprintf {{
 This is p4: \"%s\".
 It is a string with %d characters.\n
 }}, S1, strlen(S1)
 prints S2
endin
schedule(1,0,0, "Hello World !!!")
```

Note that it is also possible to process strings at performance time, since most opcodes listed above have k-rate versions for this purpose.

4.3 Spectral-Domain Signals

Spectral-domain signals represent a streaming analysis of an audio signal. They use their own rates of update, independent of the control frequency, but this is largely transparent to the end user. Users will use specific families of opcodes to analyse audio signals, process spectral-domain signals and generate new audio signals. There are two main types of these variables: f-sig and w-sig. The first one is used for standard spectral audio manipulation, while the second is mostly employed for specialist data analysis operations.

4.3.1 f-sig Variables

Standard frequency-domain, or f-sig, variables are used to hold time-ordered frames of spectral data. They are generated by certain spectral analysis opcodes and can be consumed by a variety of processing units. The data carried by an f-sig can be transformed back to the time domain via various methods of synthesis. Outside of audio-processing contexts, a frequency-domain signal may be used as is without resynthesis, such as for visualising the contents of an audio signal. Signals carried by f-sig are completely self-describing, which means that additional information on the format is carried alongside the raw data from opcode to opcode.

Frequency-domain manipulation offers additional methods to work with audio compared with time-domain processing alone. It allows time and frequencies to be worked with separately, and can be used to do things like stretching a sound over time without affecting pitch, or modifying the pitch without affecting the duration of the sound. The options for processing f-sigs will depend on the *source*. There are four sub types of f-sigs. The characteristics of the data, and their accompanying descriptive information, will depend on these:

1. PVS_AMP_FREQ: this sub type carries frames (vectors) of amplitude and frequency pairs. Each one of these refers to a given frequency band (bin), in ascending order from 0 Hz to the Nyquist frequency (inclusive). At the analysis, the spectrum is broken into a certain number of these bins, which are equally-spaced in frequency. Alongside this data, the f-sig carries the analysis frame size, the spacing between analysis points (hopsize), the window size and type used. This subtype is employed for Phase Vocoder signals, and is the most commonly used f-sig.
2. PVS_AMP_PHASE: this carries frames of amplitude and phase pairs. Similar to the above, but carrying phase information instead of frequency.
3. PVS_COMPLEX: this carries frames of real-imaginary (complex) pairs. Also similar to the previous subtype, but carrying rectangular rather than polar data.
4. PVS_TRACKS: this sub type carries partial track data, rather than a fixed set of analysis bands. This data contains amplitude, frequency and phase information, as well as an ID, which can be used to identify a given track. The ordering is by creation ('birth') time, and frequency (in the case of tracks that started at the same time). Tracks are terminated if no continuation is found, so frame sizes can vary with time.

As mentioned before, f-sigs run at their own update rate, which is independent of the instrument control frequency. In order to implement this, spectral signals carry a frame index that indicates the analysis time, so that opcodes are notified if a new frame is ready for processing. Data is produced every analysis time point, which is spaced by hopsize samples. If this spacing is smaller than ksmps, more than one analysis will be have to performed in a single k-cycle. In some cases, this will trigger the use of a sliding algorithm. In other situations, when no such processing is available, an error message will be issued. The special sliding algorithm is in general very processor-intensive and might not be suitable for real-time operations. As a rule of thumb, we should try to keep the hopsize larger than ksmps.

Usage

Frequency-domain signals of the amplitude/frequency subtype can be generated from an analysis of an input signal (eg. with pvsanal), a table (pvstanal) or by reading a sequence of pre-analysed frames from a file (pvsfread):

```
fsig pvsanal asig,ifsize,ihopsize,iwsize,iwin
fsig pvstanal ktimescal, kamp, kpitch, ktab,
    [kdetect, kwrap, ioffset,ifftsize, ihop, idbthresh]
fsig pvsfread ktime, Sname
```

Analysis will normally require the frame, hop, and window sizes, as the window type, to be given as i-time parameters. For file reading, we will require a k-rate time position from which the current frame is to be read from, and a file name. Empty f-sigs can also be created/initialised with pvsinit:

```
fsig  pvsinit  ifsize[,ihopsize,iwsize,iwin,iformat]
```

where the iformat code is 0 for amplitude-frequency, 1 for amplitude-phase, 2 for complex and 3 for tracks.

Amplitude-frequency frequency-domain signals can be resynthesised using phase vocoder synthesis (pvsynth) or using additive methods (pvsadsyn):

```
asig pvsynth fsig
ares pvsadsyn fsrc, inoscs, kfscal
```

Since an f-sig is self-describing, phase vocoder analysis does not require any parameters in addition to its input. With additive synthesis, we can determine the number of bands being resynthesised and use a frequency scaling (transposition) control. Many processing opcodes can be used between the analysis and synthesis stages. Listing 4.3 shows a simple example of a frequency shifter, which moves each analysis component above 300 Hz by 400 Hz.

Listing 4.3 Amplitude-frequency f-sig processing example

```
instr 1
 asig  in
 fs1 pvsanal asig,2048,256,2048,1
 fs2 pvshift fs1,p4,p5
 ahr pvsynth fs2
 out ahr
endin
schedule(1,0,1,300,400)
```

Track data f-sigs are generated by the partials opcode from amplitude-frequency and amplitude-phase signals, and can be further processed and resynthesised using a variety of methods. Details of spectral (frequency-domain) signal processing will be further discussed in Chapter 14.

4.3.2 w-sig Variables

In addition to f-sigs, Csound includes a non-standard frequency-domain type, the w-sig, which carries constant-Q, exponentially-spaced frequency-domain data. This signal is obtained with the spectrum opcode, and can be used in a number of analysis operations, for pitch and onset detection, display and spectral histogram. These signals are not currently used for any audio processing, and there are no resynthesis methods available for them at the moment.

4.4 Arrays

Arrays are a homogeneous set of signals where each member of the set is accessed by its index number. Arrays can be of any type, such as "an array of k-vars" or "an array of a-vars". They may also be multi-dimensional, holding rows and columns, in the case of two dimensions. The number of indices needed to select an element in an array determines its dimension.

Arrays can be used to pass around a group of signals. For example, an a-array may be used to pass around sets of audio signals. The audio signal array can then be thought of as a multichannel signal. They can also be used to hold vectors or matrices of numbers, at i-time or k-time, for various applications. In addition, it is possible to create arrays of strings and f-sigs.

Array types have their own syntax for initialisation, setting and read access, which is slightly different from that of ordinary variables.

4.4.1 Initialisation

An array is created and initialised using the `init` opcode:

```
asigs[] init 4
```

This creates a one-dimensional array of a-type signals. The variable is declared on its first use as an array using square brackets (i.e. []). The type of the array is specified using the same convention as other variables, by looking at the first letter of the variable name:

```
asigs[][] init 4, 4
```

This creates a two-dimensional array of a-type signals. It has sizes of four (rows) and four (columns), and there are 16 elements within the array ($4 \times 4 = 16$).

Note that the name of the variable is `asigs` and not `asigs[]`. The brackets are used to declare the type, but are not part of the name. After declaring the variable, subsequent uses of the variable will refer to only the name.

Creating an array and initialising it to a set of values can be verbose, especially when there are a large number of values to set in the array. An alternative for i-time and k-rate arrays is to use the `fillarray` opcode. In the following example, `fillarray` is used to create a one-dimensional, five-element array with values 0, 1, 2, 3 and 4. The resulting value is assigned to the `ivals` variable. As this is the first time `ivals` is used, the brackets are necessary to declare the variable's type. Subsequent uses only require the variable's name:

```
ivals[] fillarray 0, 1, 2, 3, 4
print lenarray(ivals)
```

The `fillarray` opcode can be used to initialise arrays after their creation as well:

```
kArr[] init 5
kArr fillarray 0, 1, 2, 3, 4
```

In this case, we will also be able to use it to initialise two-dimensional arrays. This is done in row-column order. For instance, a 2×2 matrix such as

$$\begin{pmatrix} 1 & 2 \\ 3 & 4 \end{pmatrix}$$

can be created with the following code:

```
iArr[][] init 2,2
iArr fillarray 1,2,3,4
```

4.4.2 Setting Values

To set values within an array, use the name of the array, followed by square brackets, and an *index*. You will need to use one index per dimension, and must use the same number of square brackets as there are dimensions for the array.

```
ivals[]      init 4
ivals2[][]   init 4, 4

ivals[0]   = 1
ivals[1]   = 2

ivals2[0][0] = 1
ivals2[0][1] = 2
```

In the above example, `ivals` is initialised as a one-dimensional array with size four. `ivals2` is initialised as a two-dimensional array with sizes four and four. By default, values within a newly initialised array are set to their default empty value. For example, i- and k-types default to 0, a-types default to a vector of 0's, strings default to empty strings (`""`) and so on.

Following initialisation, the first element of `ivals` is set to 1, and the second element is set to 2. The element number is determined by the index value given within the square brackets. Indexes are zero-based, meaning the first element is at index 0, and the last index is at size -1 (in this case, 3). For `ivals2`, the values of the first and second elements in the first group of elements are also set to 1 and 2.

Indexes for arrays may be constant numbers, but may also be expressions. In the following example, the variable `indx` is used to set the values 1 and 2 for the first two elements of the array:

```
ivals[] init 4
indx = 0
ivals[indx]   = 1
ivals[indx+1] = 2
```

The use of expressions for indexes is useful for programmatically setting values within an array, using programming constructs such as branching and loops. These will be introduced and discussed in Chapter 5.

4.4.3 Reading Values

Reading values from an array uses the same syntax as setting values in an array. As when writing of values, indexes for the array may be expressions. The following example prints out the value for each member of the ivals array:

```
ivals[] fillarray 1,2
print ivals[0]
print ivals[1]
```

4.4.4 Performance Time

As usual, i-time arrays are only executed at i-time, and thus their index will not change at performance time (e.g. a k-rate variable or expression for an i-time array will not force a reading/writing operation to take place). If we require arrays to be read/set, and indexes to change during performance, we need to use a k-rate array. It is important to note that even if the array contents do not change, but the indexing does, then an i-time array is not suitable. This is because indexing and unpacking (extracting the array value) are i-time operations.

Listing 4.4 k-rate array access example

```
instr 1
 kvals[] init 3
 kndx init 0
 kvals[0] = 1
 kvals[1] = 2
 kvals[2] = 3
 printk2 kvals[kndx]
 kndx = (kndx + 1) % lenarray(kvals)
endin
schedule(1,0,.1)
```

4.4.5 String and f-sig Arrays

As indicated above, arrays of strings can be created and manipulated. These allow indexing at initialisation and performance time, allowing which member is accessed to change from one k-cycle to another.

Listing 4.5 String array example

```
instr 1
 Svals[] init 2
 kval init 0
 Svals[0] = "hello"
 Svals[1] = "world"
 kval = kval % 2
 puts Svals[kval],kval+1
 kval +=1
endin
schedule(1,0,2/kr)
```

Similarly, f-sig arrays are also possible, and as they are a performance-time variable type, indexing at k-rate is allowed.

4.4.6 Arithmetic Expressions

Numeric arrays with scalar elements (i-time and k-rate) can also be combined directly into expressions using other arrays, scalar variables or constants. These can include addition (+), subtraction (-), multiplication (*), division (/) and exponentiation (^). The following example demonstrates this facility.

Listing 4.6 Arrays in arithmetic expressions

```
instr 1
 iArr[] fillarray 1,2,3
 iRes[] = (iArr * iArr) / 2
 print iRes[0]
 print iRes[1]
 print iRes[2]
endin
schedule(1,0,0)
```

Running this program, we will obtain the following printout at the console:

```
SECTION 1:
new alloc for instr 1:
instr 1:  #i2 = 0.500
instr 1:  #i3 = 2.000
instr 1:  #i4 = 4.500
```

Note that this is not extended to all types. Arrays of audio vectors (a-type) and f-sigs cannot be combined in this way, instead each member must be accessed separately via indexing:

```
nchnls = 2
instr 1
 aout[] init 2
 aout[0] diskin "fox.wav"
 aout[1] diskin "beats.wav"
 aout[0] = aout[0]*0.5
 aout[1] = aout[1]*0.5
    out aout
endin
```

Also note that the shorthand expressions +=, *= etc. are not defined for use with arrays. In this case, we need to write them in full, as in the example above (aout[0] = aout[0]*0.5).

4.4.7 Arrays and Function tables

Function tables can be considered as one-dimensional numeric arrays, that have global scope. It is possible to copy data from i-time and k-rate array variables into tables and vice versa. In the following example, we copy the contents of a function table into an array.

Listing 4.7 Copying the contents of a function table into an array

```
gifn ftgen 1,0,4,-2,1,2,3,4
instr 1
 iArr[] init ftlen(gifn)
 copyf2array iArr,gifn
 print iArr[0]
 print iArr[1]
 print iArr[2]
 print iArr[3]
endin
schedule(1,0,0)
```

Likewise, an array can be copied into a function table.

Listing 4.8 Copying the contents of an array into a function table

```
giArr[] fillarray 1,2,3,4
instr 1
 ifn ftgen 0,0,lenarray(giArr),2,0
 copya2ftab giArr,ifn
 print table(0,ifn)
```

```
 print table(1,ifn)
 print table(2,ifn)
 print table(3,ifn)
endin
schedule(1,0,0)
```

The interchangeability between the array and function table representations is explored by some opcodes. For instance, the standard oscillators (oscil, oscili, oscil3) can read from an i-time array instead of a table.

Listing 4.9 Using an array in place of a function table

```
instr 1
 iArr[] fillarray p6,p7,p8,p9
 amod oscil 1,2,iArr
 aout oscili p4,p5*amod
        out aout
endin
schedule(1,0,2,0dbfs/2,440,1,2,1.5,1.25)
schedule(1,2,2,0dbfs/2,440,1,1.5,2,1.25)
schedule(1,4,2,0dbfs/2,440,1,1.5,1.25,2)
schedule(1,6,2,0dbfs/2,440,1,1.25,1.5,2)
```

In this case, a simple arpeggiator is easily created by using a local array as the source for an oscillator, with varying patterns. Note that these opcodes have the restriction that array lengths are required to be a power-of-two value.

4.4.8 Audio Arrays

A useful application of audio arrays is to handle stereo and multichannel signals. In particular, diskin or diskin2 can write to arrays, which can then be passed directly to the output, as the opcode out can handle multiple channels held in an array:

```
nchnls=2
instr 1
 aout[] diskin2 "stereo.wav"
 out aout
endin
```

Arrays can also be used to hold various signals that will be mixed together, or further processed. In the next example, an a-sig array is initialised with one dimension of three members. Three vco2 oscillators are then used to write signals into each of the array member slots. Next their values are summed into the aout variable. The resulting signal is then sent to the out opcode.

Listing 4.10 Audio arrays handling multiple sources

```
instr 1
 asigs[] init 3
 asigs[0] = vco2(p4, p5)
 asigs[1] = vco2(p4, p6)
 asigs[2] = vco2(p4, p7)
 aout = asigs[0] + asigs[1] + asigs[2]
         out aout
endin
schedule(1,0,1,0dbfs/6,440,660,880)
```

In real-world Csound usage, we would likely use a loop to iterate over the array (which can be of varying length in that case) and the array signal source would also be separate from the processing code. For example, a source may be the `diskin` opcode, reading in multichannel audio files, and the array-processing code may be contained in a *user-defined opcode* (see Chapter 7).

4.5 Conclusions

This chapter explored advanced data types in Csound. Strings represent textual data, useful for creating diagnostic messages and file paths. Spectral-domain signals provide frequency-domain representations of audio signals, and present many new kinds of processing methods to the user. Finally, arrays provide ways of working with sets of ordered, homogenous data, simplifying the handling and processing of data meant to be processed together.

The data types discussed here open up a great number of options that were not available in the early versions of Csound. In particular, as will be explained in chapter 5, arrays enable a high level of programmability that goes hand in hand with control-flow constructs, allowing code to be made more concise and elegant. They can also be used as containers for vector and matrix operations (such as windowing and Fourier transforms) that will be very useful in a variety of sound and music computing applications.

Chapter 5
Control of Flow and Scheduling

Abstract In this chapter, we will examine the language support for program flow control in various guises. First, the text discusses conditional branching at initialisation and performance times, and the various syntactical constructs that are used for that. This will be followed by a look at loops, and their applications. The chapter continues with an examination of instrument scheduling at i-time and k-rate. Recursive instruments are explored. MIDI scheduling is introduced, and the supports for event duration control are outlined, together with an overview of the tied-note mechanism. The text is completed with a discussion of instrument reinitialisation.

5.1 Introduction

Instrument control consists of a specific set of opcodes and syntactical constructs that is essential to Csound programming. These allow us to control how and when unit generators are run, and to plan the performance of instruments to a very fine level of detail. This chapter will outline the main elements of program control offered by the language, such as branching, loops, instantiation, duration manipulation, ties and reinitialisation.

5.2 Program Flow Control

Csound has a complete set of flow control constructs, which can do various types of branching and iteration (looping). These work both at initialisation and performance times, and it is very important to keep the distinction between these two stages of execution very clear when controlling the flow of a program. As we will see, depending on the choice of opcodes, the jumps will occur only at i-time or at perf-time, or both. As before, the advice is to avoid intermixing of lines containing distinct initialisation and performance code so that the instrument can be read more

© Springer International Publishing Switzerland 2016   105
V. Lazzarini et al., *Csound*, DOI 10.1007/978-3-319-45370-5_5

easily. This is a simple means of avoiding making the mistake of assuming that the code will or will not execute at a given time.

5.2.1 Conditions

The flow of control in an instrument (or in global space) can be controlled by checking the result of conditions. These involve the checking of a variable against another, or against a constant, using a comparison operator, which yields a *true* or *false* result. They are also called *Boolean expressions*:

> a < b: true if a is smaller than b
> a <= b: true if a is smaller than or equal to b
> a > b: true if a is bigger than b
> a <= b: true if a is bigger than or equal to b
> a == b: true if a is equal to b
> a != b: true if a is not equal to b

In this case, a and b are either both scalar variables (i or k), or a variable and a constant. Furthermore, these operations can be combined into larger expressions with logical operators that work specifically with Boolean results:

> a && b: true only if a AND b are true, false otherwise
> a || b: true if a OR b is true, false if both are false

where a and b are Boolean expressions using comparison or logical operators. The execution of these at i- or k-time will depend on the type of branching employed.[1]

5.2.2 Branching

Csound allows branches to be created that depend on the value of Boolean expressions. These can be evaluated at initialisation or performance time, or both. Many of the branching constructs in Csound will use *labels*, which are names, placed anywhere in the code, that work as markers. The syntax for labels is:

```
label: ...
```

Labels are tokens composed of characters, numerals or both, and followed by a colon (:). They can be inserted at the beginning of any code line, or at a blank line.

---

[1] Also note that the conditional expression will not be evaluated at i-time at all if a k-variable is employed in it. In this case, any branching will be perf-time only. See Section 5.2.2 for details.

Initialisation-time only

The basic branching statement for init-time-only execution is

```
if  <condition>  then igoto label
```

It can also be written in an opcode form:

```
cigoto condition, label
```

These statements provide an i-time jump to a label if the condition evaluates to *true*. It allows the program to bypass a block of code, for instance

```
seed 0
ival = linrand(1)
if irnd  > 0.5 igoto second
 schedule(1,0,1)
second:
 schedule(2,0,1)
```

In this case, depending on the value of the `ival` variable, the code will schedule instruments 1 and 2 (false), or instrument 2 only (true). Another typical case is to select between two exclusive branches, for true and false conditions, respectively. In this case, we need to employ a second label, and an `igoto` statement,

```
igoto label
```

which always jumps to `label` (no condition check). So, if we were to select instrument 1 or 2, we could use the following code:

```
seed 0
ival = linrand(1)
if irnd  > 0.5 igoto second
 schedule(1,0,1)
igoto end
second:
 schedule(2,0,1)
end:
```

By inserting the `end` label after the second schedule line, we can jump to it if the condition is false. In this case, the code will instantiate instrument 1 or 2, but never both. Initialisation-time branching can be used in global space as in the example here, or inside instruments.

Note that placing perf-time statements inside these branching blocks might cause opcodes not to be initialised, which might cause an error, as the statements are ignored at performance time, and all branches are executed. For this reason, i-time-only branching has to avoid involving any perf-time code. One exception exists, which is when the *tied-note* mechanism is invoked, and we can deliberately bypass initialisation.

Initialisation and performance time

A second class of branching statements work at both i- and perf-time. In this case, we can mix code that is executed at both stages inside the code blocks, without further concerns. In this case, we can use

```
if  <condition>  then goto label
```

or

```
cggoto condition, label
```

with the limitation that the Boolean expression needs to be i-time only (no k-vars allowed). The jump with no conditions is also

```
goto label
```

The usage of these constructs is exactly the same as in the i-time-only case, but now it can involve perf-time code. An alternative to this, which sometimes can produce code that is more readable, is given by the form

```
if  <condition>  then
...
[elseif <condition> then
...]
[else
...]
endif
```

which also requires the condition to be init-time only. This form of branching syntax is closer to other languages, and it can be nested as required. In listing 5.1, we see the selection of a different sound source depending on a random condition. The branching works at i-time, where we see the printing to the console of the selected source name, and at perf-time, where one of the unit generators produces the sound.

Listing 5.1 Branching with `if – then – else`

```
seed 0
instr 1
 if linrand:i(1) < 0.5 then
  prints "oscillator \n"
  avar = oscili(p4,cpspch(p5))
 else
  prints "string model \n"
  avar = pluck(p4,cpspch(p5),cpspch(p5),0,1)
 endif
  out avar
endin
schedule(1,0,1,0dbfs/2,8.00)
```

Performance-time only

The third class of branching statements is composed of the ones that are only effective at performance time, being completely ignored at the init-pass. The basic form is

```
if  <condition>  then kgoto label
```

or

```
ckgoto condition, label
```

and

```
kgoto label
```

The alternative `if - then - endif` syntax is exactly as before, with the difference that the perf-time-only form uses k-rate conditions. In programming terms, this makes a significant difference. We can now select code branches by checking for values that can change during performance. This allows many new applications. For instance, in the next example, we monitor the level of an audio signal and mix the audio from another source if it falls below a certain threshold.

Listing 5.2 Branching with k-rate conditions

```
instr 1
 avar init 0
 avar2 = pluck(p4,cpspch(p5),cpspch(p5),0,1)
 krms rms avar2
 if krms < p4/10 then
   a1 expseg 0.001,1,1,p3-1,0.001
   avar = oscili(a1*p4,cpspch(p5))
 endif
 out avar+avar2
endin
schedule(1,0,5,0dbfs/2,8.00)
```

Note that we initialise `avar` to 0 since it always gets added to the signal, whether a source is filling it or not. The only way to guarantee that the variable does not contain garbage left over by a previous instance is to zero it. Various other uses of k-rate conditions can be devised, as this construct can be very useful in the design of instruments. When using performance-time branching it is important to pay attention to the position of any i-time statements in the code, as these will always be executed, regardless of where they are placed.

Time-based branching

Csound provides another type of performance-time branching that is not based on a conditional check, but on time elapsed. This is performed by the following opcode:

```
timout istart, idur, label
```

where istart determines the start time of the jump, idur, the duration of the branching and label is the destination of the jump. This relies on an internal clock that measures instrument time. When this reaches the start time, execution will jump to the label position, until the duration of the timeout is reached (the instrument might stop before that, depending, of course, on p3). The example below is a modification of the instrument in listing 5.2, making the addition of the second source time based. Note that the duration of the jump is for the remainder of the event.

Listing 5.3 Branching based on elapsed time

```
instr 1
 avar init 0
 avar2 = pluck(p4,cpspch(p5),cpspch(p5),0,1)
 timout 1,p3-1,sec
 kgoto end
 sec:
  a1 expseg 0.001,1,1,p3-2,0.001
  avar = oscili(a1*p4,cpspch(p5))
 end:
  out avar+avar2
endin
schedule(1,0,5,0dbfs/2,8.00)
```

Since many aspects of Csound programming are based on the passing of time, this type of branching can be very useful for a variety of applications. Note, however, that Csound offers many ways of time-based branching. So the code above could also be written using timeinsts(), to get the elapsed time of an instrument instance.

Listing 5.4 Another option for time-based branching

```
instr 1
 avar init 0
 avar2 = pluck(p4,cpspch(p5),cpspch(p5),0,1)
 if timeinsts() > 1 then
  a1 expseg 0.001,1,1,p3-3,0.001
  avar = oscili(a1*p4,cpspch(p5+.04))
 endif
 out avar+avar2
endin
schedule(1,0,5,0dbfs/2,8.00)
```

Also scheduling an instrument at a certain start time from inside another instrument is an option for time-based branching. This will be explained in Section 5.3.1.

Conditional assignment

Finally, Csound also provides a simple syntax for making assignments conditional:

```
xvar =  <condition> ?  true-expression : false-expression
```

This works at i- and perf-time:

```
avar = ivar > 0.5 ?
        pluck(p4,ifr,ifr,0,1) : oscili(p4,ifr)
```

or at perf-time only:

```
kvar = rms(avar) > 0dbfs/10 ? 1 : 0
```

This statement can also be nested freely with multiple depth levels:

```
instr 1
 kswitch line 2,p3,-1
 avar =  kswitch <= 0 ?
        oscili(p4,p5) : (kswitch <= 1 ?
        vco2(p4,p5) :  pluck(p4,p5,p5,0,1))
 out avar
endin
```

5.2.3 Loops

Csound allows the creation of iterative structures, called *loops*. As with branching, these can be i-, perf-time or both. The simplest types of loops can be created with goto etc. statements that make a jump to an earlier code line. Of course, in order to avoid eternal repetition, we also need a condition to be placed somewhere. For instance, an i-time loop to create 10 events with instrument 1 can be created with:

```
icnt init 0
loop:
if icnt == 10 igoto end
 schedule(1,icnt,2,0dbfs/2,8.00 + icnt/100)
 icnt += 1
igoto loop
end:
```

Similarly, loops on both init-pass and perf-time, or the latter alone, can be created with goto, kgoto and their associated conditional statements.

The loop_xx facility follows the same paradigm but simplifies the code, as it sets the increment and the break condition in a compact way:

```
icnt init 0
loop:
```

```
  schedule(1,icnt,2,0dbfs/2,8.00 + icnt/100)
loop_lt icnt,  1,  10,  loop
```

Until and while

Csound also offers the more common forms of `until` and `while` loops. These have the following syntax:

```
until <condition>  do
...
od
```

and

```
while <condition>  do
...
od
```

The difference between them is that `until` will loop when the condition is not true, whereas `while` keeps repeating if the condition is true. For instance, the same example given to create ten events on instrument 1 has the following form with `while`:

```
icnt init 0
while icnt < 10 do
  schedule(1,icnt,2,0dbfs/2,8.00 + icnt/100)
  icnt += 1
od
```

If we want to use `until`, then the condition test is `icnt >= 10`. These loops will work at initialisation and performance time if the condition is i-time, and at performance time only if the condition is k-rate.

Control-rate loops

The possibility of control-rate loops also opens up a number of interesting avenues, but a lot of care needs to be taken with these. Very often they can lead to code that misrepresent the underlying processes. For this reason, it is worth examining this mechanism in some detail.

Firstly, it is important to understand that a running opcode is an *object* with a *state*. In Csound programming, the syntax simplifies the process for the user, hiding the complexities of dealing with these objects. This bundles opcode creation, which involves allocating memory for its state (if needed), with the execution of its init- and perf-time subroutines in one single code entity. The following line,

```
a1 oscili  p4, p5
```

means three things:

1. Instantiate *one* (and one only) `oscili` object
2. Initialise it by running its i-time subroutine
3. Perform audio computation by calling its perf-time subroutine repeatedly, producing vectors of ksmps samples and placing them in the output variable `a1`.

While this helps to keep code tidy and objective, it can lead to a crucial misreading of the program. For instance, these lines in pseudocode

```
while kcnt < N do
 asig[kcnt] opcode kinput[kcnt]
 kcnt += 1
od
```

will instantiate one single copy of `opcode` and will run it repeatedly for `N` times each k-cycle. This is not the same as having `N` separate instances of the opcode, which are initialised and performed in succession, placing their output in an audio-rate array, which is often the intended and expected result.

There are occasions when we want to achieve the effect of running an opcode repeatedly, and loops can be used for that. To create sets of parallel opcodes, such as oscillator or filter banks, loops are not appropriate. For these applications, we should look to employ *recursion*, which will be explored later in this and other chapters.

Note that *functions* do not have state, by definition, and thus can be used without any concerns in a k-rate loop. They are invoked according to the rate of their input, and so can run exclusively at perf-time. With these, it is possible, with a bit more code, to realise designs such as a bank of sinusoidal oscillators. In the example shown in listing 5.5, we break down an oscillator into its components and recreate it from scratch so that we can generate a sound by adding sine waves of different amplitudes and frequencies. See also Section 3.5.2 for details on oscillators.

Listing 5.5 Control-rate loops for a bank of oscillators

```
instr 1
 apha init 0
 kph[] fillarray 0,0,0
 kamp[] fillarray 1,0.1,0.5
 kfr[] fillarray p5,p5*2.7,p5*3.1
 kcnt = 0
 asig = 0
 while kcnt < lenarray(kph) &&
       kcnt < lenarray(kamp) &&
       kcnt < lenarray(kfr) do
  ksmp = 0
  kpha = kph[kcnt]
  kf = kfr[kcnt]
  while ksmp < ksmps do
   kpha += (2*$M_PI*kf/sr)%(2*$M_PI)
```

```
   vaset(kpha,ksmp,apha)
    ksmp += 1
   od
   asig += kamp[kcnt]*sin(apha)
   kph[kcnt] = kpha
   kcnt += 1
  od
  k1 expseg p4,p3,0.001*p4
  out asig*k1
endin
schedule(1,0,1,0dbfs/2,300)
```

In this example, there is one loop nested inside another. The outer loop iterates over the number of oscillators needed, which is determined by the sizes of the phase, amplitude and frequency arrays. Each oscillator is made up of a phase update, a sine function lookup and amplitude scaling. The inner loop takes care of updating the phase. Since we are generating an audio signal, we need to create a phase vector `apha` sample by sample, which will contain the phases corresponding to the frequency of the oscillator. We use the opcode `vaset` to set the values of each sample in `apha`, so the loop iterates over ksmps. Once this is done, we can generate the signal by calling the `sin()` function and scaling its output. Note that we have avoided using all but one opcode in this loop. The one we employed is safe because it is actually stateless (it only copies a scalar k-var value into a position in an a-var vector), and so it works like a function. The example uses an envelope to create a decay to make it a percussive sound.

5.3 Scheduling

As we have already seen in earlier chapters, for instruments to execute and produce sound, they need to be instantiated in the engine. Csound has two main mechanisms for scheduling instances: real-time events and the numeric score. In this chapter, we will concentrate on the former, and the latter will be discussed in Chapter 8.

With regards to the output audio stream, events are scheduled at k-period boundaries. Their start time and duration are rounded to the nearest control block time. This means that there is a maximum quantisation error of $\frac{0.5}{kr}$. At the default value for kr (=4410, ksmps=10), this is equivalent to 0.11 ms, which is negligible. Even at the more common setting of ksmps = 64, the error is less than one millisecond. However, if this is not good enough, it is possible to run Csound at ksmps=1, or use the special `--sample-accurate` option (with ksmps > 1), for sample-level time event quantisation.

It is possible to issue new real-time events during performance in an instrument. When these are scheduled, their start time is relative to the next k-cycle, for the same reason as above, since Csound will sense new events in the intervals between computation. So, time 0 refers to this, even in sample-accurate mode, as the processing of a k-cycle is not interrupted when a new event is scheduled midway through it. This is also the behaviour if events are placed from an external source (e.g. a frontend through an API call).

5.3.1 Performance-Time Event Generation

So far, we have been using i-time scheduling in global space. But it is equally possible to place these schedule calls in an instrument and use them to instantiate other instruments. This can be done also at performance time, as indicated above, using the following opcode:

```
event Sop, kins, kst, kdur, [, kp4, ...]
```

Sop - a string containing a single character identifying the event type. Valid characters are i for instrument events, f for function table creation and e for program termination.

kins - instrument identifier, can be a number, i- or k-var, or a string if the instrument uses a name instead of a number for ID. This is parameter 1 (p1).

kst - start time (p2).

kdur - duration (p3).

kp4, ... - optional parameters (p4, ...)

This line will execute at every k-period, so it might need to be guarded by a conditional statement, otherwise new events will be fired very rapidly. In listing 5.6, we can see an example of an instrument that plays a sequence of events on another instrument during its performance time.

Listing 5.6 Performance-time event scheduling

```
instr 1
 ktime = timeinsts()
 k1 init 1
 if ktime%1 == 0 then
  event "i",2,0,2,p4,p5+ktime/100
 endif
endin

instr 2
 k1 expon 1,p3,0.001
 a1 oscili p4*k1,cpspch(p5)
 out a1
endin
schedule(1,0,10,0dbfs/2,8.00)
```

In addition to `event`, we also have its init-time-only version `event_i`, which follows the same syntax, but does not run during performance. Finally, another option for perf-time scheduling is given by `schedkwhen`, which responds to a trigger, and can also control the number of simultaneously running events.

5.3.2 Recursion

Since we can schedule one instrument from another, it follows that we can also make an instrument schedule itself. This is a form of *recursion* [2], and it can be used for various applications. One of these is to create multiple parallel copies of instruments that implement banks of operators. For instance, we can implement the bank of sinusoidal oscillators in listing 5.5 with recursive scheduling of instruments, using a loop.

Listing 5.7 Bank of oscillators using recursion

```
instr 1
 if p6 = 0 then
  iamp[]fillarray p4*0.1,p4*0.5
  ifr[] fillarray p5*2.7,p5*3.1
  icnt = 0
  while icnt < lenarray(iamp) &&
        icnt < lenarray(ifr) do
   schedule(1,0,p3,iamp[icnt],ifr[icnt],1)
   icnt += 1
  od
 endif
 k1 expon   1,p3,0.001
 a1 oscili p4,p5
   out a1*k1
endin
schedule(1,0,2,0dbfs/4,300,0)
```

We can also use recursion to create streams or clouds of events with very little effort, by giving a start time offset to each new recursive instance. In this case, there is no actual need for conditional checks, as long as we set the value of p2 no smaller than one control period (otherwise we would enter an infinite recursion at i-time). Here's a minimal instrument that will create a stream of sounds with random pitches and amplitudes (within a certain range).

Listing 5.8 Bank of oscillators using recursion

```
seed 0
instr 1
 k1 expon   1,p3,0.001
 a1 oscili p4,p5
```

```
 out a1*k1
 schedule(1,.1,.5,
    linrand(0dbfs/10),
    750+linrand(500))
endin
schedule(1,0,1,0dbfs/4,300)
```

Each new event comes 0.1 seconds after the previous and lasts for 0.5 seconds. The sequence is never-ending; to stop we need to either stop Csound or compile a new instrument 1 without the recursion. It is possible to create all manner of sophisticated recursive patterns with this approach.

5.3.3 MIDI Notes

Instruments can also be instantiated through MIDI NOTE ON channel messages. In this case, the event duration will be indeterminate, and the instance is killed by a corresponding NOTE OFF message. These messages carry two parameters, note number and velocity. The first one is generally used to control an instrument pitch, while the other is often mapped to amplitude. The NOTE ON - NOTE OFF pair is matched by a share note number. More details on the operation of Csound's MIDI subsystem will be discussed in Chapter 9.

5.3.4 Duration Control

Instruments can be scheduled to run indefinitely, in events of indeterminate duration that are similar to MIDI notes. We can do this by setting their duration (p3) to a negative value. In this case, to kill this instance, we need to send an event with a negative matching p1.

Listing 5.9 Indeterminate-duration event example

```
instr 1
 a1 oscili p4,cpspch(p5)
 out a1
endin
schedule(1,0,-1,0dbfs/2,8.00)
schedule(-1,5,1,0dbfs/2,8.00)
```

It is possible to use fractional instrument numbers to mark specific instances, and kill them separately by matching their p1 with a negative value. The syntax for this is num.instance:

```
schedule(2.1,0,-1,0dbfs/2,8.00)
schedule(-2.1,5,1,0dbfs/2,8.00)
```

```
schedule(2.2,1,-1,0dbfs/2,8.07)
schedule(-2.2,4,1,0dbfs/2,8.07)
```

The important thing to remember is that in these situations, p3 is invalid as a duration, and we cannot use it in an instrument (e.g. as the duration of an envelope). For this purpose, Csound provides a series of envelope generators with an associated release segment. These will hold their last-generated value until deactivation is sensed, and will enter the release phase at that time. In this case the instrument duration is *extended* by this extra time. Most opcode generators will have a release-time version, which is marked by an 'r' added to the opcode name: linsegr is the r-version of linseg, expsegr of expseg, linenr of linen etc. Note that these can also be used in definite-duration events, and will extend the p3 duration by the release period defined for the envelope.

For instance, the envelope

```
k1 expsegr 1,1,0.001,0.1,0.001
```

goes from 1 to 0.001 in 1 s, and holds that value until release time. This extends the instrument duration for 0.1 s, in which the envelope goes to its final value 0.001. The two final parameters determine release time and end value. If the instance is deactivated before 1 s, it jumps straight into its release segment. It can be used in the instrument in 5.6, to shape the sound amplitude.

Listing 5.10 Using a release-time envelope with an indefinite-duration event

```
instr 1
 k1 expsegr 1,1,0.001,0.1,0.001
 a1 oscili p4*k1,cpspch(p5)
 out a1
endin
schedule(1,0,-1,0dbfs/2,8.00)
schedule(-1,1,1,0dbfs/2,8.00)
```

Similarly, the other release time will add a final segment to the envelope, which extends the duration of the event. If multiple opcodes of this type are used, with different release times, the extra time will be equivalent to the longest of these. The Reference Manual can be consulted for the details of r-type envelope generators.

5.3.5 Ties

Together with negative (held) durations, Csound has a mechanism for *tied notes*. The principle here is that one instance can take the space of, and replace, an existing one that has been held indefinitely. This happens when an event with matching (positive) p1 follows an indefinite duration one. If this new note has a positive p3, then it will be the end of the tie, otherwise the tie can carry on to the next one.

In order to make the new event 'tie' to the previous one, it is important to avoid abrupt changes in the sound waveform. For instance, we need to stop the envelope cutting the sound, and also make the oscillator(s) continue from their previous state, without resetting their phase. Other opcodes and sound synthesis methods will also have initialisation steps that might need to be bypassed.

For this purpose, Csound provides some opcodes to detect the existence of a tied note, and to make conditional jumps in the presence of one:

```
ival tival
```

will set `ival` to one on a tied event, and zero otherwise, whereas

```
tigoto label
```

executes a conditional jump at i-time (like `igoto`) only on a tie. This is used to jump the initialisation of some opcodes, such as envelopes, which allow it. There are, however, other opcodes that cannot be run without initialisation, so this mechanism cannot be used with them. They will report an error to alert the user that they need to be initialised.

The following example in listing 5.11 demonstrates the tied-note mechanism with a very simple instrument consisting of an envelope, a sawtooth oscillator (`vco2`) and a filter (`moogladder`). The tied note is detected in `itie`, and this value is used to skip parts of the oscillator and filter initialisation. These opcodes have optional parameters that can be set to the `tival` value to control when this is to be done (check the Reference Manual for further information on this).

We also use the conditional jump to skip the envelope initialisation. This means that the duration parameters are not changed. Additionally, its internal time count is not reset and continues forward. Since we set the original envelope duration to the absolute value of the p3 in the first event, this should be used as the total duration of *all* of the tied events. We set the p3 of whatever is the final note to the remaining duration after these. If no time remains, we close the tie with an event of 0 duration.

Listing 5.11 Tied event example

```
instr 1
 itie tival
 tigoto dur
 env:
  ist = p2
  idur = abs(p3)
  k1 linen 1,0.2,idur,2
 dur:
  if itie == 0 igoto osc
   iend = idur + ist - p2
   p3 = iend > 0 ? iend : 0
 osc:
  a1 vco2 p4*k1,cpspch(p5),0,itie
  a2 moogladder a1,1000+k1*3000,.7,itie
  out a1
endin
schedule(1,0,-6,0dbfs/2,8.00)
```

```
schedule(1,2,-1,0dbfs/2,8.07)
schedule(1,3,-1,0dbfs/2,8.06)
schedule(1,4,1,0dbfs/2,8.05)
```

Note that treating ties often involves designing a number of conditional branches in the code (as in the example above). In order to make multiple concurrent tied streams the fractional form for p1 can be used to identify specific instances of a given instrument.

5.4 Reinitialisation

The initialisation pass of an instrument can be repeated more than once, through *reinitialisation*. This interrupts performance while the init-pass is executed again. This can be done for selected portions of an instrument code, or for all of it:

```
reinit label
```

will start a reinitialisation pass from label to the end of the instrument or the rireturn opcode, whichever is found first. The next example illustrates reinitialisation. During performance, timout jumps to the end for $\frac{1}{4}$ of the total duration. After this, a reinitialisation stage starts from the top, and the timout counter is reset. This makes it start jumping to the end again for the same length, followed by another reinit-pass. This is repeated yet another time:

```
instr 1
 icnt init 1
 top:
  timout 0, p3/4,end
  reinit top
 print icnt
 icnt += 1
 rireturn
end:
endin
schedule(1,0,1)
```

The printout to the console shows how the i-time variables get updated during reinitialisation:

```
SECTION 1:
new alloc for instr 1:
instr 1:   icnt = 0.000
instr 1:   icnt = 1.000
instr 1:   icnt = 2.000
instr 1:   icnt = 3.000
```

Another example shows how `reinit` can be used to reinitialise k-time variables:

```
instr 1
 puts "start", 1
 top:
 puts "reinit", 1
 kcnt init 0
 if kcnt > 10 then
  reinit top
 endif
 printk2 kcnt
 kcnt += 1
endin
schedule(1,0,0.005)
```

This will print out the following:

```
start
reinit
 i1       0.00000
 i1       1.00000
 i1       2.00000
 i1       3.00000
 i1       4.00000
 i1       5.00000
 i1       6.00000
 i1       7.00000
 i1       8.00000
 i1       9.00000
 i1      10.00000
reinit
 i1       0.00000
 i1       1.00000
 i1       2.00000
 i1       3.00000
 i1       4.00000
 i1       5.00000
 i1       6.00000
 i1       7.00000
 i1       8.00000
 i1       9.00000
 i1      10.00000
```

Csound also offers a special goto jump, active only at reinit time:

```
rigoto label
```

Any perf-time opcodes inside a reinit block will be reinitialised, as shown in the example above, with `timout`.

5.5 Compilation

In addition to being able to instantiate, initialise, perform and reinitialise, the Csound language is also capable of compiling new code on the fly. This can be done via a pair of init-time opcodes:

```
ires compilestr Sorch
ires compileorc Sfilename
```

The first opcode takes in a string containing the code to compiled, and the second the name of a plain text file containing the Csound program, returning a status code (0 means successful compilation). These opcodes allow new instruments to be added to an existing performance. Once they are compiled, they can be scheduled like any other existing instrument. If an instrument has the same number as an existing one, it will replace the previous version. New events will use the new definition, but any running instance will not be touched.

In listing 5.12, we show a simple example of how an instrument can be compiled by another. Instrument 1 contains code to run at i-time only, which will compile a string provided as an argument, and then schedule the new instrument, if successful.

Listing 5.12 Compiling code provided as a string argument

```
instr 1
 S1 = p4
 if compilestr(S1) == 0 then
  schedule(2,0,1,p5,p6)
 endif
endin
schedule(1,0,0,
{{
 instr 2
  k1 expon 1,p3,0.001
  a1 oscili p4*k1,p5
  out a1
 endin
}}, 0dbfs/2, 440)
```

Although in this case we have a more complicated way of doing something that is quite straightforward, the code above demonstrates a powerful feature of Csound. This allows new instruments, provided either as strings or in a text file, to be dynamically added to a running engine.

5.6 Conclusions

This chapter explored the many ways in which we can control instruments in a Csound program. We have explored the standard flow control structures, such as branching and looping, and saw how they are implemented within the language, in their initialisation and performance-time forms. In particular, we saw what to expect when opcodes are used inside loops, and that we need to be especially careful not to misread the code.

The various details of how instruments are scheduled were discussed. We explored how instruments can instantiate other instruments at both init and performance time. The useful device of recursion, when an instrument schedules itself, was explored with two basic examples. The text also studied how event durations can be manipulated, and how the tied-note mechanism can be used to join up sequences of events. The possibility of reinitialisation was introduced as another means of controlling the execution of Csound instruments.

Chapter 6
Signal Graphs and Busses

Abstract This chapter discusses how instruments are constructed as signal graphs, looking at how unit generators (opcodes, functions, operators) are linked together through the use of variables. We also detail the order of execution of code inside instruments, and how instances of the same or different instruments are sequenced inside a k-cycle. We introduce the notion of patch-cords and busses, as metaphors for the various types of connections that we can make in Csound program code. To conclude the chapter, the most important mechanisms for passing data between separate instances of instruments are examined in detail.

6.1 Introduction

Although Csound instruments are independent code objects, it is possible to connect them together in different ways. This is often required, if for instance we want to use audio effects that apply to all sound-generating instruments, rather than to each instance separately. This is a typical feature of music systems such as synthesis workstations and sequencing and multi-tracking programs. In these types of applications, ideally we have a single instrument instance implementing the desired effect, to which we send audio from various sources in different instruments. Connecting instances in this way creates dependencies between parts of the code, and it is important to understand how the signal flows from origin to destination, and how the execution of the various elements is ordered.

In this chapter, we will first review how unit generators are connected together to make up the instrument signal graph, looking at how variables are used to link them up and how the execution of code is sequenced. We will also look at how instances of the same and of different instruments are ordered in a k-cycle. This will be followed by a detailed discussion of various methods that can be used to set up connections between instruments.

© Springer International Publishing Switzerland 2016
V. Lazzarini et al., *Csound*, DOI 10.1007/978-3-319-45370-5_6

6.2 Signal Graphs

A Csound instrument can be described as a graph of unit generators connected together using variables, which can be thought of, conceptually, as patch-cords. This metaphor is useful to understand their operation, although it should be stressed that there are other, equally valid, interpretations of the role and behaviour of Csound program code.

In any case, whenever we connect one unit generator (opcode, function, arithmetic operator) into another, such a patch-cord will be involved, either explicitly or implicitly. We can picture an instrument as a graph whose nodes are its unit generators, and the connecting lines, its variables. For instance, we can make an oscillator modulate the amplitude of another with the following code excerpt:

```
asig oscili oscili(kndx,kfm)+p4,kfc
```

This creates the signal graph shown in Fig. 6.1. The compiler will create two *synthetic* k-rate variables, and use them to connect the first oscillator to the addition operator, and the output of this to the second oscillator amplitude. These variables, marked with a # and hidden from the user, are crucial to define the graph unambiguously. In other words, the single line of code is unwrapped internally in these three steps:

```
#k1 oscili kndx,kfm
#k2 = p4 + #k1
asig oscili #k2,kfc
```

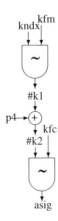

Fig. 6.1 The representation of a signal graph in Csound, which involves two synthetic variables created by the compiler to connect the unit generators defined in it

The compiler creates the structure of the signal graph from the user code. The actual executable form of this is only formed when an instance of the instrument is allocated. At that point the memory for the variables, as well as for each opcode, is reserved, and the connections to the unit generators are established. Any init-pass routine that exists in the instrument is then run. When the instance performs, the order in which the different elements are placed in the code will determine the execution sequence. In the example above, this is easy to determine:

1. The values of the variables `kndx` and `kfm` are read by the modulator oscillator, which then executes its perf-time routine, storing the result in the `#k1` memory location.
2. The value found in `p4` is added to that in `#k1` and stored in `#k2`.
3. The audio-rate oscillator takes the values in `#k2` and `kfc` as its amplitude and frequency and produces a vector of ksmps samples that is stored in `asig`.

Note that the order is determined line by line. So, if the instrument were written as

```
k1 init 0
asig oscili k1+p4,kfc
k1 oscili kndx, kfm
```

then the execution order would be different, with the audio oscillator (the carrier) producing its output before the modulator runs in the same k-cycle. Note that in order for the compiler to accept this, we need to explicitly initialise the variable `k1`. This also means that the modulator will be delayed in relation to the carrier. This is not a natural way to set up an instrument, but in some situations we might want to re-order the signal graph to realise a specific effect. For instance, if we want to create a feedback path, we need to order things in a precise way:

```
asig init 0
asig oscili asig+p4,kfc
```

In this case, we are feeding the output of the oscillator into its own amplitude input, a set up that is called *feedback amplitude modulation* (FBAM) [59]. There will always be a one-ksmps block delay in the feedback path. In Csound, it is easy to be very clear about how the signal path and execution sequence are constructed, with the language generally helping us to use the most natural order of operations without any extra effort.

6.3 Execution Order

All active instances of instruments will perform once in each k-cycle. While the execution order of unit generators in an instrument is determined by their place in the code, the performance of instruments and their instances is organised in a different way, although this is also clear and well defined. The sequence of execution is particularly important when we start connecting one instrument to another via the various types of busses.

6.3.1 Instances

When two or more instances of the same instrument are running at the same time, their order of performance is defined by the following rules:

1. if the value of their p1 is the same, then they are ordered by their instantiation time; if that is the same, then they are performed according to the event order.
2. if the value of their p1 differs (in the case of fractional values for p1, which can be used to identify specific instances), then the order is determined by ascending p1.

For instance, the following events with p4 set to 1, 2, 3 (to label their instances)

```
schedule(1,0,1,1)
schedule(1,0,2,2)
schedule(1,1,1,3)
```

will be performed in order 1, 2 (0-1 seconds), and then 2, 3 (1-2 seconds). However, the following events with different p1 values

```
schedule(1.03,0,1,1)
schedule(1.02,0,2,2)
schedule(1.01,1,1,3)
```

will execute in the order (p4 labels) 2, 1 and 3, 2. So a fractional p1 value can also be used to control the sequence in which instances are performed.

Note that these rules apply only to performance-time execution, which is normally where they matter more. The init-time sequence, in the case of events of the same instrument starting at the same time, is always determined by their instantiation order, and disregards any difference in the values of their p1.

6.3.2 Instruments

The execution sequence of instruments at performance time is determined by their number, in ascending order. So in a k-cycle, lower-number instrument instances are performed first. For this reason, if we have an instrument that is supposed to receive input from others, then we should try to give it a high number. If an instance of one instrument receives audio or control signals from a higher-order source, then there will always be a one-k-period delay in relation to it. While this might not be significant at times, it can be crucial in some applications, and so it is important to make sure the order of performance is exactly as we want it.

Again, these rules are only relevant to performance time. The order of init-pass execution is determined by the time/sequence in which events are scheduled, regardless of instrument order. In summary, within one k-cycle, opcodes, functions and operators are executed in the order they appear in an instrument (and according to specific precedence rules in the case of arithmetic); instances of the same instrument are ordered by schedule time and sequence, or by ascending p1; instances of different instruments are executed in ascending order.

Named instruments

Named instruments are transformed internally into numbers, in the order in which they appear in the orchestra. To achieve a desired sequence of execution, we must take notice of this and organise them by their line numbers. We should also avoid mixing named and numbered instruments.

6.4 Busses

An extension of the patch-cord metaphor, the *bus* is an element used to take signals from one or more sources and deliver them to one or more destinations. The end points of a bus are different instrument instances, and so they work as global patch-cords between different parts of the code. The role of a bus is very important if we want to implement things such as global effects and various types of modulation connections. Busses can be implemented in a variety of ways. In this section, we will examine the different possibilities and their applications.

6.4.1 Global Variables

Global variables are the original means of connecting instruments. Any internal type can be made global by adding a g in front of its name, including arrays:

```
gasig, gavar[]
gksig, gkvar[]
gival, givar[]
gfsig, gfsp[]
gStr, gSname[]
```

As with local variables, before global variables are used as input, they need to be explicitly declared or used as an output. In many cases, this means we need to initialise them in global space, before they can be used as busses:

```
gasig init 0
```

The usual practice is to add the source signal to the bus, possibly with a scaling factor (p6 in this case):

```
instr 1
 asrc oscili p4, p5
```

```
out asrc
gasig += asrc*p6
endin
```

where the += operator adds the right-hand side to the contents of the left-hand side. The reason for doing this is to avoid an existing source signal being overwritten by another. By summing into the bus, instead of assigning to it, we can guarantee that multiple sources can use it. At a higher-order instrument, we can read the bus, and use the signal:

```
instr 100
 arev reverb gasig, 3
 out arev
endin
```

The bus signal can feed other instruments as required. Once all destinations have read the bus, we need to clear it. This is very important, because as the signals are summed into it, the bus will accumulate and grow each k-cycle, if left uncleared. We can do this inside the last instrument using it, or in one with the specific purpose of clearing busses:

```
instr 1000
 gasig = 0
endin
```

In addition to audio variables, it is also possible to send other types of signals in global variable busses, including controls, spectral signals and strings. Arrays of global variables can also be very useful if we want to create flexible means of assigning busses. Listing 6.1 shows how we can use an array to route signals to different effects, depending on an instrument parameter (p6).

Listing 6.1 Using an array to route signals to two different effects

```
gabuss[] init 2

instr 1
 aenv expseg 0.001,0.01,1,p3,0.001
 asig oscili aenv*p4,p5
 out asig
 gabuss[p6] = asig*0.5 + gabuss[p6]
 schedule(1,0.1,0.3,
          rnd(0.1)*0dbfs,
          500+gauss(100),
          int(rnd(1.99)))
endin
schedule(1,0,0.5,
         0dbfs*0.1,
         500,0)
```

```
instr 100
 arev reverb gabuss[0],3
 out arev
 gabuss[0] = 0
endin
schedule(100,0,-1)

instr 101
 adel comb gabuss[1],4,0.6
 out adel
 gabuss[1] = 0
endin
schedule(101,0,-1)
```

The sound-generating instrument 1 calls itself recursively, randomly selecting one of the two effect busses. Note that we have to explicitly add the signal to the bus, with the line

```
gabuss[p6] = asig*0.5 + gabuss[p6]
```

as the += operator is not defined for array variables.

6.4.2 Tables

Function tables are another type of global objects that can be used to route signals from instrument to instrument. Because they can store longer chunks of data, we can also use them to hold signals for a certain amount of time before they are consumed, and this can allow the reading and writing to the bus to become decoupled, asynchronous. In listing 6.2, we have an application of this principle. A function table is allocated to have a 5-second length, rounded to a complete number of ksmps blocks. Instrument 1 writes to it, and instrument 2 reads from it, backwards. The result is a reverse-playback effect. The code uses one source and one destination, but it is possible to adapt for multiple end points.

Listing 6.2 Using a function table to pass audio signals asynchronously from one instrument to another

```
idur = 5
isamps = round(idur*sr/ksmps)*ksmps
gifn ftgen 0,0,-isamps,2,0

instr 1
 kpos init 0
 asig inch 1
 tablew asig,a(kpos),gifn
 kpos += ksmps
```

```
kpos = kpos == ftlen(gifn) ? 0 : kpos
endin
schedule(1,0,-1)

instr 2
 kpos init ftlen(gifn)
 asig table a(kpos),gifn
 kpos -= ksmps
 kpos = kpos == 0 ? ftlen(gifn) : kpos
 out asig
endin
schedule(2,0,-1)
```

Tables can also be used to route conveniently control signals, as these consist of single variables and so each table slot can hold a control value. Using `tablew` and `table` opcodes, we can write and read these signals. A more complex approach is provided by the `modmatrix` opcode, which uses function tables to create complex routings of modulation signals. It uses three tables containing the input parameters, the modulation signals and the scaling values to be applied. It writes the resulting control signals into another table:

$$R = I + M \times S \tag{6.1}$$

where I is a vector of size n containing the input parameters, M is a vector of size m containing the modulation values, and S is a scaling matrix of size $m \times n$ with the scaling gains for each modulation signal. The resulting vector R contains the combined modulation sources and inputs for each parameter n. For example, let's say we have two modulation sources, three parameters and a modulation matrix to combine these:

$$\begin{pmatrix} 0.4 & 0.6 & 0.7 \end{pmatrix} + \begin{pmatrix} 0.75 & 0.25 \end{pmatrix} \times \begin{pmatrix} 0.1 & 0.5 & 0.2 \\ 0.5 & 0.1 & 0.3 \end{pmatrix} = \begin{pmatrix} 0.6 & 1.0 & 0.925 \end{pmatrix} \tag{6.2}$$

The vectors and matrix are defined as function tables. In the case of eq. 6.2, we would have:

```
ipar ftgen 1,0,3,-2, 0.4, 0.6, 0.7
imod ftgen 2,0,2,-2, 0.75, 0.25
iscal ftgen 3,0,6,-2,0.1,0.5,0.2,0.5,0.1,0.3
ires  ftgen 4,0,3,2, 0,0,0
```

where `ipar`, `imod`, and `iscal` are the tables for input parameters, modulation sources and scaling matrix, respectively. Note that the matrix is written in row, column format. The results are held in a table of size three, which is represented by `ires`. Tables can be initialised using any GEN routine (here we used GEN 2, which just copies its parameter values to each table position). The `modmatrix` opcode itself is defined as:

```
modmatrix ires, imod, ipar, iscal, inm, inn, kupdt
```

The first four parameters are the table numbers for results, modulation, parameters and scaling matrix. The arguments `inm` and `inn` are the number of sources and parameters, respectively. The `kupdt` parameter is set to non-zero to indicate that the scaling matrix has changed. This should be zero otherwise, to make the opcode run efficiently. It is expected that the modulation signals and input parameters will be written into the table as necessary (using `tablew`), and that the results will be read using `table` at every k-cycle. Only one instance of the opcode is needed to set a modulation matrix, and the sources and destinations can be spread out through the instruments in the code. A simple example using the tables above is shown in listing 6.3

Listing 6.3 Modulation matrix example using two sources and three destinations

```
gipar ftgen 1, 0, 3, -2, 0.4, 0.6, 0.7
gimod ftgen 2, 0, 2, -2, 0, 0
giscal ftgen 3, 0, 6, -2,0.1,0.5,0.2,0.5,0.1,0.3
gires ftgen 4, 0,3,2,0,0,0

instr 1
 k1 oscili p4,p5
 tablew k1,p6,gimod
endin
schedule(1,0,-1,1,0.93,0)
schedule(1,0,-1,1,2.05,1)

instr 2
 kenv linen p4,0.01,p3,0.1
 k1 table p6,gires
 a1 oscili k1*kenv,p5
 out a1
 schedule(2,0.5,1.5,
          rnd(0.1)*0dbfs,
          500+gauss(400),
          int(rnd(2.99))))
endin
schedule(2,0,1.5,0dbfs*0.1,500,0)

instr 10
 kupdt init 1
 modmatrix gires,gimod,gipar,giscal,2,3,kupdt
 kudpt = 0
endin
schedule(10,0,-1)
```

In this example, we have two modulation sources (low-frequency oscillators, LFOs, at 0.93 and 2.05 Hz), and distribute a combination of these modulations to three independent destinations, modulating their amplitude. The sound generating instrument 2 is recursively called, randomly picking a different destination (from `gires`) number out of three choices (p6). In this example, we do not update the scaling matrix, and also keep the parameter offset values fixed (in `gipar`). Note that the modulation table values will change every k-cycle (0.75 and 0.25 were only used as sample values for eq. 6.2), as the LFOs oscillate, so we just initialise the `gimod` table with zeros.

6.4.3 Software Bus

Csound also includes a very powerful software bus, which can be used by frontends and hosts to connect with external controls and audio signals, via API calls. It is also available to be used internally by instruments to communicate with each other. In these applications, it can replace or add to the other methods discussed above. It can be used freely with i-, k-, a- and S-variables.

The software bus works with a system of named *channels*, which can be opened for reading, writing and bidirectionally. When used internally, they always work in bidirectional mode (as we will be both sending and receiving data inside a Csound orchestra). These channels can optionally be declared and initialised using the following opcodes for audio, strings and control data, respectively:

```
chn_a Sname, imode
chn_S Sname, imode
chn_k Sname, imode
```

Channels are given an `Sname`, which will identify them, and set read-only (`imode` = 1), write-only (2), or both (3). To set the value of a channel, we use

```
chnset xvar, Sname
```

where the type of `xvar` will depend on the kind of channel being set (string, audio, or control). For audio channels, we also have

```
chnmix avar, Sname
```

for summing into a channel, rather than overwriting it. Once we have consumed the audio, we need to clear it (as we have seen with global variables):

```
chnclear Sname
```

Once a channel exists, data can be read from it using

```
xvar chnget Sname
```

The software bus can be used instead of global variables. Listing 6.4 shows an example where a source instrument copies its output to a channel called `reverb`, which is then read by an instrument implementing a global reverberation effect.

Listing 6.4 Using the software bus to send signals to a reverb effect

```
instr Sound
 aenv linen p4, 0.01,p3,0.1
 asrc oscili aenv, p5
 out asrc
 chnmix asrc*p6, "reverb"
endin
schedule("Sound",0,0.5, 0dbfs/4, 400, 0.1)
schedule("Sound",1,0.5, 0dbfs/4, 500, 0.2)
schedule("Sound",2,0.5, 0dbfs/4, 300, 0.4)

instr Reverb
  asig chnget "reverb"
  arev reverb asig, 3
        out arev
  chnclear "reverb"
endin
schedule("Reverb",0,-1)
```

It is also possible to access the software bus from global variables, by linking them together through the chnexport opcode:

```
gxvar chnexport Sname, imode
```

This mechanism allows a channel to be accessible directly through a global variable.

As channels can be created dynamically, with unique names which are generated on the fly, they offer an easy way to communicate between instrument instances whose number is not yet defined at the beginning of the performance. Listing 6.5 demonstrates this by dynamically generating channel names.

Listing 6.5 Generating channel names dynamically

```
seed 0
giCount init 0
instr Create_Sender
 kCreate metro randomh:k(1,5,1,3)
 schedkwhen kCreate,0,0,"Sender",0,p3
endin
schedule("Create_Sender",0,2.5)

instr Sender
 giCount += 1
 S_chn sprintf "channel_%d", giCount
 schedule "Receiver",0,p3,giCount
 chnset randomh:k(1,100,1,3),S_chn
endin
```

```
instr Receiver
 kGet chnget sprintf("channel_%d",p4)
 if changed(kGet)==1 then
   printks "time = %.3f, channel_%d = %d\n",0,times:k(),
           p4,kGet
 endif
endin
```

The printout will show something like this:

```
time = 0.006, channel_1 = 50
time = 0.215, channel_2 = 23
time = 0.424, channel_3 = 95
time = 0.630, channel_4 = 95
time = 0.839, channel_5 = 44
time = 1.007, channel_1 = 77
time = 1.112, channel_6 = 25
time = 1.216, channel_2 = 63
time = 1.425, channel_3 = 30
time = 1.631, channel_4 = 83
time = 1.646, channel_7 = 41
time = 1.840, channel_5 = 61
time = 2.009, channel_1 = 62
time = 2.113, channel_6 = 65
time = 2.125, channel_8 = 94
time = 2.218, channel_2 = 59
time = 2.426, channel_3 = 84
time = 2.485, channel_9 = 85
time = 2.633, channel_4 = 26
time = 2.647, channel_7 = 30
time = 2.842, channel_5 = 64
time = 3.114, channel_6 = 24
time = 3.126, channel_8 = 25
time = 3.486, channel_9 = 57
time = 3.648, channel_7 = 91
time = 4.127, channel_8 = 61
time = 4.487, channel_9 = 9
```

The schedkwhen opcode is used to generate events at performance time, depending on a non-zero trigger (its first argument), which can be taken conveniently from a metro instance that produces such values periodically.

6.5 Conclusions

In this chapter, we reviewed the principles underlying the construction of signal graphs for instruments. We saw that inside each instrument, execution is sequential, line by line, and that, in some cases, operations are unwrapped by the compiler into separate steps. Variables are used as patch-cords to link unit generators together, explicitly, in the case of the ones declared in the code, and implicitly, when the compiler creates hidden, synthetic variables. We have also noted how the language coerces the user into employing the most natural connections, but there is full control of the execution sequence, and we can easily program less usual signal paths, such as feedback links.

The order of execution of instruments was also detailed. In summary, same-instrument instances will be performed by event and start time order, unless they use fractional p1 values, in which case this will be used as the means to determine the execution sequence inside a k-cycle. Different instruments are ordered by ascending p1, so higher-numbered ones will always come last in the sequence. We have also noted that this only applies to performance time, and the init-pass is always executed in event order.

The chapter concluded with a look at different ways we can connect instruments together, concentrating on global variables, function tables and the software bus. We saw how global variables can be used as busses, which can be fed from several sources and have multiple destinations. These are very flexible means of connecting instruments, but care needs to be taken to clear the variables after use. Function tables are also global code objects that can be used to connect instruments, in particular if we want to do it in a decoupled way. They can also hold control signals, and the modmatrix opcode provides an added facility to distribute modulation sources around instruments. Finally, we introduced the software bus, which can be used to connect Csound with external controls provided by hosts or frontends, and also to link up instruments internally, implementing busses to send signals from different sources to various destinations.

Chapter 7
User-Defined Opcodes

Abstract In this chapter, we will examine the different ways in which Csound code can be composed into bigger units. This is enabled by the user-defined opcode (UDO) mechanism. After looking at its syntax in detail, we will explore the various aspects of the UDO operation. We will examine the idea of local control rate and how it can be useful in signal processing applications. A very powerful programming device, *recursion*, will be introduced to the UDO context, with a discussion of its general use cases. To complete this chapter, we will look at the concept of subinstruments, which provide another means of composing Csound code.

7.1 Introduction

Although the Csound language has a large collection of opcodes, there are always situations where a new algorithm is required. In this case, there is a need for mechanisms to add new unit generators to the system. The Csound language can be expanded in two ways:

1. Plug-ins: through the addition of opcodes, written in C or C++, and compiled into dynamic libraries.
2. User-defined opcodes (UDOs): new opcodes written in the Csound language proper, which can be used in instruments in a similar way to existing unit generators.

The first method is used in situations where there is a real efficiency requirement, or in the very few occasions where some programming device not supported by the Csound language is involved. It is generally straightforward to create new plug-in opcodes, but this topic is beyond the scope of this book (an introduction to it is found in the Reference Manual). In this chapter we would like, instead, to explore the various ways in which opcodes can be created using Csound code. This is often called the *composability* element of a language, by which smaller components can be put together and used as a single unit elsewhere.

7.2 Syntax

As with instruments, UDOs are defined by a code block, which is set between two keywords `opcode` and `endop`. The general form is shown in listing 7.1.

Listing 7.1 Defining a new UDO

```
opcode <name>,<outargs>,<inargs>

endop
```

A UDO has an identification name, `<name>` in listing 7.1, which is textual. Additionally, it requires lists of output and input arguments, which follow the name after the comma. Csound allows an unspecified number of UDO definitions, and unique names are only required if the UDO has the same arguments as another existing opcode. This allows polymorphic opcodes to be defined. The keyword `endop` closes a UDO definition. Both `opcode` and `endop` need to be placed on their own separate lines. UDOs cannot be nested and need to be defined in global space. In order to perform computation, the opcode needs to be instantiated in code somewhere (instruments or global space). UDOs can call any other existing UDOs, provided of course that they have already been defined earlier in the code.

7.2.1 Arguments

The argument list for opcodes allows for all internal data types. It is composed of letters indicating the type, with added brackets (`[]`) to indicate the case of arrays. There are some extra input types to indicate optional parameters, and whether init-pass is performed on the argument:

`a`: audio-rate input.
`a[]`: audio-rate array.
`k`: control-rate, only used at perf-time.
`k[]`: control-rate array.
`J`: optional control-rate, defaults to -1.
`O`: optional control-rate, defaults to 0.
`P`: optional control-rate, defaults to 1.
`V`: optional control-rate, defaults to 0.5.
`K`: control-rate, used at both init-time and performance.
`i`: i-time.
`i[]`: i-time array.
`j`: optional i-time, defaults to -1.
`o`: optional i-time, defaults to 0.
`p`: optional i-time, defaults to 1.
`S`: string.
`S[]`: string array.

```
f: frequency-domain.
f[]: frequency-domain array.
0: no inputs.
```

For example, an input argument list with three inputs of audio, control and init-time types would be represented by the characters `aki`. Variables are copied into UDO arguments, which are accessed through the `xin` opcode:

```
var[,var2,...]  xin
```

Arguments will accept only inputs of their own type, except for k-rate parameters, which take k- and i-vars and constants, and i-time, which take i-vars and constants. For output types, the list is similar, but shorter:

```
a: audio-rate output.
a[]: audio-rate array.
k: control-rate.
k[]: control-rate array.
i: i-time.
i[]: i-time array.
S: string.
S[]: string array.
f: frequency-domain.
f[]: frequency-domain array.
0: no outputs.
```

Outputs are similarly copied from opcode variables, through the `xout` opcode:

```
xout var[,var2,...]
```

In other aspects, UDOs are very similar to instruments. It is very simple to adapt existing instruments for this. However, there are a number of specific aspects to UDO operation that are important to note. While the actual code might look the same, there are important distinctions between these two program structures. In listing 7.2, we see an example of a basic UDO that combines an envelope and an oscillator.

Listing 7.2 Envelope + oscillator combined into a UDO

```
opcode EnvOsc,a,aaiiij
 amp,afr,iri,
    idur,idec,ifn xin
 a1 oscili amp,afr,ifn
 a2 linen a1,iri,idur,idec
    xout a2
endop
```

This UDO can be used multiple times in an instrument, wherever the oscillator and envelope arrangement is required. For instance, if we want to implement an FM synthesis instrument (see Chapter 12), with double modulation, and envelopes controlling timbre and amplitude, we simply need to connect the UDOs appropriately (listing 7.3).

Listing 7.3 Double modulator FM synthesis using the `EnvOsc` UDO

```
instr 1
 amod1 EnvOsc a(p6*p5),a(p5),0.1,p3,0.1
 amod2 EnvOsc a(p7*p5),a(p5*3),0.2,p3,0.4
 asig   EnvOsc a(p4),amod1+amod2+p5,0.01,p3,0.1
     out asig
endin
schedule(1,0,1,0dbfs/2,440,2.5,1.5)
```

Note that, as discussed before, a-rate inputs in UDOs can only accept audio signals. For this reason, in listing 7.3 we needed to employ conversion for i-time parameters using `a()`. Alternatively, we could have implemented other versions of the same opcode that allowed for `ak`, `ka`, and `kk` for the first two parameters, and this polymorphic set would cover all basic input types.

The next example in listing 7.4 demonstrates how array parameters are used in UDOs. In this case, the first argument is an audio-rate array, so the local variable also has to be defined as an array. Conveniently, the UDO is able to deal with different input array sizes.

Listing 7.4 Array parameters

```
gisinetab = ftgen(0, 0, 4097, 10, 1)
opcode SinModulate, a, a[]i
 asigs[], ifreq xin
 icount = lenarray(asigs)
 iphaseAdj = 1 / icount
 aphs = phasor(ifreq)
 aout = 0
 kndx = 0
 until (kndx >= icount) do
  aphsAdj = ((kndx / icount) * aphs) % 1.0
  amod = tablei:a(aphsAdj, gisinetab, 1)
  aout += asigs[kndx] * ((amod + 1) / 3)
  kndx += 1
 od
 xout aout
endop

instr 1
 ipch = cps2pch(p4,12)
 iamp = ampdbfs(p5)
 asigs[] init 3
 aenv linsegr 0,0.01,.5,0.01,.45,2.25,0
```

```
asigs[0] vco2 iamp, ipch
asigs[1] vco2 iamp, ipch * 2
asigs[2] vco2 iamp, ipch * 3
out moogladder(SinModulate(asigs,
                    0.75)*aenv,
                    2000,.3)
endin

icnt = 0
while icnt < 6 do
  schedule(1,icnt,10-icnt,8+2*icnt/100,-12)
  icnt += 2
od
```

7.3 Instrument State and Parameters

Parameters are passed to and received from UDOs by value; this means that they cannot change an instrument variable indirectly. In other words, a UDO receives a copy of the value of its input arguments, which resides in its local variables. It passes a value or values out to its caller, which copies each one into the variable receiving it. UDOs can access global variables following the normal rules, which allow any part of the code to modify these.

However, some elements of instrument state are shared between the calling instrument and its UDO(s):

- parameters (p-variables): these are accessible by the UDO and can modify the caller instrument's duration.
- extra time: this is accessible here, affecting opcodes that are dependent on it.
- MIDI parameters: if an instrument is instantiated by a MIDI NOTE message, these are available to UDOs.

This shared state can be modified by the UDO, affecting the instrument that uses the opcode. For instance if any envelope used here uses a longer extra time duration than the one set by the calling instrument, then this will be the new duration used by the instrument instance. Likewise the total duration of an event can be modified by assigning a new duration to p3 inside the UDO. For instance, we can create an opcode that reads a soundfile from beginning to end, by checking its duration and making the event last as long as it needs to. This is shown in listing 7.5, where regardless of the original event duration, it will play the soundfile to the end.

Listing 7.5 A UDO that controls the calling event's duration

```
opcode FilePlay,a,S
 Sname xin
 p3 = filelen(Sname)
```

```
 a1 diskin Sname,1
 xout a1
endop

instr 1
 out FilePlay("fox.wav")
endin
schedule(1,0,1)
```

7.4 Local Control Rate

It is possible to set a local control rate for the UDO which is higher than the system control rate. This is done by defining a local `ksmps` value that is an even subdivision of the calling-instrument block size. All UDOs have an extra optional i-time argument in addition to the ones that are explicitly defined, which is used to set a ksmps value for any opcode instance. The following example demonstrates this (listing 7.6); we call the same counting opcode with three different ksmps values (system is set at ksmps=10, the default).

Listing 7.6 Local control rates in UDO performance

```
opcode Count,k,i
  ival xin
  kval init ival
  print ksmps
  kval += 1
  printk2 kval
  xout kval
endop

instr 1
 kval Count 0,p4
 printf "Count output = %d\n",kval,kval
endin
schedule(1,0,1/kr,1)
schedule(1,1/kr,1/kr,2)
schedule(1,2/kr,1/kr,5)
```

The console printout shows the difference in performance between these three opcode calls (each event is only one k-cycle in duration):

```
SECTION 1:
new alloc for instr 1:
instr 1:  ksmps = 1.000
 i1     1.00000
```

```
i1       2.00000
i1       3.00000
i1       4.00000
i1       5.00000
i1       6.00000
i1       7.00000
i1       8.00000
i1       9.00000
i1      10.00000
Count output = 10
  rtevent:             T   0.000 TT   0.000 M:        0.0
instr 1:  ksmps = 2.000
i1       1.00000
i1       2.00000
i1       3.00000
i1       4.00000
i1       5.00000
Count output = 5
  rtevent:             T   0.000 TT   0.000 M:        0.0
instr 1:  ksmps = 5.000
i1       1.00000
i1       2.00000
Count output = 2
```

In addition to this, it is possible to fix the control rate of an opcode explicitly by using the setksmps opcode. This can be used in UDOs and in instruments, with the same limitation that the local ksmps has to divide the calling-instrument (or system, in the case of instruments) block size. It is important to place it as the first statement in an instrument or UDO, in case any opcodes depend on a correct ksmps value for their initialisation. The following example shows this, with ksmps=1 and 2 for the UDO and instrument, respectively.

Listing 7.7 Local control rates in UDO and instrument performance

```
opcode Count,k,i
 setksmps 1
 ival xin
 kval init ival
 print ksmps
 kval += 1
 printk2 kval
 xout kval
endop

instr 1
 setksmps 2
 kval Count 0
```

```
 printf "Count output = %d\n",kval,kval
endin
schedule(1,0,1/kr)
```

In this case, in one system k-period the instrument has five local k-cycles. The opcode divides these into two k-cycles for each one of the caller's. This example prints the following messages to the console:

```
SECTION 1:
new alloc for instr 1:
instr 1:  ksmps = 1.000
 i1      1.00000
 i1      2.00000
Count output = 2
 i1      3.00000
 i1      4.00000
Count output = 4
 i1      5.00000
 i1      6.00000
Count output = 6
 i1      7.00000
 i1      8.00000
Count output = 8
 i1      9.00000
 i1     10.00000
Count output = 10
B  0.000 ..   1.000 T   1.000 TT   1.000 M:        0.0
```

The possibility of a different a-signal vector size (and different control rates) is an important aspect of UDOs. This enables users to write code that requires the control rate to be the same as the audio rate, for specific applications. This is the case when ksmps = 1. This enables Csound code to process audio sample by sample and to implement one-sample feedback delays, which are used, for instance, in filter algorithms.

For instance, we can implement a special type of process called *leaky integration*, where the current sample is the sum of the current sample and the output delayed by one sample and scaled by a factor close to 1. This requires the sample-by-sample processing mentioned above.

Listing 7.8 Sample-by-sample processing in a leaky integrator UDO

```
opcode Leaky,a,ak
 setksmps 1
 asum init 0
 asig,kfac xin
 asum = asig + asum*kfac
 xout asum
endop
```

```
instr 1
 a1 rand p4
 out Leaky(a1,0.99)
endin
schedule(1,0,1,0dbfs/20)
```

This is a very simple example of how a new process can be added to Csound by programming it from scratch. UDOs behave very like internal or plug-in opcodes. Once defined in the orchestra code, or included from a separate file via a #include statement, they can be used liberally.

7.5 Recursion

A powerful programming device provided by UDOs is *recursion* [2], which we have already explored in Chapter 5 in the context of instruments. It allows an opcode to instantiate itself an arbitrary number of times, so that a certain process can be dynamically spawned. This is very useful for a variety of applications that are based on the repetition of a certain operation. Recursion is done in a UDO by a call to itself, which is controlled by conditional execution. A minimal example, which prints out the instance number, is show below.

Listing 7.9 Minimal recursion example

```
opcode Recurse,0,io
iN,icnt xin
if icnt >= iN-1 igoto cont
   Recurse iN,icnt+1
cont:
  printf_i "Instance no: %d\n",1,icnt+1
endop

instr 1
 Recurse 5
endin
schedule(1,0,0)
```

Note that as the o argument type is set by default to 0, so icnt starts from 0 and goes up to 4. This will create five recursive copies of Recurse. Note that they execute in the reverse order, with the topmost instance running first:

```
SECTION 1:
new alloc for instr 1:
Instance no: 5
Instance no: 4
Instance no: 3
```

```
Instance no: 2
Instance no: 1
B  0.000 ..  1.000 T  1.000 TT  1.000 M:         0.0
```

This example is completely empty, as it does not do anything but print the instances. A classic example of recursion is the calculation of a factorial, which can be defined as $N! = N * (N - 1)!$ (and $0! = 1, 1! = 1$). It can be implemented in a UDO as follows:

Listing 7.10 Factorial calculation by recursion

```
opcode Factorial,i,ip
 iN,icnt xin
 ival = iN
 if icnt >= iN-1 igoto cont
   ival Factorial iN,icnt+1
 cont:
   xout   ival*icnt
endop

instr 1
 print Factorial(5)
endin
schedule(1,0,0)
```

this prints

```
SECTION 1:
new alloc for instr 1:
instr 1:  #i0 = 120.000
```

Recursion is a useful device for many different applications. One of the most important of these, as far as Csound is concerned, is the spawning of unit generators for audio processing. There are two general cases of this, which we will explore here. The first of these is for sources/generators, where the audio signal is created in the UDO. The second is when a signal is processed by the opcode. In the first case, we need to add the generated signal to the output, which will be a mix of the sound from all recursive instances.

For instance, let's consider the common case where we want to produce a sum of oscillator outputs. In order to have individual control of amplitudes and frequencies, we can use arrays for these parameters, and look up values according to instance number. The UDO structure follows the simple example of listing 7.9 very closely. All we need is to add the audio signals, control parameters and oscillators:

Listing 7.11 Spawning oscillators with recursion

```
opcode SumOsc,a,i[]i[]ijo
 iam[],ifr[],iN,ifn,icnt xin
 if icnt >= iN-1 goto syn
   asig SumOsc iam,ifr,iN,ifn,icnt+1
```

```
 syn:
    xout asig + oscili(iam[icnt],ifr[icnt],ifn)
 endop

 gifn ftgen 1,0,16384,10,1,1/2,1/3,
                      1/4,1/5,1/7,1/8,1/9
 instr 1
  ifr[] fillarray 1,1.001,0.999,1.002,0.998
  iam[] fillarray 1,0.5,0.5,0.25,0.25
  a1 SumOsc iam,ifr*p5,lenarray(iam),gifn
  out a1*p4/lenarray(iam) * transeg:k(1,p3,-3,0)
 endin
 schedule(1,0,10,0dbfs/2,440)
```

The principle here is that the audio is passed from the lowermost to the topmost instance, to which we add all intermediary signals, until it appears at the output of the UDO in the instrument. In the case of audio processing instruments, the signal has to originate outside the instrument, and be passed to the various instances in turn. For instance, when implementing higher-order filters from a second-order section (see Chapters 12 and 13), we need to connect these in series, the output of the first into the input of the second, and so on. A small modification to the layout of the previous example allows us to do that.

Listing 7.12 Higher-order filters with recursive UDO

```
opcode ButtBP,a,akkio
 asig,kf,kbw,iN,icnt xin
 if icnt >= iN-1 goto cont
   asig ButtBP asig,kf,kbw,iN,icnt+1
 cont:
   xout butterbp(asig,kf,kbw)
endop

instr 1
 a1 rand p4
 a2 ButtBP a1,1500,100,4
 out a2
endin
schedule(1,0,1,0dbfs/2)
```

We can characterise these two general cases as *parallel* and *cascade* schemes. In listing 7.11, we have a series of oscillators side by side, whose output is mixed together. In the case of listing 7.12, we have a serial connection of opcodes, with the signal flowing from one to the next. To make a parallel connection of processors receiving the same signal, we need to pass the input from one instance to another, and then mix the filter signal with the output, as shown in listing 7.13.

Listing 7.13 Parallel filter bank with recursive UDO

```
opcode BPBnk,a,ai[]kio
 asig,ifc[],kQ,iN,icnt xin
 if icnt >= iN-1 goto cont
    afil BPBnk asig,ifc,kQ,iN,icnt+1
 cont:
    xout afil +
         butterbp(asig,ifc[icnt],ifc[icnt]/kQ)
endop

instr 1
 a1 rand p4
 ifc[] fillarray 700, 1500, 5100, 8000
 a2 BPBnk a1,ifc,10,lenarray(ifc)
 out a2
endin
schedule(1,0,10,0dbfs/2)
```

7.6 Subinstruments

Another way of composing Csound code is via the *subinstr* mechanism. This allows instruments to call other instruments directly as if they were opcodes. The syntax is

```
a1[,...]  subinstr id[, p4, p5, ...]
```

where id is the instrument number or name (as a string). This shares many of the attributes of a UDO, but is much more limited. We are limited to i-time arguments, which are passed to the instrument as parameters (p-vars). An advantage is that there is no need to change our code to accommodate the use of an instrument in this manner. In listing 7.14, we see an example of instrument 1 calling instrument 2, and using its audio output.

Listing 7.14 A subinstrument example

```
instr 1
iamp = p4
ifreq = p5
a1 subinstr 2,iamp,ifreq,0.1,p3,0.9,-1
    out   a1
endin
schedule(1,0,1,0dbfs/2,440)

instr 2
 a1 oscili p4,p5,p9
 a2 linen a1,p6,p7,p8
```

```
    out a2
endin
```

In addition to `subinstr`, which runs at init- and perf-time (depending on the called instrument code), we can also call an instrument to run *only* at the initialisation pass. The syntax is:

```
subinstrinit id[, p4, p5, ...]
```

In this case, no performance takes place and no audio output is present. The subinstrument is better understood as another way of scheduling instruments, rather than as a regular opcode call. In this way, it can be a useful means of reusing code in certain applications.

7.7 Conclusions

This chapter explored a number of ways in which Csound can be extended, and code can be composed into larger blocks. The UDO mechanism allows users to implement new processing algorithms from scratch or to package existing opcodes that can be used in instrument-like ordinary unit generators. With these possibilities, it is simpler to organise our code into components that can be combined later in different ways. UDOs can be saved in separate files and included in code that uses them. They can also be very handy for showing the implementation of synthesis and transformation processes, and we will be using them widely for this purpose in later chapters of this book.

Part III
Interaction

Chapter 8
The Numeric Score

Abstract This chapter discusses the standard numeric score, which can be used to control events in Csound. Following a general introduction to its syntax and integration with the system, the text explores the basic statement types that compose it, and how they are organised. The score processor and its functionality are explored, followed by a look at loops and playback control. The chapter concludes with a discussion of score generation by external programs and scripting languages.

8.1 Introduction

The standard numeric score was the original means of controlling Csound, and for many composers, it is still seen as the fundamental environment for creating music with the software. Its concept and form goes a long way back to MUSIC III and IV, which influenced its direct predecessors in the MUSIC 360 and MUSIC 11 systems. It was created to respond to a need to specify instances of compiled instruments to run at certain times, as well as create function tables as needed. Scores were also designed to go through a certain amount of processing so that tempo and event sorting could be applied. A traditional separation of signal processing and event scheduling was behind the idea of score and orchestra. The lines between these two were blurred as Csound developed, and now the score is one of many ways we can use to control instruments.

The score can be seen as a data format rather than a proper programming language, even though it has many scripting capabilities. An important aspect to understand is that its syntax is very different from the Csound language proper (the 'orchestra'). There are reasons for this: firstly, the score serves a different purpose; secondly, it developed separately, and independently from the orchestra. Thus, we need to be very careful not to mix the two. In particular, the score is much simpler, and has very straightforward syntax rules. These can be summarised as:

1. The score is a list of statements.

© Springer International Publishing Switzerland 2016
V. Lazzarini et al., *Csound*, DOI 10.1007/978-3-319-45370-5_8

2. Statements are separated by line breaks.
3. All statements start with a single character identifying its purpose, sometimes called an *opcode* (not to be confused with Csound language opcodes, or unit generators).
4. Statement parameters are separated by blank spaces, and are also called *p-fields*. The space between the opcode and the first p-field is optional.
5. Parameters are mostly numeric, but strings in double quotes can be used wherever appropriate.

A couple of important things to remember always are that commas are not used to separate parameters (as in the orchestra), but spaces; and that statements do not necessarily execute in the order they are placed, also unlike the orchestra language, where they do. The score is read at the first compilation of Csound and stored in memory, and its playback can be controlled by the Csound language. Further single events, or full scores can be sent to Csound, but these will be taken as *real-time events*, which are treated slightly differently as some score features such as tempo warping are not available to them.

Score comments are allowed in the same format used in the Csound language. Single-line comments, running until the end of the line, are started by a semicolon (;). Multiple-line comments are enclosed by /* and */, and cannot be nested. Scores are generally provided to Csound in a CSD file, inside the `<CsScore>` section (see Chapter 2).

Ultimately, the score is concerned with event scheduling. These events are mostly to do with instantiating instruments, but can also target function table generation. Before the appearance of function table generator opcodes and real-time events, all of these could only be executed through the score. While modern usage can bypass and ignore it completely, the score is still an important system resource. Many classic examples of Csound code rely heavily on the score. This chapter will provide a full description of its features.

8.2 Basic Statements

As outlined above, the score is a list of statements, on separate lines, each one with the following format:

```
op [p1 p2 ...]
```

where op is one of a, b, e, f, i, m, n, q, r, s, t, v, x, y, {, and }. The statement parameters (p-fields) are p1, p2 etc. separated by spaces. Depending on op, these might be optional.

The main score statement is indicated by the i opcode, and it schedules an event on a given instrument:

```
i p1 p2 p3 [p4 ...]
```

p1 is the instrument number, or name (a string in double quotes). If using a non-integral number, the fractional part identifies a given instance, which can be used for tied notes. A negative p1 turns a corresponding indeterminate-length instance off.

p2 is starting time.

p3 is duration time: if negative, an indeterminate-length is set, which starts a held note.

p4 ... are extra parameters to instruments.

The i-statement format follows the form already shown for scheduling instances. Time is set in arbitrary units called *beats*, which are later translated in tempo processing (defaulting to 1 second). Events can be placed in the score in any order (within a section, see Section 8.3), as they are sorted before performance. As usual, any number of concurrent i-statements are allowed for a given instrument (disregarding any computation time requirements). Parameters are usually numeric, but strings can also be used in double quotes, if an instrument expects them.

A second statement affecting events is defined by q. This can be used to mute i-statements, but only affects events before they get started. The form of the statement is:

```
q p1 p2 p3
```

p1 is the instrument number, or name (in double quotes).

p2 is the action time; the statement will affect instruments with start time equal to or later than it.

p3 is 0 to mute an instrument and 1 to un-mute it.

This can be used to listen to selected parts of the score by muting others.

Finally, we can create function tables from the score with an f-statement. This has the following general form:

```
f p1 p2 p3 p4 p5 [p6 ...]
```

p1 is the table number; a negative number destroys (deallocates) the table.

p2 is the creation or destruction time.

p3 is the table size.

p4 is the GEN routine code; negative values cause re-scaling to be skipped.

p5 ... are the relevant GEN parameters.

The form of the f-statement is very similar to the `ftgen` opcode, and all the aspects of table creation, such as GEN codes, re-scaling and sizes are the same here. A special form of the f-statement has a different, very specific, purpose. It is used to create a creation time with no associated table, which will extend the duration of the score by a given amount of time:

```
f 0 p2
```

In this case, Csound will run for p2 beats. This is useful to keep Csound running for as long as necessary, in the presence of a score. If no score is provided, Csound will stay open until it is stopped by a frontend (or killed).

8.3 Sections

The score is organised in sections. By default there is one single section, but further ones can be set by using an s-statement, which sets the end of a section:

```
s [p1]
```

A single optional opcode can be used to define an *end time*, which is used to extend the section (creating a pause) before the start of the next. For this, the p1 value needs to be beyond the end of the last event in the ordered list, otherwise it has no effect. Extending a section can be useful to avoid cutting off the release of events that employ extra time.

The score is read section by section, and events are sorted with respect to the section in which they are placed. Time is relative to the start of the section, and gets reset at the end of each one. Sections also have an important control role in Csound, as memory for instrument instances is also freed/recovered at the end of a section. This was more significant in older platforms with limited resources, but it is still part of the system.

The final section of a score can optionally be terminated by an e-statement:

```
e [p1]
```

This also includes an optional extended time, which works in the same way as in the s-statement. Note that any statements after this one will be ignored. Once the last section of a score finishes, Csound will exit unless an f 0 statement has been used. An example of a score with two sections is given in listing 8.1, where four statements are performed in each section. If p4 and p5 are interpreted as amplitude and frequency, respectively, we have an upward arpeggio in the first section, and a downward one in the second. Note that the order of statements does not necessarily indicate the order of the performed events.

Listing 8.1 Score example

```
f 1 0 16384 10 1 0.5 0.33 0.25 0.2
; first section
i1 0 5 1000 440
i1 1 1 1000 550
i1 2 1 1000 660
i1 3 1 1000 880
s
; second section
i1 3 1 1000 440
i1 2 1 1000 550
i1 1 1 1000 660
i1 1 5 1000 880
e
```

The x-statement can be used to skip the subsequent statements in a section:

```
x
```

It takes no arguments. Reading of the score moves to the next section.

8.4 Preprocessing

Scores are processed before being sent to the Csound engine for performance. The three basic operations applied to a score's events are carry, tempo, and sort, executed in this order. Following these, two other processing steps are optionally performed, to interpret next-p/previous-p and ramping symbols.

8.4.1 Carry

Carry works on groups of consecutive i-statements. It is used to reduce the need to type repeated values. Its basic rules are:

1. Carry works only within a consecutive group of i-statements whose integral p1 values are the same (fractional parts indicating specific instances are ignored).
2. Any empty p-field takes the value of the previous line. An empty p-field can be marked with a dot (.), or left blank if there are no further ones in the same line.
3. For p2 only: the symbol + is interpreted as the sum of the previous line's p2 and p3. This symbol can be carried itself.
4. For p2 only: the symbols ^+ or ^- followed by a value x set p2 to a new value by adding or subtracting x to/from p2 on the previous line.
5. The symbol ! will block the implicit carry of blank p-fields in a statement and subsequent ones.

In listing 8.2 we see these features being employed and their interpretation in the comments.

Listing 8.2 Carry examples and their interpretation

```
i 1 0 5 10000 440
i . + 2 . 330      ;i1 5 2 1000 440
i . + 1            ;i1 7 1 1000 330
i 2 0 1 55.4 22.1
i . ^+1 . 67.1     ;i2 1 1 67.1 22.1
i . ^+2 . !        ;i2 3 1 [no p4 or p5 are carried]
```

8.4.2 Tempo

A powerful feature of the score is tempo processing. Each section can have an in-dependent tempo setting, which can also be time-varying. Tempo is controlled by a t-statement:

```
t   p1   p2   [p3   p4   ...]
```

p1 should be 0 to indicate beat 0
p2 beat 0 tempo
p3, p5, ... odd p-fields indicate optional times in beats
p4, p6, ... even p-fields are the tempi for each beat in the preceding p-field.

Tempo processing can create accelerandos, ritardandos and abrupt tempo changes. Each pair of p-fields, indicating beats and tempi, will determine how the p2 and p3 fields in each i-statement are interpreted. Tempo values are defined in beats per minute. For instance, a ritardando is created with the following tempo pattern:

```
t  0  60  20  30
```

Tempo at the start is set at 60 bpm and by beat 20 it is 30 bpm. The tempo remains at the last-defined value until the end of the section. Equally, an accelerando can be determined by a t-statement such as:

```
t  0  120  20  180
```

Any number of time-tempo pairs can be used. A single pair determines a fixed tempo for the section; if no tempo statements are provided, beats are interpreted as seconds (60 bpm). Only one such statement is valid per section. The $-t$ N option can be used to override the score tempi of all sections, which will be set to N bpm.

The v-statement can be used as a *local* time warp of one or more i-statements:

```
v  p1
```

where p1 is a time warp factor that will affect only the subsequent lines, multiplying their start times (p2) and durations (p3). It can be cancelled or modified by another v-statement. This has no effect on carried p2 and p3 values.

8.4.3 Sort

Following carry and tempo processing, the section i-statements are sorted into as-cending time order according to their p2 value. If two statements have the same p2 value, they are sorted in ascending order according to p1. Likewise, if both p2 and p1 coincide, sorting is performed in p3 order. Sorting affects both i- and f-statements, and if these two have the same creation time, the f-statement will take precedence.

8.4.4 Next-p and Previous-p

The score also supports the shorthand symbols npN and ppN, where N is an integer indicating a p-field. These are used to copy values from other p-fields: the first indicates the source of the p-field to the next i-statement, and the second refers to the previous i-statement. Since these are interpreted *after* carry, tempo and sort, they can be carried and will affect the final sorted list. Listing 8.3 shows some examples with their interpretation in the comments. Note if there is no previous or next line, the result is 0.

Listing 8.3 Next-p and previous-p examples

```
i 1 0 5 10000 440   pp5 np4 ; i 1 0 5 10000 0 30000
i 1 1 2 30000 330   pp4   ; i 1 1 2 30000 330 10000 20000
i 1 2 1 20000 55 . pp5 ; i 1 2 1 20000 55 30000 330
```

These symbols are recursive and can reference other existing np and pp symbols. Next-p and previous-p cannot be used in p1, p2 and p3, but can refer to these. References do not cross section boundaries.

8.4.5 Ramping

A final preprocessing offered by a score for i-statements can make values ramp up or down from an origin to a given target. This is p-field-oriented manipulation. The symbol < is used for a linear ramp between two values, and (or) can be used to create an exponential curve. A tilde symbol (∼) can be used to generate random values in a uniform distribution between two values. Ramping is not allowed in p1, p2 or p3.

Listing 8.4 Ramping example

```
i 1 0 1   100 ; 100
i 1 0 1   <   ; 200
i 1 0 1   <   ; 300
i 1 0 1   <   ; 400
i 1 0 1   500 ; 500
```

8.4.6 Expressions

Ordinarily, p-fields accept only constant values. However, using square brackets ([and]), we can invoke a special preprocessing mode that can evaluate arithmetic expressions and place the result in the corresponding p-field. The common operators used in the Csound languages can be used here (+, −, *, / , ^, and %). The usual rules of precedence apply, and parentheses can be used to group expressions. Bitwise logical operators (& for AND, | for OR and # for exclusive-OR) can also be used. The ~ symbol can be used to mean a random number between 0 and 1. A y-statement can be used to seed the random number generator with a starting value:

```
y [p1]
```

The value of p1 is used as the seed, but it can be omitted, in which case the system clock is used. The following examples use expressions in p-field 4:

```
i 1 0 1   [2*(4+5)] ; 18
i 1 0 1   [10 * ~] ; random (0 - 10)
i 1 0 1   [ 5 ^ 2 ] ; 25
```

Expressions also support a special @N operator that evaluates to the next power-of-two greater or equal to N. The @@N operator similarly yields the next power of two plus one.

8.4.7 Macros

The preprocessor is responsible for macro substitution in the score; it allows text to be replaced all over the score. As in the Csound language, we can use it in two ways, with or without arguments. The simplest form is

```
#define  NAME  #replacement text#
```

where $NAME will be substituted by the replacement text, wherever it is found. This can be used, for instance, to set constants that will be used throughout the code.

Listing 8.5 Score macro example

```
#define FREQ1 #440#
#define FREQ2  #660#
i1 0 1 1000 $FREQ1 ; i1 0 1 1000 440
i1 0 1 1500 $FREQ2 ; i1 0 1 1000 660
i1 0 1 1200 $FREQ1
i1 0 1 1000 $FREQ2
```

When a space does not terminate the macro call, we can use a full stop for this purpose.

```
#define  OCTAVE  #8.#
i1 0 1 1000  $OCTAVE.09  ; 8.09
i1 2 1 1000  $OCTAVE.00  ; 8.00
```

Another form of #define specifies arguments:

```
#define  NAME(a'b)  #replacement text#
```

where the parameters a and b are referred to in the macro as $a and $b. More arguments can be used. In listing 8.6 we use arguments to replace a whole score line.

Listing 8.6 Macro example, with arguments

```
#define   FLUTE(a'b'c'd)  #i1 $a $b $c $d#
#define   OBOE(a'b'c'd)   #i2 $a $b $c $d#

FLUTE(0 '  0.5 '  10000 '  9.00)
OBOE(0 '  0.7 '  8000 '  8.07)
```

Score macros can also be undefined with:

```
#undef   NAME
```

8.4.8 Include

The preprocessor also accepts an include statement, which copies the contents of an external text file into the score. For instance, the text

```
#include "score.inc"
```

where the file *score.inc* contains a series of statements, will include these to the score in the point where that line is found. The file name can be in double quotes or any other suitable delimiter character. If the included file is in the same directory as the code, or the same working directory as a running Csound, then its name will be sufficient. If not, the user will need to pass the full path to it (in the usual way used by the OS platform in which Csound is running). A file can be included multiple times in a score.

8.5 Repeated Execution and Loops

The execution of sections can be repeated by the use of the r-statement

```
r p1   p2
```

where p1 is the number of times the current section is to be repeated, and p2 is the name of a macro that will be substituted by the repetition number, starting from 1. For example:

```
r   3   REPEAT
i1 0 1 1000 $REPEAT
i1 1 1 1500 $REPEAT
s
```

will be expanded into:

```
i1  0  1  1000  1
i1  1  1  1500  1
s
i1  0  1  1000  2
i1  1  1  1500  2
s
i1  0  1  1000  3
i1  1  1  1500  3
s
```

Another way to repeat a section is to place a named marker on it, which can then be referred to later. This is done with the m-statement

```
m  p1
```

where p1 is a unique identifier naming the marker, which can contain numerals and/or letters. This can then be referenced by an n-statement to repeat the section once:

```
n  p1
```

The n-statement uses p1 as the name of the marker whose section is to be played again.

Listing 8.7 Section repeats in score

```
m  arpeg
i1  0  1  10000  8.00
i1  1  1  10000  8.04
i1  2  1  10000  8.07
i1  3  1  10000  9.00
s
m  chord
i1  0  4  10000  7.11
i1  0  4  10000  8.02
i1  0  4  10000  8.07
i1  0  4  10000  9.02
s
n  arpeg
n  chord
n  arpeg
n  chord
n  chord
n  arpeg
```

The example in listing 8.7 will alternate the first two sections twice, then repeat the second and finish with the first.

Finally, the score supports a non-sectional loop structure to repeat any statements defined inside it. It uses the following syntax:

```
{ p1 p2
...
}
```

where p1 is the number of repeats required, and p2 is the name of a macro that will be replaced by the iteration number, this time starting from 0. This is followed by the score statements of the loop body, which is closed by a } placed on its own line. Multiple loops are possible, and they can be nested to a depth of 39 levels.

The following differences between these loops and section repeats are important to note:

- Start times and durations (p2, p3) are left untouched. In particular, if p3 does not reference the loop repeat macro, events will be stacked in time.
- Loops can be intermixed with other concurrent score statements or loops.

The following example creates a sequence of 12 events whose p5 goes from 8.00 to 8.12 in 0.01 steps. The event start times are set from 0 to 11.

Listing 8.8 Score loop example

```
{ 12 N
i1 $N 1 10000 [8. + $N/100]
}
```

8.6 Performance Control

Two statements can be used to control playback of a score section. The first one is the a-statement:

```
a p1 p2 p3
```

This advances the score playback by p3 beats, beginning at time p2 (p1 is ignored). It can be used to skip over a portion of a section. These statements are also subject to sorting and tempo modifications similarly to i-statements.

The final performance control statement is:

```
b p1
```

This specifies a clock reset. It affects how subsequent event start times are interpreted. If p1 is positive, it will add this time to the following i-statements p2 times, making them appear later; if negative, it will make the events happen earlier. To set the clock back to normal, we need to set p1 to 0. If the clock effect on the next i-statement makes p2 negative, then that event will be skipped.

8.6.1 Extract

Csound also allows a portion of the score to be selected for performance using its *extract* feature. This is controlled by a text file containing three simple commands that will determine the playback:

1. `i <num>` – selects the required instruments (default: all)
2. `f <section>:<beat>` – sets the start (*from*; default: beginning of score)
3. `t <section>:<beat>` – sets the end (*to*; default: end of score)

If a command is not given, its default value is used. An example extract file looks like this:

```
i 2  f 1:0  t 1:5
```

This will select instrument 2 from section 1 beat 0 to section 1 beat 5. The extract feature is enabled by the option

```
-x fname
```

where `fname` is the name of the extract file.

8.6.2 Orchestra Control of Score Playback

In the Csound language, three unit generators can issue score i-statements, which will be sent to the engine as real-time events. These are the opcodes `scoreline`, `scoreline_i` and `readscore`. The first two accept strings with i- and f-statements (multiple lines are allowed) at performance and initialisation time, respectively. The third also works only at i-time, but allows carry, next-p, previous-p, ramping, expressions and loops to be preprocessed (although tempo warping and section-based processing are not allowed).

In addition to these, the loaded score can be controlled with two init-time opcodes:

```
rewindscore
```

which takes no parameters and rewinds the score to the beginning, and

```
setscorepos ipos
```

This can be used as a general transport control of the score, moving forwards or backwards from the current position. The parameter `ipos` sets the requested score position in seconds. These two opcodes can be very useful, but care must be taken not to create problematic skips, when using them in instruments that can themselves be called by a score. It is possible, for instance, to create a never-ending loop playback by calling `rewindscore` in an instrument as the last event of a score.

8.6.3 Real-Time Events

Score events can be sent to Csound in real-time from external sources. The most common use of this is via the API, where hosts or frontends will provide various means of interacting with Csound to instantiate instruments. In addition to this, Csound has a built-in facility to take events from a file or a device by using the -L dname option, where dname is the name of a source. We can, for instance, take events from the standard input (which is normally the terminal), by setting dname to stdin:

```
<CsoundSynthesizer>
<CsOptions>
-L stdin -o dac
</CsOptions>
<CsInstruments>
instr 1
 out oscili(p4*expon(1,p3,0.001),p5)
endin
</CsInstruments>
</CsoundSynthesizer>
```

Similarly, It is also possible to take input from a text file, by setting dname to the file name. Finally, in Unix-like systems, it is possible to pipe the output of another program that generates events. For instance the option

```
-L " | python score.py"
```

will run a Python program (score.py) and take its output as real-time events for Csound. It is also possible to run a language interpreter, such as the Python command, in an interactive mode, outputting events to Csound. In this case we would take the input from stdin, and pipe the output of the interpreter into Csound. For instance, the terminal command

```
python -u | csound -L stdin -o dac ...
```

will open up the Python interpreter, which can be used to generate real-time events interactively.

8.7 External Score Generators

A full numeric score is very easily generated by external programs or scripts. Its straightforward format allows composers and developers to create customised ways to manipulate scores. Csound supports this very strongly. A powerful feature is the possibility to invoke external score generators from inside a CSD file. These can be given parameters or a complete code, which they can use to write a score text file that is then read directly by Csound.

This is provided as an optional attribute in the `<CsScore>` tag of a CSD file:

```
<CsScore bin="prog" >
...
</CsScore>
```

where `prog` is an external program that will be used to generate or preprocess the score. The contents of the `<CsScore>` section in this case will not be interpreted as a numeric score, but as input to `prog`. They will be written to a file, which is then passed as the first argument to `prog`. The external generator will write to a text file whose name is passed to it as the second argument. This is then loaded by Csound as the score. So this program needs to be able to take two arguments as input and output files, respectively. Any such generator capable of producing valid score files can be used.

As an example, let's consider using the Python language to generate a minimal score. This will create 12 events, with increasing p2, p4 and p5 (similar to the loop example in listing 8.8).

Listing 8.9 External score generator using Python

```
<CsScore bin="python">
import sys
f = open(sys.argv[1], "w")
stm = "i 1 %f 1 %f %f \n"
for i in range(0,12):
    f.write(stm % (i,1000+i*1000,8+i/100.))
</CsScore>
```

In listing 8.9 we see the relevant CSD section. The command `python` is selected as the external binary. A simple script opens up the output file (`argv[1]`, since `argv[0]` is the input, the actual program file), and writes the twelve statements using a loop. If we ran this example on the terminal, using the Python command, the output file would contain the following text:

```
i 1 0.000000 1 1000.000000 8.000000
i 1 1.000000 1 2000.000000 8.010000
i 1 2.000000 1 3000.000000 8.020000
i 1 3.000000 1 4000.000000 8.030000
i 1 4.000000 1 5000.000000 8.040000
i 1 5.000000 1 6000.000000 8.050000
i 1 6.000000 1 7000.000000 8.060000
i 1 7.000000 1 8000.000000 8.070000
i 1 8.000000 1 9000.000000 8.080000
i 1 9.000000 1 10000.000000 8.090000
i 1 10.000000 1 11000.000000 8.100000
i 1 11.000000 1 12000.000000 8.110000
```

When Csound runs this script as an external generator, it uses temporary files that get deleted after they are used. This mechanism can be used with a variety of scripting languages, just by observing the simple calling convention set by the system.

8.8 Alternatives to the Numeric Score

As we have pointed out at the outset, the score is but one of the many optional ways we can control Csound. The score-processing features discussed in this chapter can be replaced, alternatively, by code written in the Csound orchestra language. We can demonstrate this with a simple example, in which two score-processing elements, tempo statement (ritardando) and ramping, are integrated directly in the program.

The score calls an instrument from beat 0 to beat 10. The tempo decreases from metronome 240 to 60, whilst the pitch decreases in equal steps from octave 9 (C5) to octave 8 (C4).

Listing 8.10 Tempo and ramping score

```
t  0  240  10  60
i  2  0  1  9
i  .  +  .  <
i  .  +  .  <
i  .  +  .  <
i  .  +  .  <
i  .  +  .  <
i  .  +  .  <
i  .  +  .  <
i  .  +  .  <
i  .  +  .  <
i  .  +  .  8
```

The scoreless version in listing 8.11 uses a UDO that calculates an interpolation value. This is used for both pitch ramping and tempo (time multiplier) modification. The events themselves are generated by a loop.

Listing 8.11 Csound language realisation of score in listing 8.10

```
opcode LinVals,i,iiii
 iFirst,iLast,iNumSteps,iThisStep xin
 xout iFirst-(iFirst-iLast)/iNumSteps*iThisStep
endop

instr 1
 iCnt,iStart init 0
 until iCnt > 10 do
  iOct = LinVals(9,8,10,iCnt)
  schedule 2,iStart,1,iOct
  iCnt += 1
```

```
 iStart += LinVals(1/4,1,10,iCnt+0.5)
 od
endin
schedule(1,0,0)

instr 2
 out mode(mpulse(1,p3),cpsoct(p4),random:i(50,100))
endin
```

8.9 Conclusions

The Csound numeric score was the original method for controlling the software, prior to the introduction of real-time and interactive modes. It has been used widely in the composition of computer music works, and, for many users, it provides an invaluable resource. Although the score is not a complete programming language in a strict sense, it has a number of programmable features.

As we have seen, the score is best regarded as a data format, that is simple and compact, and can be very expressive. It allows the use of external software and languages to construct compositions for algorithmic music, and this feature is well integrated into the Csound system, via the CSD <CsScore> tag bin attribute. Scores can also be included from external files, and their use can be integrated with real-time controls and events.

Chapter 9
MIDI Input and Output

Abstract In this chapter, we will explore the implementation of the MIDI protocol in Csound. Following an introductory discussion of the protocol and its relevant parts, we will explore how it is integrated into the software. We will show how MIDI input is designed to fit straight into the instrument-instance model of operation. The dedicated opcodes for input and output will be introduced. The chapter will also examine the different backend options that are currently supported by the system.

9.1 Introduction

The Musical Instrument Digital Interface (MIDI) protocol is a very widespread means of communication for musical devices. It establishes a very straightforward set of rules for sending various types of control data from a source to a destination. It supports a point-to-point unidirectional connection, which can be made between hardware equipment, software or a combination of both.

Although MIDI is significantly limited when compared to other modern forms of communication (for instance, IP messages), or data formats (e.g. the Csound score), it is ubiquitous, having been adopted by all major music-making platforms. Due to its limitations, it is also simple, and therefore cheap to implement in hardware. While its death has been predicted many times, it has nevertheless lived on as a useful means of connecting musical devices. It is however important to understand the shortcomings of MIDI, and its narrow interpretation of music performance, so that these do not become an impediment to flexible music making.

9.2 MIDI Messages

The MIDI protocol incorporates three types of messages [87]: channel messages, system common messages and system real-time messages. The first type is used to send control information to music devices, e.g. when to play a sound, to change a control value etc. The second type includes system exclusive messages, which is used for custom message types, defining various types of data transfer (e.g. for synthesiser editors, bulk dumps etc.), MIDI time code frames, song selection and positioning. System real-time messages are used mainly for synchronisation through timing clocks.

The MIDI implementation in Csound mostly addresses the receiving and sending of channel messages. Such messages are formatted in a specific way, with two components:

1. **status**: this comprises a code that defines the message type and a channel number. There are seven message types and sixteen independent channels.
2. **data**: this is the data that is interpreted according to the message type. It can be made up of one or two values, in the range of 0-127.

At the low level, MIDI messages are transmitted as a sequence of eight-bit binary numbers, or bytes. The status part occupies a single byte, the first four bits containing the message type and the second the channel (thus the limited number of channels and message types allowed). The data can comprise one or two bytes, depending on the message types (each byte carrying a single value). Status bytes always start with a 1, and data bytes with 0, thus limiting the useful bits in each byte to seven (hence the fundamental MIDI value range of 0-127).

9.2.1 Channel Message Types

The seven message types are:

1. **NOTE ON**: whose numeric code is 144, and which contains two data values. The first is a note number, and the second a note velocity (e.g. telling how hard a particular key is pressed).
2. **NOTE OFF**: with code 128, and also containing two values, a note number and velocity. This message can also be emulated with a NOTE ON containing zero velocity.
3. **aftertouch**: using code 208 and containing a single value, the amount of aftertouch. This is typically generated by a keyboard whose keys are pressed down while being played.
4. **polyphonic aftertouch**: whose code is 166, and provides a note-by-note aftertouch control with two data values (note number and amount).
5. **program change**: code 192, carries one single value, the program number. This is often used to change stored parameter sets in music devices.
6. **control change**: using code 176, carries two values, a control number and a value. It is used for interactive parameter adjustments.
7. **pitch bend**: code 224, also carrying two values, a coarse and a fine pitch bend amount, which can be combined to control parameters (such as frequency).

Note that while these messages were originally intended for particular purposes, they can be mapped to any function. There is nothing stopping, for instance, pitch bend being mapped to filter frequency, or any other parameter in a given instrument. Even NOTE messages, which are normally associated with starting and stopping sounds, can be reassigned to other ends.

9.3 The Csound MIDI System

Csound has a very straightforward MIDI input implementation. By default, NOTE ON messages are sensed and trigger instrument instances. These instances are allocated as usual, but have a particular characteristic: as discussed in Section 5.3.3, they will have indeterminate duration (i.e. a negative p3 value). NOTE OFF messages with a matching note number will stop them playing. So, for this standard type of MIDI operation, we need to make sure our instruments do not depend on p3. Many of the Opcodes that do depend on p3 have versions designed to be independent of a set duration (the 'r' ones introduced in Section 5.3.4).

Also by default, channels are tied in to their respective instrument numbers: channel 1 to instr 1 through channel 16 to instr 16. If one of these is not defined, instrument 1 or the lowest-order instrument will be used. It is possible to modify this using the massign opcode:

```
massign ichnl, instr
```

where ichnl is the channel to be assign to instr instr. It is also possible to link program numbers with instruments using pgmassign:

```
pgmassign ipgm, instr[, ichn]
```

where ipgm is the program to be assigned to instr instr.

Note that Csound has unlimited polyphony by default, so it will respond to any number of simultaneous NOTE ON messages. However, given that computers have various finite levels of resources, the practical number of notes played together will not be unlimited. While Csound does not have any built-in voice-stealing or reallocation mechanism, such things can be constructed in user code. The maxalloc and active opcodes can be used to limit the allocation for a given instrument and find the number of currently active instances, respectively. In additiion, cpuprc or cpumeter can be used to check for CPU use (although these two opcodes are currently only available on Linux).

9.3.1 Input

MIDI input opcodes can be separated into two categories:

1. For instruments triggered directly: these are opcodes that expect the instrument
to have been instantiated by a NOTE ON message. They assume certain data to
be there: note number, velocity and channel.
2. For generic use: these opcodes will normally have access to all MIDI messages
coming into Csound.

In the first category, we have

`veloc`: note velocity.
`notnum`: note number.
`aftouch`: aftertouch amount.
`polyaft`: aftertouch amount for a given note.
`pchbend`: pitch-bend amount.
`ampmidi`: takes the note velocity and translates it into amplitude.
`cpsmidi` and `cpsmidib`: translates note number into Hz, in the second case
incorporating pitch-bend data.
`midictrl`, `midic7`, `midic14`, and `midic21`: access any control change
data in the instrument input MIDI channel. The '14' and '21' opcodes can use
multiple messages together to make up higher-resolution data.

All of these opcodes are only applicable to MIDI-triggered instruments. For more
generic use we have

`chanctrl`, `ctrl7`, `ctrl14` and `ctrl21`: control change data for a given
channel.
`midiin`: raw MIDI data input.

In order to access the MIDI device, it is necessary to include the `-M dev` option,
where `dev` is a device identifier that will be dependent on the MIDI backend used,
and the devices available in the system.

MIDI files can also be used, in addition to real-time/device input (including simultaneously). All we need to do is to supply the filename with the `-F` option.

MIDI input examples

The following example shows a simple MIDI subtractive synthesiser, with support
for pitch bend and modulation wheel (controller 1). It takes in the velocity and
maps it to amplitude, with note number converted to Hz and used as the oscillator
frequency. The filter cut-off is controlled by the modulation wheel input and note
frequency.

Listing 9.1 Simple MIDI subtractive synthesiser

```
instr 1
 kcps cpsmidib 2
 iamp ampmidi 0dbfs
 kcf midictrl 1,2,5
```

```
out linenr(moogladder(
      vco2(iamp,kcps,10),
      kcf*(kcps +
      linenr(kcps,0.1,0.1,0.01)),
      0.7), 0.01,0.1,0.01)
endin
```

The next example shows the use of a generic MIDI input to report the messages received, a type of MIDI monitor. Note that we use `massign` with parameters set to zero, which disables triggering of instruments (NOTE ON messages do not create instances of any instrument). This is because we want to run only a single copy of instrument 1 that will monitor incoming MIDI data. This approach can be used in case we want to parse MIDI messages before dispatching them to other parts of the code.

Listing 9.2 MIDI monitor to print out input messages

```
massign 0,0
instr 1
 k1,k2,k3,k4 midiin
 if k1 == 144 then
  S1 strcpyk "NOTE ON"
 elseif k1 == 128 then
  S1 strcpyk  "NOTE OFF"
 elseif k1 == 166 then
  S1 strcpyk  "POLY AFTERTOUCH"
 elseif k1 == 208 then
  S1 strcpyk  "AFTERTOUCH"
 elseif k1 == 192 then
  S1 strcpyk  "PROGRAM CHANGE"
 elseif k1 == 176 then
  S1 strcpyk "CONTROL CHANGE"
 elseif k1 == 224 then
  S1 strcpyk  "PITCH BEND"
 else
  S1 strcpyk  "UNDEFINED"
 endif
 printf "%s chn:%d data1:%d data2:%d\n",
         k1,S1,k2,k3,k4
endin
schedule(1,0,-1)
```

Mapping to instrument parameters

It is also possible to map NOTE ON/OFF data to given instrument parameter fields (p4, p5 etc.). This allows us to make instruments ready for MIDI without needing to modify them too much, or use some of the MIDI opcodes above. In order to do this, we can use the following options:

--midi-key=N: route MIDI note on message key number to p-field N as MIDI value.

--midi-key-cps=N: route MIDI note on message key number to p-field N as cycles per second (Hz).

--midi-key-oct=N: route MIDI note on message key number to p-field N as linear octave.

--midi-key-pch=N: route MIDI note on message key number to p-field N as octave pitch-class.

--midi-velocity=N: route MIDI note on message velocity number to p-field N as MIDI value.

--midi-velocity-amp=N: route MIDI note on message velocity number to p-field N as amplitude change data for a given channel.

For instance, the options

```
--midi-key-cps=5 --midi-velocity-amp=4
```

will map key numbers to Hz in p5 and velocities to p4 (amplitudes in the range of 0-0dbfs). In this case, the following minimal instrument will respond directly to MIDI:

```
instr 1
 out(linenr(oscili(p4,p5),0.01,0.1,0.01))
endin
```

The only thing we need to worry about is that the instruments we are hoping to use with MIDI do not depend on the event duration (p3). If we use envelopes with built-in release times as in the previous example, then we will not have any problems reusing code for MIDI real-time performance.

9.3.2 Output

MIDI output is enabled with the -Q dev option, where, as before, dev is the destination device. It is also possible to write MIDI files using --midioutfile=.... Similarly to input, Csound provides a selection of opcodes that can be used for this purpose:

midion, midion2, noteon, noteoff, notendur, and noteondur2 – NOTE channel messages.
outiat and outkat – aftertouch.
outipat and outkpat – polyphonic aftertouch.
outic and outkc – control change.
outic14 and outkc14 – control change in two-message packages for extra precision.

outipat and outkpat – program change.
outipb and outkpb – pitch bend.
midiout – generic MIDI message output.

Some of the NOTE opcodes will put out matching ON - OFF messages, so there is no particular problem with hanging notes. However, when using something more raw (such as noteon and midiout), care needs to be taken so that all notes are killed off correctly.

MIDI output example

The following example demonstrates MIDI output. We use the midion opcode, which can be run at k-rate. This opcode sends NOTE ON messages when one of its data parameters (note number, velocity) changes. Before a new message is sent, a NOTE OFF cancelling the previous note is also sent. The metro opcode is used to trigger new messages every second (60 bpm), and the note parameters are drawn from a pseudo-random sequence. For each note, we print its data values to the screen.

Listing 9.3 Simple MIDI output example

```
instr 1
 k1 metro 1
 if k1 > 0 then
  kn = 60+rnd:k(12)
  kv = 60+birnd:k(40)
  printf "NOTE %d %d\n", kn, kn, kv
  midion 1,kn,kv
 endif
endin
```

9.3.3 MIDI Backends

Depending on the operating system, various MIDI backends can be used. These can be selected with the option -+rtmidi=.... The default backend for all platforms is portmidi. Some frontends can also implement their own MIDI IO, in which case there is no particular need to choose a different backend.

PortMidi

PortMidi is a cross-platform MIDI library that works with whatever implementation is offered by the host OS. It can be selected by setting -+rtmidi=pmidi, but it is also the default backend. Devices are listed numerically, so in order to select inputs or outputs, we should provide the device number (e.g. `-M 0`, `-Q 0`). Devices can be listed by running Csound with the sole option `--midi-devices`. The option `-M a` enables all available input devices together.

CoreMIDI

CoreMIDI is the underlying OSX MIDI implementation, which can be accessed using `-+rtmidi=coremidi`. Devices are also listed numerically, as with PortMidi.

ALSA raw MIDI

ALSA raw MIDI is the basic Linux MIDI implementation, accessed using `-+rtmidi =alsa`. Devices are listed and set by name, e.g. `hw:1,0`.

ALSA sequencer

The Alsa sequencer is a higher-level MIDI implementation, accessed using `-+rtmidi =alsaseq`. Devices are also listed and set by name.

Jack

The Jack connection kit MIDI implementation works in a similar way to its audio counterpart. We can set it up with `-+rtmidi=jack`, and it will connect by default to the system MIDI sources and/or destinations. It is also possible to define the connections by passing port names to `-M` and `-Q`. The Jack patchbay can be used to set other links. Jack is particularly useful for inter-application MIDI IO.

9.4 Conclusions

In this chapter we have examined another means of controlling Csound, through the use of the MIDI protocol. This can be employed for real-time device control, as well as in offline scenarios with file input. The use of MIDI can be intermingled with other forms of control, score, and orchestra code. It provides a flexible way to integrate Csound with other software and external hardware.

It was pointed out that MIDI has its limitations in that it supports a certain approach to music making that lacks some flexibility. In addition, the protocol has limited precision, as most of the data is in the seven-bit range. Alternatives to it have been shown elsewhere in this book, and in particular, we will see in the next chapter a more modern communications protocol that might eventually supersede MIDI. However, at the time of writing, this is still the most widespread method of connecting musical devices together for control purposes.

Chapter 10
Open Sound Control and Networking

Abstract In this chapter, we examine the networking capabilities of Csound. Through the use of Open Sound Control, the network opcodes, and/or the server, users can interact with external software, as well as with distributed processes in separate machines. The chapter explores these ideas, introducing key concepts and providing some examples of applications.

10.1 Introduction

Csound has extensive support for interaction and control via the network. This allows processes running on separate machines (or even on the same one) to communicate with each other. It also provides extra means of allowing the system to interact with other software that can use the common networking protocols. There are three areas of functionality within Csound that use networking for their operation: the Open Sound Control (OSC) opcodes; the all-purpose network opcodes; and the Csound server.

10.2 Open Sound Control

Open Sound Control was developed in the mid-1990s at CNMAT [133, 134], and is now widely used. Its goal was to create a more flexible, dynamic alternative to MIDI. It uses modern network communications, usually based on the user datagram transport layer protocol (UDP), and allows not only communication between synthesisers but also between applications and remote computers.

© Springer International Publishing Switzerland 2016   181
V. Lazzarini et al., *Csound*, DOI 10.1007/978-3-319-45370-5_10

10.2.1 The OSC Protocol

The basic unit of OSC data is a message. This is sent to an address which follows the
UNIX path convention, starting with a slash and creating branches at every following slash. The names inside this structure are free, but the convention is that the name
should fit the content, for instance /voice/3/freq or /ITL/table/point0.
So, in contrast to MIDI, the address space is not pre-defined and can be changed
dynamically.

An OSC message must specify the type(s) of its argument(s). The basic types
supported by Csound are:

- integer 32-bit (type specifier: "i")
- long integer 64-bit ("h")
- float ("f")
- double ("d")
- character ("c")
- string ("s").

Once data types are declared, messages can be sent and received. In OSC termi-
nology, anything that sends a message is a client, and anything that receives it is a
server. Csound can be both, in various ways, as it can

- send a message and receive it in another part of the same program;
- receive a message which is sent by any other application on this computer (localhost) or anywhere in the network;
- send a message to another application anywhere in the network.

10.2.2 Csound Implementation

The basic OSC opcodes in Csound are OSCsend and OSCreceive. As their
names suggest, the former sends a message from Csound to anywhere, and the latter
receives a message from anywhere in Csound:

```
OSCsend kwhen, ihost, iport, idestination,
            itype [, kdata1, kdata2, ...]
```

where kwhen sends a message whenever it changes, ihost contains an IP address,
iport specifies a port number, and idestination is a string with the address
space. The itype string contains one or more of the above-mentioned type speci-
fiers which will then occur as kdata1, kdata2 and so on.

```
ihandle OSCinit iport
kans OSClisten ihandle, idest, itype
            [, xdata1, xdata2, ...]
```

The opcode outputs 1 (to `kans`) whenever a new OSC message is received, otherwise zero. The `ihandle` argument contains the port number which has been opened by `OSCinit`; `idest` and `itype` have the same meaning as in `OSCsend`. The arguments `xdata1`, `xdata2` etc. must correspond to the data types which have been specified in the `itype` string. These will receive the contents of the incoming message, and must have been declared in the code beforehand, usually with an `init` statement to avoid an undefined-variable error.

The following code sends one integer, one float, one string, and one more float once a second via port 8756 to the localhost (`"127.0.0.1"`), and receives them by another Csound instance.

Listing 10.1 Client (sender) program

```
instr send_OSC
 kSend = int(times:k())+1
 kInt  = int(random:k(1,10))
 String = "Hello anywhere"
 kFloat1 = random:k(0,1)
 kFloat2 = random:k(-1,0)
 OSCsend kSend, "127.0.0.1", 8756,
       "/test/1", "ifsf",
      kInt, kFloat1,   String, kFloat2
 printf {{
OSC message %d sent at time %f!
int = %d, float1 = %f,
String = '%s', float2 = %f\n
}},
 kSend, kSend, date:k(),
 kInt,
 kFloat1, String,
 kFloat2
endin
schedule("send_OSC",0,1000)
```

Listing 10.2 Server (receiver) program

```
giRecPort OSCinit 8756
instr receive_OSC
 kI, kF1, kF2, kCount init 0
 Str = ""
 kGotOne OSClisten giRecPort, "/test/1", "ifsf",
                     kI, kF1, Str, kF2
 if kGotOne == 1 then
    kCount += 1
    printf {{
OSC message %d received at time %f!
int = %d, float1 = %f,
```

```
String = '%s', float2 = %f\n
      }},kCount, kCount, date:k(),
         kI, kF1, Str, kF2
 endif
endin
schedule("receive_OSC", 0, 1000)
```

Running the two programs on two separate processes (different terminals/shells) shows this printout for the sender:

```
OSC message 1 sent at time 1449094584.194864!
int = 8, float1 = 0.291342,
String = 'Hello anywhere', float2 = -0.074289

OSC message 2 sent at time 1449094585.167630!
int = 5, float1 = 0.680684,
String = 'Hello anywhere', float2 = -0.749450

OSC message 3 sent at time 1449094586.166411!
int = 1, float1 = 0.871381,
String = 'Hello anywhere', float2 = -0.070356

OSC message 4 sent at time 1449094587.168919!
int = 3, float1 = 0.615197,
String = 'Hello anywhere', float2 = -0.863861
```

And this one for the receiver:

```
OSC message 1 received at time 1449094584.195330!
int = 8, float1 = 0.291342,
String = 'Hello anywhere', float2 = -0.074289

OSC message 2 received at time 1449094585.172991!
int = 5, float1 = 0.680684,
String = 'Hello anywhere', float2 = -0.749450

OSC message 3 received at time 1449094586.171918!
int = 1, float1 = 0.871381,
String = 'Hello anywhere', float2 = -0.070356

OSC message 4 received at time 1449094587.169059!
int = 3, float1 = 0.615197,
String = 'Hello anywhere', float2 = -0.863861
```

So in this case the messages are received with a time delay of about 5 milliseconds.

10.2.3 *Inter-application Examples*

Processing using Csound as an audio engine

Processing is a Java-based programming language for visual arts[1]. Its visual engine can interact easily with Csound's audio engine via OSC. A simple example uses the "Distance1D" code from the built-in Processing examples. Four lines (one thick, one thin, on two levels) move depending on the mouse position (Fig. 10.1).

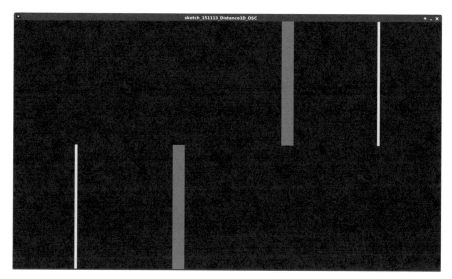

Fig. 10.1 Processing user interface

To send the four x-positions as OSC messages, we use the code shown in listing 10.3 in the Processing sketchbook.

Listing 10.3 Processing program for OSC messaging

```
import oscP5.*;
import netP5.*;

OscP5 oscP5;
NetAddress myRemoteLocation;
float xpos1;
float xpos2;
float xpos3;
float xpos4;
...
```

---

[1] http://processing.org

```
void setup()
{
  ...
  oscP5 = new OscP5(this,12001);
  myRemoteLocation = new NetAddress("127.0.0.1",12002);
  ...
}

void draw()
{
  ...
  OscMessage xposMessage =
      new OscMessage("/Proc/xpos");

  xposMessage.add(xpos1);
  xposMessage.add(xpos2);
  xposMessage.add(xpos3);
  xposMessage.add(xpos4);

  oscP5.send(xposMessage, myRemoteLocation);
}
```

The Processing sketch sends the four locations as x-positions in one OSC message with the address /Proc/xpos on port 12,002. The message consists of four floating point numbers which represent xpos1, xpos2, xpos3, and xpos4. The Csound code receives these messages and scales the pixel range (0, ..., 640) to the octave C-C (MIDI 72-84) for the upper two lines, and to the octave F#-F# (MIDI 66-78) for the lower two lines. As the lines move at different speeds, the distance changes all the time. We use this to increase the volume of the two lines on the same level when they become closer to each other, and vice versa. The movement of the mouse leads to different chords and different changes between chords (listing 10.4).

Listing 10.4 OSC-controlled Csound synthesis code

```
opcode Scale, k, kkkkk
 kVal, kInMin, kInMax, kOutMin, kOutMax xin
 kValOut = (((kOutMax - kOutMin) / (kInMax - kInMin)) *
                    (kVal - kInMin)) + kOutMin
 xout kValOut
endop

giPort OSCinit 12002

instr 1

 ;initialize variables and receive OSC
 kx1, kx2, kx3, kx4 init 0
```

```
kPing OSClisten giPort, "/Proc/xpos", "ffff",  kx1,
                              kx2, kx3, kx4

;scale x-values to MIDI note numbers
km1 Scale kx1, 0, 640, 72, 84
km2 Scale kx2, 0, 640, 72, 84
km3 Scale kx3, 0, 640, 66, 78
km4 Scale kx4, 0, 640, 66, 78

;change volume according to distance 1-2 and 3-4
kdb12 = -abs(km1-km2)*2 - 16
kdb12 port kdb12, .1
kdb34 = -abs(km3-km4)*2 - 16
kdb34 port kdb34, .1

;produce sound and output
ax1 poscil ampdb(kdb12), cpsmidinn(km1)
ax2 poscil ampdb(kdb12), cpsmidinn(km2)
ax3 poscil ampdb(kdb34), cpsmidinn(km3)
ax4 poscil ampdb(kdb34), cpsmidinn(km4)

ax = ax1 + ax2 + ax3 + ax4

kFadeIn linseg 0, 1, 1

out ax*kFadeIn, ax*kFadeIn

endin
schedule(1,0,1000)
```

Csound using INScore as an intelligent display

Open Sound Control can not only be used for real-time sonification of visual data
by Csound, as demonstrated in the previous section. It can also be used the other
way round: Csound produces sounding events which are then visualised. INScore[2],
developed at GRAME[3], is one of the many applications which are capable of doing
this.

The approach here is slightly different from the one using Processing: INScore
does not need any code. Once it is launched, it listens to OSC messages (by default
on port 7,000). These messages can create, modify and remove any graphical representation. For instance, the message /ITL/csound new will create a new panel

---

[2] http://inscore.sourceforge.net

[3] http://grame.fr

called `csound`[4]. The message `/ITL/csound/point0 set ellipse 0.1 0.2` will create an ellipse with address `point0` and size `(0.1,0.2)` (x,y) in this panel. In Csound code, both messages look like this:

```
OSCsend 1, "", 7000, "/ITL/csound", "s", "new"
OSCsend 1, "", 7000, "/ITL/csound/point0",
                "ssff", "set", "ellipse", 0.1, 0.2
```

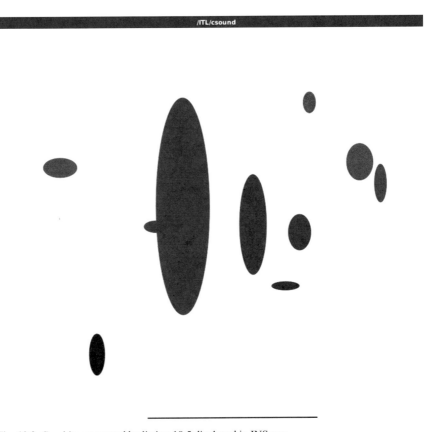

Fig. 10.2 Graphics generated by listing 10.5 displayed in INScore

The example code in listing 10.5 uses this flexibility in various ways. Ten instances of `OneTone` are created (Fig. 10.2). Each instance carries a unique ID, as it is called with a p4 from zero to nine. This ID via p4 is used to link one instance to one of the ten points in the INScore panel. Each instance will create a successor with the same p4. The ten points are modified by the sounds in many ways:

- "High" and "low" pitches are placed high and low in the panel.

---

[4] `/ITL` is the address space for the INScore application.

- Left/right position is determined by the panning.
- The form of the ellipse depends on shorter/duller (more horizontal) or longer/-more resonant (more vertical) sounds.
- The colour is a mixture of red (depending on pitch) and some blue (depending on filter quality).
- The size of an ellipse depends on the volume. Each point disappears slowly when the sound gets softer.

Except for the continuous decrement of the size, it is sufficient to send all messages only once, by using kwhen=1. For the transformation of the size, a rate of 15 Hz is applied for sending OSC messages, instead of sending on every k-cycle.

Listing 10.5 OSC-generating Csound code

```
ksmps = 32
nchnls = 2
0dbfs = 1
seed 0

;inscore default port for receiving OSC
giOscPort = 7000

opcode Scale, i, iiiii
 iVal, iInMin, iInMax, iOutMin, iOutMax xin
 iValOut = (((iOutMax - iOutMin) / (iInMax - iInMin))
                 * (iVal - iInMin)) + iOutMin
 xout iValOut
endop

instr Init
 OSCsend 1, "", giOscPort, "/ITL/csound", "s", "new"
 OSCsend 1, "", giOscPort, "/ITL/csound/*", "s", "del"
 gkSend metro 15
 indx = 0
 while indx < 10 do
  schedule "OneTone", 0, 1, indx
  indx += 1
 od
 schedule "Reverb", 0, p3
endin

instr OneTone
 ;generate tone and send to reverb
 iOct random 7,10
 iDb random -20,0
 iQ random 100,1000
 p3 = iQ/100
```

```
aStrike butlp mpulse(ampdb(iDb), p3), cpsoct(iOct)
aTone linen mode(aStrike, cpsoct(iOct), iQ), 0, p3, p3/2
iPan random 0,1
aL,aR pan2 aTone, iPan
chnmix aL, "left"
chnmix aR, "right"
;send OSC messages to Inscore
S_address sprintf "/ITL/csound/point%d", p4
iSizeX Scale iDb, -20, 0, .1, .3
iY_rel Scale iQ, 100, 1000, .1, 2
OSCsend 1, "", giOscPort, S_address, "ssff", "set",
                    "ellipse", iSizeX, iSizeX*iY_rel^2
OSCsend 1, "", giOscPort, S_address, "si", "red",
                    Scale(iOct,7,10,0,256)
OSCsend 1, "", giOscPort, S_address, "si", "blue",
                    Scale(iQ,100,1000,100,0)
OSCsend 1, "", giOscPort, S_address, "sf", "y",
                    Scale(iOct,7,10,.7,-.7)
OSCsend 1, "", giOscPort, S_address, "sf", "x",
                    Scale(iPan,0,1,-1,1)
OSCsend gkSend, "", giOscPort, S_address, "sf",
                    "scale", line:k(1,p3,0)
;call a new instance of this ID
 schedule "OneTone", p3, 1, p4
endin

instr Reverb
 aL chnget "left"
 aR chnget "right"
 aLrv, aRrv reverbsc aL, aR, .7, sr/3
 out aL*.8+aLrv*.2, aR*.8+aRrv*.2
 chnclear "left"
 chnclear "right"
endin

schedule("Init",0,9999)
```

Note that OSCsend will default to the localhost if passed an empty string as its destination network address. When using OSC for inter-application connections in the same machine, this is the normal procedure.

10.3 Network Opcodes

In addition to the specific OSC opcodes, which are designed to read this particular protocol, Csound has more general network opcodes that can be used to send and receive data via UDP or TCP (Transport Control Protocol) messages. These opcodes can be used to work with a-variables, mostly, but can also be used to send and receive control data.

The UDP opcodes are

```
asig sockrecv iport,ilen
ksig sockrecv iport,ilen
socksend asig,Sadrr,iport,ilen
socksend ksig,Sadrr,iport,ilen
```

where the network address is given by `Sadrr`, the port used by `iport`. It is also necessary to set the length of the individual packets in UDP transmission: `ilen`, which needs to be small enough to fit a single maximum transmission unit (MTU, 1,456 bytes). While UDP signals can theoretically carry audio signals, in practice it is difficult to have a reliable connection. However, it can carry control or modulation data well. In the following examples, we have a sender producing a glissando signal that is applied to an oscillator frequency at the receiver.

The send instrument produces an exponential control signal, and sends it to the localhost, port 7,708:

```
instr 1
 socksend expon(1,10,2),
          "127.0.0.1",7708,200
endin
schedule(1,0,-1)
```

The receiver, running in a separate process, takes the UDP data and uses it as a frequency scaler.

```
instr 1
 k1 sockrecv 7708,200
 out oscili(0dbfs/2, k1*440)
endin
schedule(1,0,-1)
```

In addition to these, Csound has a pair of TCP opcodes. These work differently, requiring the two ends of the connection to handshake, so the receiver will have to look for a specific sender address, instead of receiving data from any sender:

```
stsend asig,Sadrr,iport
asig strecv Sadrr,iport
```

Messages are send in a stream, and the connection is more reliable. It works better for sending audio, as in the example below:

```
instr 1
 stsend oscili(0dbfs/2,440),  "127.0.0.1",8000
endin
schedule(1,0,-1)

instr 1
 out strecv("127.0.0.1",8000)
endin
schedule(1,0,-1)
```

10.4 Csound UDP Server

In addition to its OSC and network opcodes, Csound can also be set up as a server
that will listen to UDP connections containing code text. As soon as it is received,
a text string is compiled by Csound. In this way, Csound can work as lightweight
audio server. If Csound is passed the --port=N option, it will listen to messages
on port N, until it is closed (by sending it a kill signal, ctrl-c).

Messages can be sent to Csound from any UDP source, locally or on the network.
For instance, we can start Csound with these options:

```
$ csound --port=40000 -odac
```

And we would see the following messages (among others):

```
0dBFS level = 32768.0
UDP server started on port 40000
orch now loaded
audio buffered in 1024 sample-frame blocks
PortAudio V19-devel (built Sep  4 2014 22:30:30)
   0: dac0 (Built-in Output)
   1: dac1 (Soundflower (2ch))
   2: dac2 (Soundflower (64ch))
   3: dac3 (Aggregate Device)
PortAudio: selected output device 'Built-in Output'
writing 1024 sample blks of 64-bit floats to dac
SECTION 1:
```

Csound is running, with no instruments. We can use the command nc (netcat) to
send a UDP message to it, using the heredoc facility, from another shell:

```
$ nc -uw 1  127.0.0.1 40000 <<end
> instr 1
> a1 oscili p4,p5
> out a1
> endin
```

```
> schedule(1,0,2,0dbfs/2,440)
> end
```

and all the text from the top to the keyword `end`, but excluding it, will be sent to Csound via UDP. The `nc` command takes in the network address (127.0.0.1, the localhost) and a port (40,000) to send the message to. This demonstrates the simplicity of the system, which can be used in a variety of set-ups.

As shown with OSC, third-party networking programs can be used to interface with the system. In particular, text editors (such as emacs or vim) can be configured to send selections of text to Csound, allowing them to implement interactive frontend facilities. Note that just by adding the `--port=N` option, we can start the network server, which works even alongside other input methods, scores, MIDI etc. The server itself can be closed by sending the string `"##close##"` to it.

10.5 Conclusions

Open Sound Control gives Csound a fast and flexible communication with any other application which is able to use OSC. It offers a wide range of use cases in sonification and visualisation. It makes it possible to control Csound remotely via any OSC-capable software, as explored in the examples based on Processing and In-score.

In complement to the OSC implementation, we have the generic network opcodes. These allow the user to send raw data over UDP and TCP connections, for both control and audio applications.

Finally, we introduced the built-in UDP server that is provided as part of the audio engine. Once enabled, it is possible to use it to send strings containing Csound code to be compiled and run. This provides a third way in which users can interact with the system over network communications infrastructure.

Chapter 11
Scripting Csound

Abstract In this chapter, we will discuss controlling Csound from applications written in general-purpose scripting languages. This allows users to develop music programs that employ Csound as an audio engine, for specific uses.

11.1 Introduction

Throughout this book we've discussed what can be done within the Csound system itself. In this chapter, we will discuss embedding Csound into applications, including how to externally control Csound, as well as communicate between the host application and Csound. The examples provided will be shown using the Python programming language. However, Csound is available for use in C, C++, Java (including JVM languages, such as Scala and Clojure), C#, Common Lisp and others.

11.2 Csound API

As discussed in Part I, Csound provides an Application Programming Interface (API) for creating, running, controlling, and communicating with a Csound engine instance. *Host applications*, written in a general-purpose programming language, use the API to embed Csound within the program. Such applications can range from a fully fledged frontend to small, dedicated programs designed for specific uses (e.g. an installation, a composition, prototyping, research).

The core Csound API is written in C. In addition, a C++ wrapper is available, and bindings for other programming languages (Python, Java, Lua) are generated using SWIG[1]. Each version of the API differs slightly to accommodate differences in each programming language, but they all share the same basic design and functionality.

---

[1] http://www.swig.org

© Springer International Publishing Switzerland 2016
V. Lazzarini et al., *Csound*, DOI 10.1007/978-3-319-45370-5_11

In this chapter, we will provide an overview of the use of Csound in a scripting-language environment, Python. Such scripts can be constructed to provide various means of interaction with the software system.

11.3 Managing an Instance of Csound

At the heart of a script that uses Csound through the API lies an instance of the whole system. We can think of it as an *object* that encapsulates all the elements that are involved in the operation of the software, e.g. compilation, performance, interaction etc. A number of basic actions are required in this process: initialising, compiling, running the engine, managing threads, stopping and cleaning up. In this section, we will look at each one of these.

11.3.1 Initialisation

There are two initialisation steps necessary for running Csound in Python. The first one of these is to load the Python module, so that the functionality is made available to the interpreter. There are two components in the `csnd6` module:

1. the Python wrapper code (`csnd6.py`)
2. the binary library module (`_csnd6.so`).

   In any regular installation of Csound, these would be placed in the default locations for Python. The first one contains all the Python-side API code that is needed for the running of the system, and the second contains the binary counterparts to it that talk directly to the Csound C library.

   When using Csound from Python, it is possible either to write the code into a text file and then run it using the interpreter, or else to type all commands interactively at the interpreter prompt. Csound can also be run from specialised Python shells such as IPython and its Notebook interface.

   In order to load Csound, the `import` command is used, e.g.

from csnd6 **import** Csound

This will load specifically the `Csound` class of the API, whereas

import csnd6

will load all the classes, including some other utility ones that allow the user to access some parts of the original C API that are not idiomatically adapted to the Python language. In this text, we will use this full-package import style. If you are using Python interactively you can use the following commands to get a listing of everything available in the API:

```
>> import csnd6
>> help(csnd6)
```

The next step is to create in memory a Csound object. After using the full-package form of the import command shown above, we can instantiate it with the following line:

```
cs = csnd6.Csound()
```

Once this is done, we should see the following lines printed to the console (but the reported version and build date will probably be different):

```
time resolution is 1000.000 ns
virtual_keyboard real time MIDI plugin for Csound
0dBFS level = 32768.0
Csound version 6.07 beta (double samples) Dec 12 2015
libsndfile-1.0.25
```

We can then configure options for Csound using the SetOption method. This allows us to set flags for Csound. It employs the same flags as one would use when executing Csound from the command line directly, or in the CsOptions tags within a CSD project file. For example

```
cs.SetOption('-odac')
```

will set the output to the digital-to-analogue converter (i.e., real-time audio output). Note that SetOption allows for setting only one option at a time. Options can also be set at first compilation, as shown in the next section.

11.3.2 First Compilation

In order to run Csound, the next required stage is to compile the Csound code, which we can do by passing a string containing the name of a CSD file to the Compile() method of the Csound class:

```
err = cs.Compile('test.csd')
```

Up to four separate options can be passed to Csound as string arguments to Compile(), in addition to the CSD file name:

```
err = cs.Compile('test.csd', '-odac', '-iadc','-dm0')
```

In this example, the input and output are set to real-time, and both displays and messages are suppressed. Alternatively, we can use the utility class CsoundArgV-List() to compose a variable list of arguments and pass it to Compile(), using its argc() and argv() methods which hold the number and array of arguments, respectively:

```
args = csnd6.CsoundArgVList()
args.Insert(0, 'test.csd')
args.Insert(0, '-odac')
args.Insert(0, '-idac')
args.Insert(0, '-dm0')
err = cs.Compile(args.argc(), args.argv())
```

In this case, any number of options can be passed to the compilation by inserting these as strings in the argument list.

Once the CSD is compiled with no errors (`err == 0`), it is possible to start making sound. Note that by compiling a first CSD in this way, the engine is started and placed in a ready-to-go state. Additional code can be further compiled by Csound at any point with `CompileOrc()`. Note that `Compile()` can only be called on a clean instance (either a newly created one, or one that has been reset, as will be discussed later).

11.3.3 Performing

At this stage, we can start performance. A high-level method can be used to run the Csound engine:

```
cs.Perform()
```

Note that this will block until the performance ends (e.g. at the end of an existing score) but if Csound is set to run in real-time with no score, this stage might never be reached in our lifetime. In that case, we would need to find a different way of performing Csound.

The API also allows us to perform one k-cycle at a time by calling the `Perform Ksmps()` method of the `Csound` class. Since this performs only a single `ksmps` period, we need to call it continuously, i.e. in a loop:

```
while err == 0:
  err = cs.PerformKsmps()
```

At the end of performance, `err` becomes non-zero. However, this will still block our interaction with the code, unless we place some other event-checking function inside the loop.

Performance thread

Another alternative is to run Csound in a separate thread. This is cumbersome in pure Python as the global interpreter lock (GIL) will interfere with performance. The best way to do it is to use a Csound performance thread object, which uses a separate native thread for performance. We can access it from the `csnd6` package to create a thread object by passing our Csound engine (`cs`) to it:

```
t = csnd6.CsoundPerfomanceThread(cs)
```

At this point, we can manipulate the performance simply by calling `Play()`, `Pause()`, and `Stop()` on the thread object, e.g.

```
t.Play()
```

Note that each one of these methods does not block, so they work very well in interactive mode. When using these in a script, we will have to make sure some sort of event loop or a sleep function is used, otherwise the interpreter will fall through these calls and close before we are able to hear any sound. Also in order to clear up the separate thread properly, on exiting we should always call `Join()` after `Stop()`.

Real-time performance and file rendering

As with other modes of interaction with Csound, it is possible to run Csound as a real-time performance or to render its output to a file. There is nothing specific to Python or to scripting in general with regards to these modes of operation. They can be selected as described in Part I, by employing the `-o` option. As discussed before, by default, Csound renders to file (called test.wav or test.aif depending on the platform), and to enable real-time output, the `-o dac` option is required.

11.3.4 Score Playback Control and Clean-up

If we are using the numeric score, performance will stop at the end of this score. In order to reuse the Csound instance, we need to either rewind it or set its position to some other starting point. There are two methods that can be used for this:

```
cs.SetScoreOffsetSeconds(time)
cs.RewindScore()
```

Note that these can be used at any point during performance to control the score playback. We can also get the current score position with:

```
cs.GetScoreTime()
```

Alternatively, we can clean up the engine so that it becomes ready for another initial compilation. This can be accomplished by a reset:

```
cs.Reset()
```

This can also be done, optionally, by recreating the object. In this case, Python's garbage collection will dispose of the old Csound instance later, and no resetting is needed.

11.4 Sending Events

One of the main applications for Python scripts that run Csound is to enable a high level of interactivity. The API allows external triggering of events (i.e. instantiation of instruments), in a flexible way. While the engine is performing, we can call methods in the `Csound` class that will schedule instruments for us. This is thread-safe, which means that it can be called from a different thread to the one where the engine is running, and calls can be made at any time.

There are two main methods for this, `InputMessage()` and `ReadScore`. The first one of these dispatches real-time events with no further processing, whereas the second one can take advantage of most of the score-processing capabilities (except for tempo warping) as expected for a score file (or section of a CSD file). The syntax of both methods is the same; they are passed strings defining events in the standard score format:

```
cs.InputMessage('i 1 0 0.5 8.02')
```

Events with start time (p2) set to 0 are played immediately, whereas a non-zero p2 would schedule instances to start sometime in the future, relative to the current time in the engine. Multiple events in multi-line strings are allowed, and the Python string conventions (single, double and triple quoting) apply.

11.5 The Software Bus

Csound provides a complete software bus to pass data between the host and the engine. It allows users to construct various means of interaction between their software and running instruments. The bus uses the channel system as discussed in Chapter 6, and it provides means of sending numeric control and audio data through its named channels. In the specific case of Python, given the nature of the language, we will tend to avoid passing audio data in strict real-time situations, as the system is not designed for efficient number crunching. However, in other situations, it is perfectly feasible to access an audio stream via the bus.

Data sent or received to Csound is processed once per k-cycle, so if two or more requests to a bus channel are sent inside one single k-period, the final one of these will supersede all others. All calls to the software bus are thread-safe, and can be made asynchronously to the engine.

11.5.1 Control Data

The most common use of the software bus is to pass numeric control data to and from Csound. The format is a scalar floating-point value, which can be manipulated via two methods:

```
var = cs.GetChannel('name')
cs.SetChannel('name', var)
```

The first parameter is a string identifying the name of the channel. The counterparts for these functions in Csound code are

```
kvar chnget  "name"
chnset kvar, "name"
```

These opcodes will work with either i- or k-rate variables for control data. Thus, when passing data from Python to Csound, we use `SetChannel()` and `chnget`, and when going in the other direction, `chnset` and `GetChannel()`. These functions are also thread-safe, and will set/get the current value of a given channel.

11.5.2 Audio Channels

As we have learned before, audio data in Csound is held in vectors, which will have `ksmps` length. So, if we want to access audio data, we will need to use a special Python object to handle it. The `csnd6` package offers a utility class to handle this, `CsoundMYFLTArray`, which can then be used in conjunction with the `GetAudioChannel()` and `SetChannel()` methods of the `Csound` class. For instance, in order to get the data from a channel named `"audio"`, we can use this code:

```
aud = csnd6.CsoundMYFLTArray(cs.GetKsmps())
cs.GetAudioChannel('audio', aud.GetPtr(0))
```

Each one of the elements in the vector can be accessed using `GetValue()`:

```
for i in range(0,cs.GetKsmps()):
  print aud.GetValue(i)
```

To pass data in the other direction, using the same object, we can do

```
cs.SetChannel('audio', aud.GetPtr(0))
```

The `GetPtr()` method is needed to allow the channel to access the vector memory created in the Python object. Its parameter indicates which starting position in the vector we will pass to the channel (0, the beginning, is the usual value).

11.5.2.1 Input and Output Buffers

A similar method used for audio channels can be used to access the input and/or output buffers of Csound, which hold the data used by the `in` and `out` (and related) opcodes.

As discussed earlier in Section 2.8.2, Csound has internal buffers that hold a vector of ksmps audio frames. These are called `spin` and `spout`. We can access them from the API using a `CsoundMYFLTArray` object:

```
spout = csnd6.CsoundMYFLTArray()
spout.SetPtr(cs.GetSpout())
spin = csnd6.CsoundMYFLTArray()
spin.SetPtr(cs.GetSpin())
```

As discussed above, the `SetValue()` and `GetValue()` methods of the `CsoundMYFLTArray` class can then be used to manipulate the audio data. The size of these vectors will be equivalent to `ksmps*nchnls` (if `nchnls_i` has been defined, then the size of `spin` is `ksmps*nchnls_i`).

Likewise, we can access the actual input and output software buffers, which are larger, and whose size in frames is set by the `-b N` option:

```
inbuffer = csnd6.CsoundMYFLTArray()
inbuffer.SetPtr(cs.GetInputBuffer())
outbuffer = csnd6.CsoundMYFLTArray()
outbuffer.SetPtr(cs.GetOutputBuffer())
```

Note that all of these buffers are only available after the first compilation, when the engine has started.

11.6 Manipulating Tables

Function tables are also accessible through the API. It is possible to read and/or write data to existing tables. The following methods can be used:

- `TableLength(tab)`: returns the length of a function table, `tab` is the table number.
- `TableGet(tab, index)`: reads a single value from a function table, `tab` is the table number and `index` is the position requested.
- `TableSet(tab, index, value)`: writes a single value of a function table, `tab` is the table number and `index` is the position to be set with `value`.

- `TableCopyIn(tab, src)`: copies all values from a vector into a table. The `src` parameter is given by a `GetPtr()` from a `CsoundMYFLTArray` created with the same size as the requested function table. It holds the values to be copied into the table.
- `TableCopyOut(tab, dest)`: copies all values from a table into a vector. The `dest` parameter is given by a `GetPtr()` from a `CsoundMYFLTArray` created with the same size as the requested function table. It will hold a copy of the values of the function table after the function call.

The `CsoundMYFLTArray` object used in the data copy should have a least the same size as the requested table. Trying to copy data to/from a destination or source with insufficient memory could lead to a crash.

11.7 Compiling Orchestra Code

It is possible to compile orchestra code straight from a Python string using the
`CompileOrc()` method of the `Csound` class. This can be done at any time, and
it can even replace the `Compile()` method that reads a CSD from file. However,
in this case, we will need to start the engine explicitly with a call to `Start()`.
Consider the following example (listing 11.1).

Listing 11.1 Compiling directly from a string containing Csound code

```
cs = csnd6.Csound()
cs.SetOption('-odac')
if cs.CompileOrc('''
instr 1
 a1 oscili p4, p5
 out a1
endin
schedule(1,0,2,0dbfs,440)
''') == 0:
  cs.Start()
  t = csnd6.CsoundPerformanceThread(cs)
  t.Play()
  while(t.isRunning() == 1):
    pass
```

Here, we create a Csound engine and set it to real-time audio output. Then we
send it the initial code to be compiled. Following this, if the compilation is successful, we start the engine, set up a performance thread and play it. The infinite loop at
the end makes sure the script does not fall through; it can be ignored if the code is
typed at the Python shell, and it should be replaced by an event listener loop in an
interactive application.

Any number of calls to `CompileOrc()` can be issued before or after the engine
has started. New instruments can be added, and old ones replaced using this method.
This method is thread-safe and may be called at any time.

11.8 A Complete Example

To complement this chapter, we will show how Csound can be combined with a
graphical user interface package to create a simple program that will play a sound
in response to a button click. There are several Python GUI packages that can be
used for this purpose. We will provide an example using the Tkinter module, which
is present in most platforms under which Python can be run.

The application code, in listing 11.2, can be outlined as follows:

1. A class defining the GUI interface is created.

2. A Csound object is placed inside this application class and code is compiled.
3. The engine is run in a separate thread provided by
   `CsoundPerformanceThread`.
4. Callbacks are defined to react to each button press and for quitting Csound.
5. The Application class is instantiated and its main loop is run.

The application consists of a main window (Fig. 11.1) with a single button
('play'), which will trigger an event lasting for 2 seconds.

Listing 11.2 Python GUI application example

```python
#!/usr/bin/python
import Tkinter as tk
import csnd6

class Application(tk.Frame):
 def __init__(self, master=None):
   # setup Csound
   self.cs = csnd6.Csound()
   self.cs.SetOption('-odac')
   if self.cs.CompileOrc('''
       instr 1
         a1 oscili p4, p5
         k1 expseg 1,p3,0.001
         out a1*k1*0dbfs
       endin
       ''') == 0:
    self.cs.Start()
    self.t = csnd6.CsoundPerformanceThread(self.cs)
    self.t.Play()
    # setup GUI
    tk.Frame.__init__(self, master)
    tk.Frame.config(self, height=200, width=200)
    self.grid(ipadx=50, ipady=25)
    self.Button = tk.Button(self, text='play',
     command=self.playSound)
    self.Button.pack(padx=50, pady=50, fill='both')
    self.master.protocol("WM_DELETE_WINDOW",
     self.quit)
    self.master.title('Beep')

  # called on quit
 def quit(self):
  self.master.destroy()
  self.t.Stop()
  self.t.Join()
```

```
    # called on button press
  def playSound(self):
    self.cs.InputMessage('i 1 0 2 0.5 440')

app = Application()
app.mainloop()
```

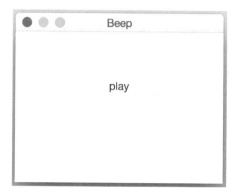

Fig. 11.1 Main window of Python application (as defined in listing 11.2)

11.9 Conclusions

This chapter set about introducing a scripting dimension to the interactive side of Csound. We have looked at how the engine can be instantiated and its performance controlled in a very precise and flexible way through the Python language. A simple, yet complete, example was given at the end, providing pointers to the types of applications this functionality can have. In addition to the ideas discussed here, we should mention that Csound can be fully integrated into the workflow of a programming environment such as Python, providing a highly efficient and configurable audio engine. Applications in music performance, composition, sonification and in signal-processing research are enabled by the system.

Part IV
Instrument Development

Chapter 12
Classic Synthesis

Abstract In this chapter, we will explore a variety of classic synthesis designs, which have been developed throughout the history of computer music. These are organised into three *families*: source-modifier methods (also known as *subtractive*), distortion techniques, and additive synthesis. The text begins by introducing some fundamental concepts that underpin all classic approaches, relating to the duality of waveform and spectrum representations. Following this, source-modifier methods are explored from their basic components, and two design cases are offered as examples. Distortion synthesis is presented from the perspective of the various related techniques that make up this family, with ready-to-use UDOs for each one of them. Finally, we introduce the principle of additive synthesis, with examples that will link up to the exploration of spectral techniques in later chapters.

12.1 Introduction

Synthesis techniques have been developed over a long period since the appearance of the first electronic instruments, and then through the history of computer music. Of these, a number of classic instrument designs have emerged, which tackled the problem of making new sounds from different angles. In this chapter, we will explore these in detail, providing code examples in the Csound language that can be readily used or modified, in the form of user-defined opcodes (UDOs). The three *families* of classic sound synthesis will be explored here: source-modifier methods, distortion and additive synthesis. Our approach will be mostly descriptive, although we will provide the relevant mathematical formulae for completeness, wherever this is relevant.

© Springer International Publishing Switzerland 2016
V. Lazzarini et al., *Csound*, DOI 10.1007/978-3-319-45370-5_12

12.1.1 Waveforms and Spectra

The principle that joins up all classic synthesis techniques is the idea that in order to realise a given waveform, we need to specify, or at least approximate, its *spectrum*. This is another way to represent sound. While audio waves give us a picture of how a given parameter (e.g. air pressure at a point) varies over time, a spectral plot shows the components of the signal at a point in time. These components, also known as partials, are very simple waveforms, which can be described by a *sinusoidal* function (such as sin() and cos()) with three parameters: amplitude, frequency and phase. They can be thought of as the building blocks of a given sound, sometimes called, informally, *pure frequencies*. Theoretically, any waveform can be broken down into such components, although in some cases this might involve an infinite number of them.

Mathematically stated, we can represent a target signal $s(t)$ as [63]

$$s(t) = \sum_{k=1}^{N} A_k \cos(\omega_k + \theta_k) \tag{12.1}$$

where A_k is the amplitude of partial k, $\omega_k = 2\pi f_k t$, with f_k its frequency in Hz, and θ_k, its phase. This representation allows us to look for the best way to generate a given sound. Depending on how large the number of components N is, and how complicated the configuration of the amplitudes, frequencies and phases is, we will prefer certain methods to others. For instance, with small numbers of components, it is simpler to create the sound by generating the components separately and mixing them up (this is the principle of additive synthesis). As N gets larger, there are more efficient ways, through distortion techniques or source-modifier methods. In the case of noisy sounds, which might have a (statistically) infinite number of components, distributed across large frequency bands, adding components together is not feasible. The idea that complex functions can be decomposed into sinusoidal partials was originally explored by Joseph Fourier [43], and methods employing this principle often bear his name (e.g. *Fourier series, Fourier transform*).

Once we can describe the components of a signal in terms of its parameters, we might then choose one of the classic methods to synthesise it. For some sounds, it is possible to have a clear picture of what the spectrum looks like. For instance, Fig. 12.1 shows a waveform and its corresponding amplitude spectrum. Each component in the waveform is shown as a vertical line, whose height indicates its (relative) amplitude (or *weight*). The horizontal axis determines the frequency of the component, from 0 Hz upwards (phase is not plotted). The spectral plot refers to a given moment in time, but as the wave is fixed, it is a valid description for this particular shape. For each line in the spectral plot, we have a sinusoidal wave in the signal, and if we start with this recipe, we could reproduce the sound by adding these simples sounds together.

When partial frequencies are connected by a simple integer relationship (e.g. 1, 2, 3, 4,...), they form a *harmonic* spectrum. In the example of Figure 12.1, this is the case, as we can see that the partials are spaced evenly apart. They are multiples

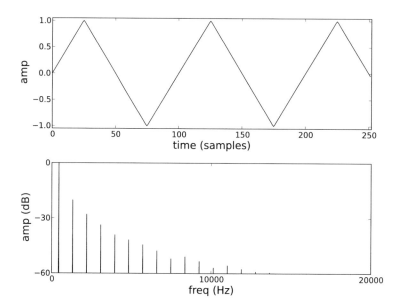

Fig. 12.1 Waveform (top) and amplitude spectrum (bottom) of an audio signal (sr = 40 kHz)

of a *fundamental frequency*, which determines the perceived pitch of the sound. In this case we can call these partials by the name of *harmonics*. Fig. 12.2 shows these components as sinusoidal waves that add up to make a complex waveform. Harmonic spectra can have gaps in it (missing harmonics), as long as the existing partials are closely aligned to a series of integers of a fundamental frequency. This is also called the *harmonic series*. If the partials are not closely related to this series, then the spectrum is *inharmonic*, and the sense of pitch less defined (as there is no fundamental frequency to speak of). In this case we have *inharmonic partials*. However, it is important to note that small deviations from the harmonic series can still evoke the feeling of a fundamental (and pitch). In fact, some instruments often have partials that deviate slightly from the harmonic series, but produce recognisable pitched notes.

More generally, spectra of different sounds are significantly more complex than the example in Fig. 12.1. In particular, interesting sounds tend to vary in time, and generating time-varying spectra is particularly important for this. In this case, the amplitude A_k and frequency f_k parameters of each partial can change over time, and the chosen synthesis method will need to try and emulate this. In the following sections, we will explore the three classic methods that can be used for creating different types of spectra.

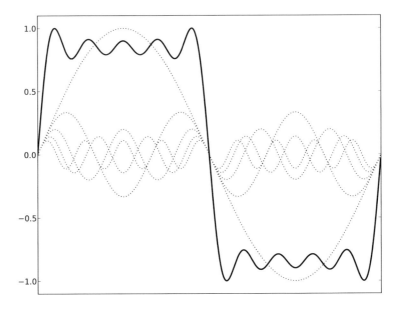

Fig. 12.2 One cycle of a complex periodic waveform (solid) and its sinusoidal components (harmonics, in dots)

12.2 Source-Modifier Methods

Source-modifier is the most common of all the classic synthesis methods, which is ubiquitously present in hardware synthesisers, their software emulations, and music programming systems. It is commonly called *subtractive synthesis*, which does not accurately describe it, since there is no actual subtraction involved in it. That term is used with reference to the fact that this method is complementary to additive synthesis, which involves actual summation of single-source signals. Here, we start with component-rich sounds, and modify them. However, this process is not necessarily a subtraction, although in some cases it involves removal of parts of the original spectrum.

12.2.1 Sources

In order to allow for further modification, a source needs to be complex, i.e. it needs to have a good number of partials in its spectrum. This means that in this case, a sinusoidal oscillator, which contains a single component, is not a very good choice. Waveforms of various other shapes can be used, and also recordings of real instruments, voices and other component-rich sounds. Non-pitched sources can also be used, and a variety of noise generators are available in Csound for this. In this section, we will examine these possibilities in detail.

Oscillators

The basic wavetable oscillators in Csound provide the most immediate types of sound sources for further modification. There are two types of simple oscillators: the `oscil` and `poscil` families. The former employs an indexing algorithm written using fixed-point (or integer) mathematics, whereas the latter uses floating-point (decimal-point) arithmetic. In earlier computing platforms, an integral index was significantly faster to calculate, making `oscil` and derived opcodes much more efficient. The limitations were twofold: function tables were required to be power-of-two size, and with large table sizes indexing errors would occur, which resulted in loss of tuning accuracy at lower frequencies. In modern platforms, very little difference exists in performance between fixed- and floating-point operation. Also, in 64-bit systems, the indexing error in the former method has been minimised significantly. So the difference between these two oscillator families is very small. In practical terms, it mainly involves the fact that `poscil` and derived opcodes do not require tables to have a power-of-two size, whereas the others do.

Both families of oscillators have a full complement of table-lookup methods, from truncation (`oscil`), to linear (`oscili`, `poscil`) and cubic interpolation (`oscil3`, `poscil3`) [67]. They can work with tables containing one or more waveform cycles (but the most common application is to have a single period of a wave stored in the table). Csound provides the GEN9, GEN10, and GEN19 routines to create tables based on the additive method. GEN10 implements a simple Fourier Series using sine waves,

$$ftable[n] = \sum_{k=1}^{N} P_k \sin(2\pi kn/L) \tag{12.2}$$

where L is the table length, N the number of harmonics (the number of GEN parameters), P_k is parameter k and n the table index. GEN9 and GEN19 have more generic parameters that include phase and relative frequencies, as well as amplitudes. For instance, with GEN10, a wave containing ten harmonics can be created with the following code:

```
gifun   ftgen   0, 0, 16384, 10, 1, 1/2, 1/3, 1/4, 1/5,
                        1/6, 1/7, 1/8, 1/9
```

The required parameters are the different amplitudes (weights) of each harmonic partial, starting from the first. In this case, the harmonics have decreasing amplitudes, 1 through to $\frac{1}{10}$ (Fig. 12.3 shows a plot of the function table). The main disadvantage of using such simple oscillators is that the bandwidth of the source cannot be controlled directly, and is dependent on the fundamental frequency. For instance, the above waveform at 100 Hz will contain components up to 900 Hz, whereas an-

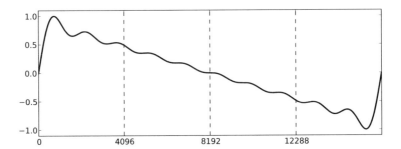

Fig. 12.3 Function table created with GEN10, containing ten harmonics with amplitudes $\frac{1}{n}$, where n is the harmonic number

other one at 1000 Hz will extend until 9000 Hz. So we will have be careful with our range of fundamentals, so that they do not go beyond a certain limit that would push the 10th harmonic ($9 \times f_0$) over the Nyquist frequency. The other difficulty is that low-frequency sounds will lack brightness, as they are severely bandlimited.

A solutions to this can be to supply a variety of tables, which would be selected depending on the oscillator fundamental. This is a good way of implementing a general table-lookup system, but can be complex to realise in code. An alternative is to use a ready-made solution provided by bandlimited oscillators. In Csound, there are four basic opcodes for this: `buzz`, `gbuzz`, `vco`, and `vco2`. The first two implement different forms of bandlimited waveforms (see Section 12.3.1 for an idea of how these algorithms work), with `buzz` providing all harmonics with the same amplitude, and `gbuzz` allowing for some control of partial weights.

The other two offer two different flavours of virtual analogue modelling, producing various types of classic waveforms. The typical shapes produced by these oscillators are

- **sawtooth**: contains all harmonics up to the Nyquist frequency with weights determined by $\frac{1}{n}$ (where n is the harmonic number).
- **square**: odd harmonics only, with weights defined by $\frac{1}{n}$.
- **pulse**: all harmonics with the same weight.
- **triangle**: odd harmonics only, with weights defined by $\frac{1}{n^2}$.

Plots of the waveforms and spectra of two classic waves generated by `vco2` are shown in Figs. 12.4 and 12.5, where we see that their spectra are completed all the way up to the Nyquist frequency. In addition to the waveshapes listed above, it is also possible to produce intermediary forms between pulse and square by modifying the width of a pulse wave (its *duty cycle*). This can be changed from a narrow impulse to a square wave (50% duty cycle).

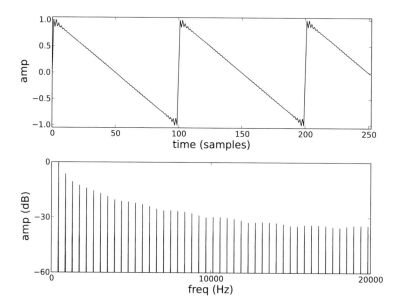

Fig. 12.4 A plot of the waveform (top) and spectrum (bottom) of a sawtooth wave generated by vco2, with $sr = 44,100$ Hz

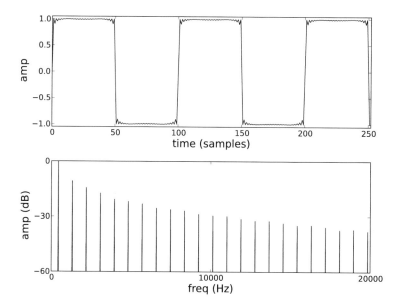

Fig. 12.5 A plot of the waveform (top) and spectrum (bottom) of a square wave generated by vco2, with $sr = 44,100$ Hz

Sampling

Yet another set of sources for modification can be provided by *sampling*. This can be achieved by recording sounds to a soundfile and then loading these into a table for playback. These tables can be read by any of the simple oscillators (with some limitations if power-of-two sizes are required), or a specialised sample-playback opcode can be used. The GEN 1 routine is used to load soundfiles into tables. Any of the file formats supported by Csound are allowed here (see Section 2.8.4). For sounds that are not mono, it is possible to read each channel of the soundfile separately, or to load the whole file with all the channels into a single table. In this case, samples are *interleaved* (see Section 2.4.3 for a detailed explanation), and ordinary oscillators cannot be used to read these. Here is a code example showing GEN1 in use to load a mono file ("fox.wav") into a function table:

```
gifun   ftgen   0, 0, 0, 1,"fox.wav", 0, 0, 0
```

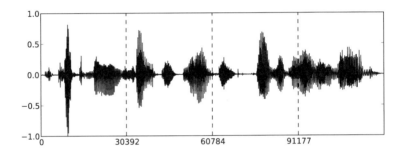

Fig. 12.6 Function table created with GEN1, containing the samples of a soundfile

Note that the table size is set to 0, which means that the soundfile size will determine how long the table will be. A plot of this function table can be seen in Fig. 12.6. When using a simple oscillator to read this, we will need to adjust the frequency to depend on the table size and the sampling rate. If we want to read the table at the original speed, we set the sampling (or phase) increment (si) (see Section 3.5.2) to 1. The relationship between oscillator frequency f, si, table size l and sampling rate sr is:

$$f = si \times \frac{sr}{l} \tag{12.3}$$

We can use si in eq. 12.3 to control the playback speed, which can be negative for backwards play. Listing 12.1 shows the code for a simple mono sample playback instrument using the `poscil` oscillator. It has parameters for amplitude, playback speed (faster, $kspd > 1$; slower, $0 < kspd < 1$; backwards, $kspd < 0$) and function table.

Listing 12.1 Mono sampler UDO

```
/**************************************
asig MonoSampler kamp, kspd, ifn
kamp - amplitude
kspd - playback speed
ifn - source function table
***************************************/
opcode MonoSampler,a,kki
 kamp,kspd,ifun xin
 xout poscil(kamp,kspd*sr/ftlen(ifun),ifun)
endop
```

This simple method will not allow control over the loop points (the oscillator will loop over the whole table). More complex instruments can be created with table-reading opcodes. Alternatively, we can employ a specialised opcode for this. There are two options for this: looping points can be taken from the soundfile header (RIFF-Wave and AIFF formats allow for this), or we can set them directly in Csound. Some soundfile-editing programs allow the user to mark looping points and then save these in the file. GEN1 will load this information as part of the function table, and Csound provides the `loscil` opcode family to read these tables, which can also be multichannel. The opcode will loop the sound as specified in the saved data. It is also possible to save the *base frequency* of the recorded sample in the soundfile header, which is also read by GEN1. The frequency control in `loscil` will be able to use this information to transpose the file correctly according to this information.

If it is not convenient to edit and save loop points with external software, we can use the `flooper` opcode family to define looping points. The advantage here is that these opcodes can crossfade the loop points. This means that the end of the loop overlaps with the beginning, generally creating a smooth join. The loop points and cross-fade time can be changed during performances with `flooper2`, which also allows different types of looping (normal, backwards, and back-and-forth). These opcodes feature a playback speed control (which is equivalent to the *si* parameter discussed above), and can play tables forwards or backwards (negative *si*).

Noise generators

In order to generate non-pitched sounds, we can avail of a set of noise generators. There are three classic opcodes that implement this functionality: `rand`, `randh`, and `randi`. These use a basic pseudo-random number generator to output white and bandlimited noise. White noise, generated by `rand`, has a theoretical flat frequency response, containing an infinite number of components with the same weight from 0 Hz to the Nyquist frequency. The bandlimited generators `randh` and `randi`, will periodically draw a new random number, according to a frequency control, holding it or interpolating it linearly to the next value, respectively. Both of them have less energy in the higher part of the spectrum. The result is slightly different, however: the interpolating version will suppress high frequencies more audibly than the sample-and-hold one (Fig. 12.7).

```
xres rand xamp [, iseed] [, isel] [, ioffset]
xres randh xamp, xcps [, iseed] [, isize] [, ioffset]
xres randi xamp, xcps [, iseed] [, isize] [, ioffset]
```

The actual numbers output by these generators are always the same sequence, starting from the same point. If we want a different set of values, then we need to *seed* the generators. This can be done by passing a different number to the `iseed` parameter, or by making the opcode use the current system time for this. These opcodes can use either the default 16-bit noise generator (`isel` = 0) or a more precise 31-bit algorithm (`isel` = 1). An offset can optionally be given (`ioffset`), which will be added to the random value output.

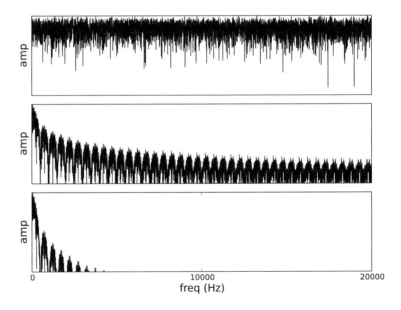

Fig. 12.7 The output spectra of the three classic noise generators: `rand` (top), `randh` (middle), and `randi` (bottom)

In addition to white noise, it is possible to generate approximations to *pink* noise. This is a type of signal whose spectral envelope follows a $\frac{1}{f}$ pattern, inversely proportional to frequency. Two opcodes provide this functionality: `pinkish` and `pinker`. The first one operates in a similar way to the classic noise generators in that it uses a pseudo-random sequence that can be seeded, and it provides an option of three methods of operation. The other opcode is an implementation of Stefan Wenzel's algorithm, and it does not have any parameters (a normalised signal, -1 to

1, is generated). `pinker` is very close to having a $\frac{1}{f}$, and does not suffer from a periodicity which can happen with `pinkish`.

If a higher level of confidence that the sequence is random is required it is possible to use the `urandom` opcode, which uses other timed activities on the host computer, such as key strokes, network packets and interrupts, to increase the randomness.

12.2.2 Modifiers

The typical modifier for an arbitrary source is called a *filter*. These are processing units designed to amplify or suppress parts of the spectrum. Digital filters work by delaying signals by small amounts, and combining these with different scaling amounts. This process can involve the re-injection of the filter output back into its input (with a certain delay), in which case we have a *feedback* filter. If there is no such signal path, then we call it a *feedforward* filter. The amount of delay used can vary, from a single to several samples, and this is one of the aspects that will determine the filter characteristics. The other is the scaling values associated with each delayed signal. Typically, a feedback filter will have delays of one or two samples, and feedforward filters will need longer delays.

The operation of a filter can be quite complex to describe, but at its heart, it is controlled by a simple principle: signals will sum up constructively or destructively when in or out of phase. Two sound waves of the same frequency are *in phase* if their starting point is aligned, otherwise they are out of phase. In the simple case of a sinusoid, if the signals are completely out of phase, they will destructively interfere with each other. This is the case when one of them is delayed by $\frac{1}{2}$ cycle: they will have opposite signs. If the waves have the same amplitude, they will cancel each other out. Conversely, if they are in phase, they will add up to twice their amplitude. In between these two extremes, out-of-phase signals will interfere with each other by varying amounts. By using delays and scaling, filters can introduce areas of amplification and suppression at different frequencies in the spectrum.

This is simple to observe in feedforward filters. Consider the simple case where we combine one signal with itself delayed by one sample (Fig. 12.8). In this case, certain parts of the spectrum will add up constructively, while others will be suppressed. To assess this, we can use sinusoidal waves of different frequencies. Take a cosine wave of frequency 0 and amplitude 1: its digital representation consists of the sequence of samples $\{1, 1, 1, 1, 1, ...\}$. If we pass it through our filter, the resulting sequence will be $\{1, 2, 2, 2, 2, 2, ...\}$; the signal is amplified. At the other side of the spectrum, a cosine wave at the Nyquist frequency is an alternation of positive and negative samples, $\{1, -1, 1, -1, 1, -1, ...\}$. In this case, the output will be always 0, as the signal is completely out of phase with its delayed path. In between these two extremes, there is an increasing amount of attenuation, from 0 Hz upwards. This principle can be extended to feedback filters, but in that case it is not possible to provide a simple analysis such as the one in Fig. 12.8.

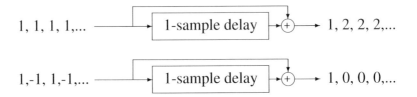

Fig. 12.8 Two digital cosine waves, at 0 Hz (top) and Nyquist (bottom), being fed to a first-order feedforward filter. The filter lets all of the 0 Hz signal through, but blocks the one at the Nyquist frequency

Frequency response, filter types and parameters

Filters can be described by their *frequency response*, which indicates how the input signal is modified at various frequencies in the spectrum [63]. In addition to the amplitude changes imposed, a filter also affects the phase of its input, delaying it by various amounts. The frequency response of a filter is made up of its amplitude and phase responses. We can classify filters in general terms by the shape of their response. In terms of amplitude, we have

- **low-pass**: low-frequency components are passed with little alteration, and high-frequency components are attenuated.
- **high-pass**: low-frequency components are suppressed, high-frequency components are passed.
- **band-pass**: low- and high-frequency components outside a certain band are cut.
- **band-reject**: components inside a certain band are attenuated.
- **all-pass**: all components are passed with the same amplitude.

A plot of low-, high- and band-pass filter amplitude responses is shown in Fig. 12.9. The phase response of a filter can be of two types:

- **linear**: the same delay is imposed at all frequencies (phases are linearly shifted).
- **non-linear**: frequencies are delayed by different amounts.

Linear phase responses are a feature of some feedforward filters, whereas feedback types are non-linear with respect to phase. Generally, in musical applications, we will be interested in the amplitude response of filters, and less so in their phase response. An exception is the case of all-pass filters, where the amplitude response does not change with frequency. In this case, we will be interested in the delays imposed at different frequencies. Also, for specific applications, we might need to ensure that the phase response is linear, in which case we have to be careful when designing a filter for them.

Filters can also be classified by their order. In the case of digital filters, this is measured by the delays used. So, for instance, if we only use a one-sample delay (as

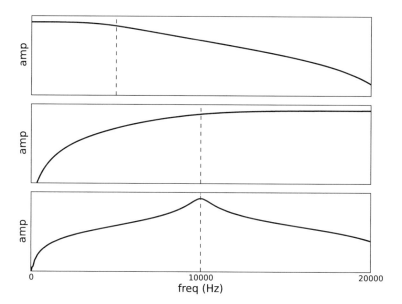

Fig. 12.9 Amplitude response plots of low- (top), high- (middle) and band-pass (bottom) filters. The dash lines indicate the cut-off/centre frequencies.

in the example above), the filter is first-order; a second-order filter has delays of up to two samples, and so on. This is also called the filter length, especially in the case of feedforward filters. For feedback filters, in general the effect of increasing the filter order is to make it have a steeper *roll-off* in its amplitude response curve. The roll-off describes the transition between the pass band (the region in the spectrum where all frequency components are not modified) and the stop band (where the spectrum is attenuated).

Feedback filters are the most common types used in musical applications, as they have controls that can be varied over time, and are very intuitive to use. They come in different forms, which tend to have specific characteristics, determining the shape of their frequency response, which is also controlled by the filter parameters. The most common of these are

- **tone control**: first-order low-pass or high-pass with a gentle amplitude response curve. The sole filter parameter is cutoff frequency. This determines the point in the spectrum between the pass band and the stop band. Examples of these in Csound are `tone` and `atone`.
- **resonators**: second-order band-pass, which can have sharp resonances depending on parameters. These filters have centre frequency and bandwidth controls, which define the characteristics of the amplitude response. In Csound, the opcodes `reson`, `resonr` and `resonz` implement this design.

- **low-pass resonant**: second- or fourth-order filters with a resonant peak. They feature cut-off frequency and resonance controls. In Csound, we have `moogladder` `moogvcf`, and `lpf18`, among others, as examples of this type.
- **Butterworth**: classic design with a maximally flat pass band, in low-, high-, band-pass and band-reject forms. Their controls are cut-off or centre frequency, and bandwidth (in the case of band-pass and band-reject types). The `butter*` family of filter opcodes in Csound implements this type as second-order sections.
- **equalisers**: equaliser filters are generally band-pass/band-reject designs with a characteristic all-pass behaviour outside their region of boost/attenuation. They often have a gain control, in addition to centre frequency and bandwidth, which controls whether the filter cuts or amplifies a particular portion of the spectrum. In Csound, this type of filter is implemented in `eqfil`.

Some filters will have a Q control in place of bandwidth or resonance. This parameter can be defined as

$$Q = \frac{f}{B} \tag{12.4}$$

where f is the filter frequency (centre or cutoff, depending on the type), and B is the bandwidth of the pass- or reject-band, or resonance region. The Q parameter is inversely proportional to the width of the filter band or resonance region. By keeping the Q constant, the bandwidth will vary with frequency, which can have significant importance in musical applications.

Impulse response

A filter can also be classified by its impulse response (IR), which is its output when fed with a single sample impulse (a full amplitude sample followed by zeros). Feedforward types are called Finite Impulse Response (FIR), because the IR will eventually be zero after the impulse has circulated through all the delays. On the other hand, feedback makes the filter feature an infinite impulse response, and so these are called IIR filters. This is because the impulse will keep being re-injected into the process, and, theoretically, the output will never be zero. The behaviour of the IR of feedback filters is very important: if it never decays, the filter is considered unstable as its output will continue growing until it exceeds the numerical range of the system. A decaying IR will eventually approach zero.

More generally, IRs can be used to describe a linear time-invariant (LTI) system, and are used to implement feedforward filters that represent it. For instance, if we record the IR of a room, we can use it to impart its characteristics into an arbitrary sound via an FIR filter. In this case, the filter length is equivalent to the IR size, and each sample of the IR is used as a scaling factor for each filter delay. This process is called *convolution*, and it can be very useful for reverberation and other applications where a fixed-parameter ("time-invariant") filter is required.

Multiple filters

Multiple filters can be connected in two ways: serially (cascade) or in parallel. In the first form, the output of one filter is sent to the input of the next, making the combination a higher-order filter. For instance, we can make a fourth-order filter, with its associated steeper roll-off, from a series connection of two identical second-order sections. Other applications of serial filters include equaliser filterbanks, where each filter is set to cover a specific area of the spectrum. Parallel filter connections are made by feeding the same input to various filters, and mixing their output. These can also be used to construct filterbanks that implement a more complex amplitude response with various peaks in the spectrum, with band-pass filter types.

12.2.3 Design Example 1: Analogue Modelling

Analogue electronic musical instruments have attracted a lot of interest in recent years. In particular, the possibility of recreating their sound in the digital domain has been a topic of research in musical signal processing. These instruments have been, for the most part, constructed with source-modifier designs, following the pattern of connecting oscillators and noise generators to filters, whose parameters can be controlled by envelopes. The classic Minimoog synthesiser is based on this: two or three oscillators, noise and external sources are sent into a mixer, whose output is shaped by a low-pass resonant filter. Two envelope generators are used, one for the filter frequency, and another for the amplitude.

Such instruments can be easily modelled with Csound. Virtual analogue oscillators exist in the form of the `vco` and `vco2` opcodes. Models of classic filters such as the Moog ladder and state variable (`moogvcf`, `moogladder`, `svfilter` and `statevar`) are also available. In addition to these, other components can be modelled from scratch or using unit generators provided by the system.

For instance, we would like to design an envelope following the classic attack-decay-sustain-release (ADSR) design, which closely models how it might be implemented in an analogue synthesiser. In these instruments, the envelope is triggered by a voltage gate signal: when this is high, the envelope runs; and when this is low, it releases. We can use the reference values 1 and 0 for high and low, and use logic to move from one stage to another (from attack to decay, and to sustain). The actual envelope can be constructed by using a portamento opcode `portk`, which goes smoothly from one value to another over a given time. This is in fact a first-order low-pass filter (like `tone`), which acts on control signals. In order to allow an instrument to trigger the release stage on note-off, we use the opcode `xtratim` to extend the duration for the release time. With these ideas, we can construct a UDO, shown in listing 12.2.

Listing 12.2 Analogue modelling ADSR envelope UDO

```
/ * * * * * * * * * * * * * * * * * * * * * * * * * * * * * * * * * * * * * * *
```

```
ksig ADSR kmax,iatt,idec,ksus,irel,ktrig
kmax - max amplitude after attack
iatt - attack time
idec - decay time
ksus - sustain amplitude
irel - release time
ktrig - trigger signal
*******************************************/
opcode ADSR,k,kiikik
 kmax,iatt,idec,ksus,irel,ktrig xin
 xtratim irel
 ktime init 0
 kv init 0
 iper = 1/kr
 if (ktrig == 1) then
   ktime = ktime + iper
   if ktime < iatt then
     kt  = iatt
     kv = kmax
   else
     kt = idec
     kv = ksus
   endif
 else
   kt = irel/8
   kv = 0
   ktime = 0
 endif
 kenv   portk  kv, kt
 xout   kenv
endop
```

A basic analogue modelling instrument following the Minimoog structure can now be realised with this envelope generator, the `vco2` opcode and the `moogladder` filter. Listing 12.3 shows the code for this instrument, which plays a chromatic scale.

Listing 12.3 Analogue modelling instrument example

```
instr 1
 ktrig = (release() == 1 ? 0 : 1)
 iatt1 = 0.01
 idec1 = 0.1
 isus1 = 0.8
 irel1 = 0.1
 kenv1 ADSR p4,iatt1,idec1,isus1,irel1,ktrig
 iatt2 = 0.1
 idec2 = 0.2
```

```
  isus2 = 0.4
  irel2 = 0.05
  kenv2 ADSR 1,iatt2,idec2,isus2,irel2,ktrig
  a1 vco2 0dbfs/3, p5
  a2 vco2 0dbfs/3, p5*1.005
  a3 vco2 0dbfs/3, p5*.995
  a4 moogladder a1+a2+a3, p5*6*(1+kenv2), 0.7
     out a4*kenv1
endin

i1 = 0
while i1 < 12 do
 schedule(1,i1,1,i1/12,cpspch(8+i1/100))
 i1 += 1
od
```

12.2.4 Design Example 2: Channel Vocoder

The channel vocoder is a very interesting music instrument design, which was originally developed by the telecommunications industry as means of encoding speech
[36]. It was the first breakthrough in the reproduction of speech using electronic
means. The vocoder operates by first carrying out an analysis of signal amplitude in
separate bands of the spectrum. The information obtained by this analysis is used
to resynthesise the signal by applying it to matched synthesis channels that operate in the same bands. The excitation source for the synthesis component is generally based on broad-bandwidth oscillators, but this can be substituted by arbitrary
spectrum-rich sounds. This arrangement performs reasonably well for vocal signals of pitched nature (the so-called *voiced* signals), mostly represented by vowels
and semivowels [48]. It is possible to use an added noise generator for non-pitched
sounds, represented by some of the consonants, to improve the overall quality of the
reproduced sound. However, as a speech spectrum analyser, the channel vocoder
was superseded for engineering applications. It did find, however, a distinct place
in music applications. The vocoder still features as an instrument in various types
of electronic music. It has an ability to blend vocal and instrumental sounds (when
these replace the excitation oscillators) into something of an other-worldly quality.

There are various ways to realise the vocoder in Csound. The simplest is to decide
on an analysis/synthesis channel (or band) design and then apply as many of these as
we need across the spectrum. There are three elements to a vocoder band: a filter for
the analysis signal, another one for the excitation signal, and a means of controlling
the amplitude of the band. This is done by the `balance` opcode, which controls
the level of one signal by comparing it to another, matching the two. It is a type
of *envelope follower* that can apply the energy detected to the signal being output
at each one of the vocoder bands. The filter needs to be reasonably selective, so

that it can be focused on a given frequency. For this job, fourth-order Butterworth filters are a good choice. The code for a vocoder band is shown in listing 12.4. It employs two `butterbp` filters in cascade for each of the excitation and analysis signal paths, and the balancing unit is used to match the output level to the energy detected at that band (Fig. 12.10).

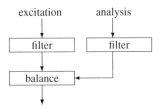

Fig. 12.10 A single band (channel) of a vocoder

Listing 12.4 Vocoder band UDO

```
/************************
asig VocBand as,an,kfreq,kbw
as - excitation signal
an - analysis signal
kfreq - band frequency
kbw - bandwidth
************************/
opcode VocBand,a,aakk
 as,an,kf,kbw xin
 xout(balance(butterbp(
              butterbp(as,kf,kbw),kf,kbw),
              butterbp(
              butterbp(an,kf,kbw),kf,kbw)))
endop
```

With this in hand, we have the choice of employing these opcodes directly in an instrument, which would fix the number of bands used, or we could possibly go one step further to make this variable. In this case, we can use recursion as a programming strategy to build up a bank of vocoder bands, whose number and spacing can be modified by the user. These start at a minimum frequency and extend up to a maximum, and are spaced at even musical intervals. For this, the expression to calculate the band spacing is

$$f_n = f_{min} \left(\frac{f_{max}}{f_{min}} \right)^{\frac{N-1}{n-1}}$$

(12.5)

where n is the band number, f_{min}, f_{max}, and f_n are the max, min and band frequencies, and N is the number of bands. We use a fixed Q parameter to make sure each band is perceptually the same size as well. The full code is shown in listing 12.5.

Listing 12.5 Channel vocoder UDO

```
/ * * * * * * * * * * * * * * * * * * * * * * * * * * * * * * * * * * * * * * *
asig Vocoder as,an,kmin,kmax,kq,ibnd
as - excitation signal
an - analysis signal
kmin - lowest band frequency
kmax - highest band frequency
kq - vocoder band Q
ibnd - number of bands
* * * * * * * * * * * * * * * * * * * * * * * * * * * * * * * * * * * * * * * * /
opcode Vocoder, a, aakkkpp
 as,an,kmin,kmax,kq,ibnd,icnt   xin
 if kmax < kmin then
    ktmp = kmin
    kmin = kmax
    kmax = ktmp
 endif
 if kmin == 0 then
    kmin = 1
 endif
 if (icnt >= ibnd) goto bank
    abnd   Vocoder as,an,kmin,kmax,kq,ibnd,icnt+1
 bank:
    kfreq = kmin*(kmax/kmin)^((icnt-1)/(ibnd-1))
    kbw = kfreq/kq
    ao VocBand as,an,kfreq,kbw
    amix = ao + abnd
    xout amix
endop
```

The vocoder UDO can be used in a variety of settings. For instance, we can use it to resynthesis speech, as originally designed. In this case, we should use an appropriate excitation source, such as a classic sawtooth wave (from vco or vco2), and we can also track the pitch of the voice and its envelope to control the oscillator fundamental frequency and amplitude, respectively. In listing 12.6, we show a simple instrument example that does this, using plltrack to estimate the pitch and a combination of rms and port to control the overall envelope of the sound. This instrument can be modified to use other types of excitation sources, to create a cross-synthesis blend with the analysis sound.

Listing 12.6 Vocoder instrument example

```
instr 1
```

```
S1 = "fox.wav"
imin = 100
imax = 10000
asig diskin2 S1,1, 0,1
ap,aloc plltrack asig, 0.1
krms port rms(asig), 0.01
anoi vco krms,ap,1,0
aout Vocoder anoi,asig,imin,imax,25,32
outs aout,aout
endin
```

12.3 Distortion Synthesis Methods

Distortion techniques are based on the non-linear modification of a basic sinusoid waveform. They try to address the key question of how to generate complex time-evolving spectra, composed of discrete components (eq. 12.1). An elegant solution to this problem is to find a way of combining a few simple sources (i.e. sine wave oscillators) to generate many partials. This is the approach taken by distortion synthesis. In this section, we will survey these techniques with their associated Csound examples.

12.3.1 Summation Formulae

Closed-form summation formulae provide several possibilities for generating complex spectra. They take advantage of well-known expressions that can represent arithmetic series, such as the harmonic series, in a compact way. In fact, it is fair to say that all distortion techniques implement specific closed-form summation formulae. We will be concentrating here on those that stem directly from simple closed-form solutions to the harmonic series.

Bandlimited pulse

The general case of eq. 12.1 can be simplified considerably, if we constrain to certain conditions. An example of this is a spectrum made up of harmonic components added up with the same weight [131]:

$$s(t) = \frac{1}{N} \sum_{k=1}^{N} \cos(k\omega_0) \tag{12.6}$$

where $\omega_0 = 2\pi f t$. This produces what we call a bandlimited pulse.

We can synthesise this spectrum by taking into account one of the best known closed-forms of an arithmetic series:

$$\sum_{k=-N}^{N} r^k = r^{-N} \frac{1 - r^{2N+1}}{1 - r} \tag{12.7}$$

Using this we get [32]:

$$\frac{1}{N} \sum_{k=1}^{N} \cos(k\omega_0) = \frac{1}{2N} \times \left[\frac{\sin((2N+1)\frac{\omega_0}{2})}{\sin(\frac{\omega_0}{2})} - 1 \right] \tag{12.8}$$

with $\omega_0 = 2\pi f_0 t$, and f_0 the fundamental frequency in Hz. The parameter N determines the number of harmonics. In this, and in the subsequent formulae in this section, the relevant synthesis expression is on the right-hand side; the left-hand expansion shows the resulting spectrum.

The only issue here is the possible zero in the denominator, in which case, we can substitute 1 for the whole expression (as a simple measure; a full solution to the problem requires more complicated logic). A sample Csound example from first principles is shown below. We will use a phasor to provide an index so we can look up a sine wave table. Then we just apply the expression above.

Listing 12.7 bandlimited pulse UDO

```
/********************************
asig Blp kamp,kfreq
kamp  - amplitude
kfreq - fundamental frequency
*******************************/
opcode Blp,a,kki
  setksmps 1
  kamp,kf  xin
  kn = int(sr/(2*kf))
  kph phasor kf/2
  kden tablei kph,-1,1
  if kden != 0 then
      knum tablei kph*(2*kn+1),-1,1,0,1
      asig = (kamp/(2*kn))*(knum/kden - 1)
   else
    asig = kamp
   endif
  xout asig
endop
```

Because of the extra check, we need to run the code with a ksmps block of 1, so we can check every sample (to avoid a possible division by zero). This code can be modified to provide time-varying spectra (by changing the value of `kn`). With this, we can emulate the effect of a low-pass filter with variable cut-off frequency. The algorithm used in `Blp` is implemented internally in Csound by the `buzz` opcode, which is, as such, more efficient than the UDO version.

Generalised Summation Formulae

Another set of closed-form summation formulae allows for a more flexible control of synthesis [91]. In particular, these give us a way to control spectral roll-off and to generate inharmonic partials. The synthesis expression for bandlimited signals is

$$
\sum_{k=1}^{N} a^k \sin(\omega + k\theta) =
$$
$$
\frac{\sin(\omega) - a\sin(\omega - \theta) - a^{N+1}(\sin(\omega + (N+1)\theta) - a\sin(\omega + N\theta))}{1 - 2*a\cos(\theta) + a^2}
\tag{12.9}
$$

where $\omega = 2\pi f_1 t$ and $\theta = 2\pi f_2 t$, with f_1 and f_2 as two independent frequency values in Hz. N is the number of components, as before, and a is an independent parameter controlling the amplitudes of the partials.

Here, by modifying a and N, we can alter the spectral roll-off and bandwidth, respectively. Time-varying these parameters allows the emulation of a low-pass filter behaviour. By choosing various ω to θ ($f_1 : f_2$) ratios, we can generate various types of harmonic and inharmonic spectra. The only extra requirement is a normalising expression, since this method will produce a signal whose gain varies with the values of a and N:

$$
\sqrt{\frac{1 - a^2}{1 - a^{2N+2}}}
\tag{12.10}
$$

Using the synthesis equation 12.9 and its corresponding scaling expression, the following Csound opcode can be created, which produces components up to the Nyquist frequency by setting N accordingly (listing 12.8).

Listing 12.8 Generalised summation-formulae UDO

```
/********************************
asig Blsum kamp,kfr1,kfr2,ka
kamp - amplitude
kfr1 - frequency 1 (omega)
kfr2 - frequency 2 (theta)
ka - distortion amount
******************************/
opcode Blsum,a,kkkki
 kamp,kw,kt,ka xin
 kn = int(((sr/2) - kw)/kt)
 aphw phasor kw
```

```
apht phasor kt
a1 tablei aphw,-1,1
a2 tablei aphw - apht,-1,1,0,1
a3 tablei aphw + (kn+1)*apht,-1,1,0,1
a4 tablei aphw + kn*apht,-1,1,0,1
acos tablei apht,-1,1,0.25,1
kpw pow ka,kn+1
ksq = ka*ka
aden = (1 - 2*ka*acos + ksq)
asig = (a1 - ka*a2 - kpw*(a3 - ka*a4))/aden
knorm = sqrt((1-ksq)/(1 - kpw*kpw))
xout asig*kamp*knorm
endop
```

If we are careful with the spectral roll-off, a much simpler non-bandlimited expression is available:

$$\sum_{k=1}^{\infty} a^k \sin(\omega + k\theta) = \frac{\sin(\omega) - a\sin(\omega - \theta)}{1 - 2*a\cos(\theta) + a^2} \qquad (12.11)$$

In this case, we will not have a direct bandwidth control. However, if we want to know what, for instance, our -60 dB bandwidth would be, we just need to know the maximum value of k for $a^k > 0.001$. The normalising expression is also simpler:

$$\sqrt{1 - a^2} \qquad (12.12)$$

The modified Csound code to match this expression is shown in listing 12.9.

Listing 12.9 Non-bandlimited generalised summation-formulae UDO

```
/*******************************
asig NBlsum kamp,kfr1,kfr2,ka
kamp - amplitude
kfr1 - frequency 1 (omega)
kfr2 - frequency 2 (theta)
ka - distortion amount
*******************************/
opcode NBlsum,a,kkkk
 kamp,kw,kt,ka  xin
 aphw phasor kw
 apht phasor kt
 a1 tablei aphw,-1,1
 a2 tablei aphw - apht,itb,1,0,1
 acos tablei apht,-1,1,0.25,1
 ksq = ka*ka
 asig = (a1 - ka*a2)/(1 - 2*ka*acos + ksq)
 knorm = sqrt(1-ksq)
 xout asig*kamp*knorm
```

endop

12.3.2 Waveshaping

The technique of waveshaping is based on the non-linear distortion of the amplitude of a signal [77, 5]. This is achieved by mapping an input, generally a simple sinusoidal one, using a function that will shape it into a desired output waveform. The amount of distortion can be controlled by a distortion index, which controls the amplitude of the input signal.

Traditionally, the most common method of finding such a function (the so-called transfer function) has been through polynomial spectral matching. The main advantage of this approach is that polynomial functions will precisely produce a bandlimited matching spectrum for a given sinusoid at a certain amplitude.

However, the disadvantage is that polynomials also have a tendency to produce unnatural-sounding changes in partial amplitudes, if we require time-varying spectra. More recently, research has shown that for certain waveshapes, we can take advantage of other common functions, from the trigonometric, hyperbolic etc. repertoire. Some of these can provide smooth spectral changes. We will look at the use of hyperbolic tangent transfer functions to generate nearly bandlimited square and sawtooth waves. A useful application of these ideas is in the modelling of analogue synthesiser oscillators.

Hyperbolic tangent waveshaping

A simple way of generating a (non-bandlimited) square wave is through the use of the *signum*() function, mapping a varying bipolar input. This piecewise function outputs 1 for all non-zero positive input values, 0 for a null input and -1 for all negative arguments. In other words, it clips the signal, but in doing so, it generates lots of components above the Nyquist frequency, which are duly aliased. The main cause of this is the discontinuity at 0, where the output moves from fully negative to fully positive. If we can smooth this transition, we are in business.

The hyperbolic tangent is one such function that can be used instead of *signum*() [71], as it has a smooth transition at that point but it also preserves some of its clipping properties (see Fig. 12.11). If we drive this function with a sinusoidal input, we will be able to produce a nearly bandlimited signal. How bandlimited will depend on how hard we drive it, as higher input amplitudes will produce more and more harmonics, and take less advantage of its smoothing properties. As with all types of waveshaping, the amplitude of the input signal will determine signal bandwidth, in a proportional way. If we want to keep a steady output amplitude, but vary the spectrum, we will need to apply an amplitude-dependent scaling. This is generally done by using a scaling function that takes the input amplitude as its argument and produces a gain that can be applied to the output signal. We can then refer to the amplitude of the input sinusoid as the distortion index.

The code for general-purpose waveshaping is based on standard function mapping. It takes an input sinusoid, maps it using table lookup and then applies the required gain, also obtained through table lookup.

Listing 12.10 Waveshaping UDO

```
/*******************************************
asig  Waveshape kamp,kfreq,kndx,ifn1,ifn2
kamp - amplitude
kfreq - frequency
kndx - distortion index
ifn1 - transfer function
ifn2 - scaling function
*******************************************/
opcode Waveshape,a,kkkiii
 kamp,kf,kndx,itf,igf xin
 asin oscili 0.5*kndx,kf
 awsh tablei asin,itf,1,0.5
 kscl tablei kndx,igf,1
 xout awsh*kamp*kscl
endop
```

For hyperbolic waveshaping, we will need to provide two function tables, for the transfer (tanh) and the scaling functions, shown in listing 12.11. The first Csound GEN draws $tanh(x)$, over $\pm\frac{P}{50}$, and the second automatically generates a scaling function based on the previous table. This is necessary to keep the overall amplitude level, as the input gain changes with the distortion index.

To keep aliasing at bay, the index of distortion (kndx) can be estimated roughly as kndx = 10000/(kf*log10(kf)).

Listing 12.11 Hyperbolic tangent waveshaping function tables

```
f2 0 16385 "tanh" -157 157
f3 0 8193 4 2 1
```

If we would like to generate a sawtooth wave instead, we could take our square signal and apply the following expression:

$$saw(t) = square(\omega)(\cos(\omega) + 1) \qquad (12.13)$$

By heterodyning it with a cosine wave, we can easily obtain the missing even components that make up the sawtooth signal. There will be a slight disparity in the amplitude of the second harmonic (about 2.5 dB), but the higher harmonics will be nearly as expected. This is an efficient way of producing a sawtooth wave (listing 12.12) [71].

Listing 12.12 Sawtooth wave oscillator based on waveshaping

```
/*******************************************
```

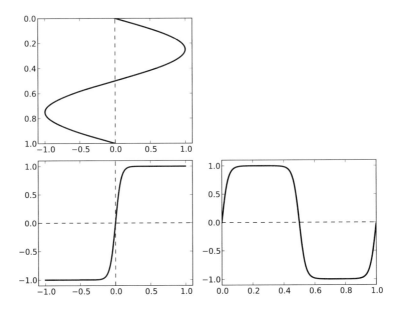

Fig. 12.11 Hyperbolic tangent waveshaping. The plot on the top shows the input to the waveshaper, whose transfer function is shown on the bottom left, and the resulting output on the right

```
asig  Sawtooth kamp,kfreq,kndx,ifn1,ifn2
kamp - amplitude
kfreq - frequency
kndx - distortion index
ifn1 - transfer function
ifn2 - scaling function
*****************************************************/
opcode Sawtooth,a,kkkii
 kamp,kf,kndx,itf,igf xin
 amod oscili 1,kf,-1,0.25
 asq Waveshape kamp*0.5,kf,kndx,-1,itf,igf
       xout asq*(amod + 1)
endop
```

12.3.3 Frequency and Phase Modulation

Frequency and Phase Modulation (FM/PM), pioneered by John Chowning in the early 1970s [26], are classic distortion techniques that have been used extensively in music. They are a special case of summation formulae that have a very straightforward implementation with two oscillators.

We will concentrate here on implementing PM, as it is the more flexible of the two, and allows a number of combinations and variations of the basic principle. FM/PM works by using one or more oscillators, the modulator(s), to modulate the frequency or the phase of one or more carriers. Sinusoidal oscillators are generally used for this.

The expression for the basic form of this technique, with one carrier, and one modulator, is given by.

$$\sum_{n=-\infty}^{\infty} J_n(k)\cos(\omega_c + n\omega_m) = \cos(\omega_c + k\sin(\omega_m)) \qquad (12.14)$$

where $J_n(k)$ are called Bessel functions of the first kind, which will determine the amplitude of each component in the spectrum; $\omega_c = 2\pi f_c t$ and $\omega_m = 2\pi f_m t$ are the carrier and modulator frequencies, respectively.

The value of these functions will vary with the k argument, which is called the *index* of modulation. For low k, high-order $J_n(k)$ are zero or close to zero. As the index rises, these Bessel functions tend to increase, and then fluctuate between positive and negative values (see Fig. 12.12). This 'wobble' is the cause of the complex spectral evolution observed in PM.

The spectrum of PM will be made up of the sums and differences of the carrier and modulator frequencies (plus the carrier itself), scaled by the Bessel functions. Each one of these $f_c \pm f_m$ frequencies is called a *sideband*, lying as it does on each side of the carrier. Any cosine component on the negative side of the spectrum is 'reflected' back on the positive side. The $f_c : f_m$ ratio will determine whether the spectrum is harmonic or inharmonic. The rule of thumb is that if it involves small whole numbers, we will have harmonic partials, otherwise the spectral components will not fuse to make an audible periodic waveform. The exact amplitude of each component in the spectrum can be worked out from equation 12.14 above and the values of $J_n(k)$ for a given k. The spectrum is non-bandlimited, but most of the energy will be concentrated in the first $k + 1$ sidebands.

An implementation of the PM algorithm is seen in listing 12.13. For time-varying spectra, as with any of the other distortion techniques, we have to change the modulation amount by manipulating the index.

Listing 12.13 PM synthesis UDO

```
/*************************************************
asig  PM kamp,kfc,kfm,kndx
kamp - amplitude
kfc - carrier frequency
kfm - modulation frequency
kndx - distortion index
*************************************************/
opcode PM,a,kkkk
 kamp,kfc,kfm,kndx xin
```

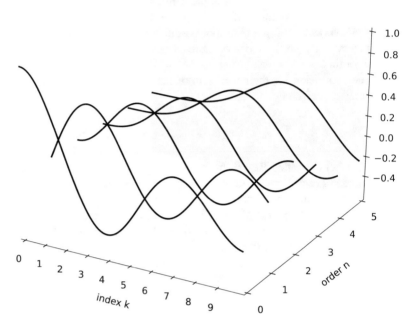

Fig. 12.12 Bessel functions of orders 0 to 5. Note how they oscillate between positive and negative values as the index of modulation increases

```
acar  phasor kfc
amod  oscili kndx/(2*$M_PI),kfm
apm   tablei acar+amod,-1,1,0.25,1
      xout apm*kamp
endop
```

Another way we could package PM synthesis is by using the Yamaha DX series principle of operators. These are a combination of an oscillator, whose phase can be modulated, with an ADSR envelope. In this case, we design a UDO that models the operator, and then we can connect these together in another opcode or instrument.

Listing 12.14 PM operator UDO

```
/*******************************************
asig PMOp kamp,kfr,apm,iatt,idec,isus,irel[,ifn]
kamp - amplitude
kfr - frequency
apm - phase modulation input
iatt - attack
idec - decay
```

```
isus - sustain
irel - release
ifn - optional wave function table (defaults to sine)
*********************************************/
opcode PMOp,a,kkaiiiij
 kmp,kfr,apm,
    iatt,idec,
    isus,irel,ifn xin
 aph phasor kfr
 a1  tablei aph+apm/(2*$M_PI),ifn,1,0,1
 a2  madsr iatt,idec,isus,irel
    xout  a2*a1*kmp
endop
```

This operator form of PM is extremely versatile, as we can arrange these opcodes in any modulation combination, including complex [114, 24] and feedback [123] modulation (ksmps=1 is needed for this). Each operator has its own envelope, so we can make time-varying spectra very easily. Modulator envelopes will control timbre, while carrier envelopes control amplitude. An equivalent process to the simple PM in the PM UDO of listing 12.13 can easily be implemented with two operators. This has the added bonus that we have envelopes included, as we demonstrate in listing 12.15.

Listing 12.15 PM operator UDO example

```
instr 1
 amod PmOp p6,p5,a(0),0.1,0.1,0.5,0.1
 acar PmOp p4,p5,amod,0.01,0.1,0.9,0.1
    out acar
endin
schedule(1,0,1,0dbfs/2,440,7)
```

Asymmetrical PM Synthesis

An interesting variation on FM/PM synthesis is Asymmetrical PM [98]. In this formulation, the original PM model is ring-modulated by an exponential signal. This has the effect of introducing a new parameter controlling spectral symmetry that allows the peaks to be dislocated above or below the carrier frequency. The expression for this technique (excluding a normalisation factor) is

$$
\sum_{n=-\infty}^{\infty} r^n J_n(k) \sin(\omega_c + n\omega_m) =
$$
$$
\exp(0.5k(r - \frac{1}{r})\cos(\omega_m)) \times \sin(\omega_c + 0.5k(r + \frac{1}{r})\sin(\omega_m))
$$

(12.15)

where, as before, $J_n(k)$ are called Bessel functions of the first kind; $\omega_c = 2\pi f_c t$ and $\omega_m = 2\pi f_m t$ are the carrier and modulator frequencies, respectively.

The new parameter r is the symmetry control, $r < 1$ pulling the spectral peak below the carrier frequency ωc and $r > 1$ pushing it above. It is a very nice feature which can be added at the expense of a few multiplies and a couple of extra table lookups (for the cosine and the exponential). Note that what we have here is actually the ring modulation of a waveshaper output (using an exponential transfer function) and the PM signal. This is a nice way of tying up two distortion techniques together.

Implementing this is not too complicated. The exponential expression needs normalisation, which can be achieved by dividing it by $exp(0.5k[r - \frac{1}{r}])$. When coding this, we will draw up an exponential table from 0 to an arbitrary negative value (say -50) and then look it up with a sign reversal ($exp(-x)$). This allows us to use the limiting table-lookup mechanism in case we have an overflow. Since the values of $exp()$ tend to have little variation for large negative values, limiting will not be problematic.

Listing 12.16 Asymmetric FM UDO

```
/*********************************************
asig Asfm kamp,kfc,kfm,kndx,kR,ifn,imax
kamp - amplitude
kfc - carrier frequency
kfm - modulation frequency
kndx - distortion index
ifn - exp func between 0 and -imax
imax - max absolute value of exp function
*********************************************/
opcode Asfm,a,kkkkkii
 kamp,kfc,kfm,knx,kR,ifn,imax
 kndx = knx*(kR+1/kR)*0.5
 kndx2 = knx*(kR-1/kR)*0.5
 afm oscili kndx/(2*$M_PI),kfm
 aph phasor kfc
 afc tablei aph+afm,ifn,1,0,1
 amod oscili kndx2, kfm, -1, 0.25
 aexp tablei -(amod-abs(kndx2))/imx, ifn, 1
        xout kamp*afc*aexp
endop
```

with the exponential function table (ifn) drawn from 0 to -imx (-50):

```
f5 0 131072 "exp" 0 -50 1
```

12.3.4 Phase-Aligned Formant Synthesis

One of the most recent new methods of distortion synthesis is the Phased-Aligned Formant (PAF) algorithm [102]. Here, we start with a desired spectral description and then work it out as a ring modulation of a sinusoid carrier and a complex spectrum (with low-pass characteristics). The interest is in creating formant regions, so we will use the sinusoid to tune a spectral bump around a target centre frequency.

The shape of the spectrum will be determined by its modulator signal, which in turn is generated by waveshaping using an exponentially shaped transfer function. So we have PAF, in its simplest formulation, as

$$\sum_{n=-\infty}^{\infty} g^{|n|} \cos(\omega_c + n\omega_m) = \frac{1+g}{1-g} \times f\left(\frac{2\sqrt{g}}{1-g} \sin\left(\frac{\omega_m}{2}\right)\right) \cos(\omega_c) \qquad (12.16)$$

$$f(x) = \frac{1}{1+x^2} \qquad (12.17)$$

$$g = \exp(\frac{f_c}{B}) \qquad (12.18)$$

The waveshaper transfer function is $f(x)$. The signal has bandwidth B, fundamental frequency $\omega_m = 2\pi f_m t$ and formant centre frequency $\omega_c = 2\pi f_c t$. To this basic formulation, where we expect f_c to be an integer multiple of f_m, a means of setting an arbitrary centre frequency is added (basically by using a pair of modulators). In addition, the complete PAF algorithm provides a *frequency shift* parameter, which, if non-zero, allows for inharmonic spectra.

The complete Csound code of a more or less literal implementation of PAF is shown in listing 12.17.

Listing 12.17 PAF UDO

```
opcode Func,a,a
 asig xin
       xout 1/(1+asig^2)
endop

/*****************************************
asig PAF kamp,kfun,kcf,kfshift,kbw
kamp - amplitude
kfun - fundamental freq
kcf - centre freq
kfshift - shift freq
kbw - bandwidth
******************************************/
opcode PAF,a,kkkkki
  kamp,kfo,kfc,kfsh,kbw   xin
```

```
kn = int(kfc/kfo)
ka = (kfc - kfsh - kn*kfo)/kfo
kg = exp(-kfo/kbw)
afsh phasor kfsh
aphs phasor kfo/2
a1 tablei 2*aphs*kn+afsh,-1,1,0.25,1
a2 tablei 2*aphs*(kn+1)+afsh,-1,1,0.25,1
asin tablei aphs, 1, 1, 0, 1
amod Func 2*sqrt(kg)*asin/(1-kg)
kscl = (1+kg)/(1-kg)
acar = ka*a2+(1-ka)*a1
asig = kscl*amod*acar
      xout asig*kamp
endop
```

The waveshaping here is performed by directly applying the function, since there are no GENs in Csound which can directly generate such a table. This is of course not as efficient as lookup, so there are two alternatives: write a code fragment to fill a table with the transfer function, to be run before synthesis; or, considering that the resulting distorted signal is very close to a Gaussian shape, use GEN 20 to create one such wavetable. A useful exercise would be to reimplement the PAF generator above with table-lookup waveshaping.

12.3.5 Modified FM Synthesis

Modified FM synthesis (ModFM) is based on a slight change in the FM/PM algorithm, with some important consequences [72]. One form of the PM equation, when cast in complex exponential terms, can look like this:

$$\Re\{e^{i\omega_c + iz\cos(\omega_m)}\} \tag{12.19}$$

where $\Re\{x\}$ is the real part of x.

If we apply a change of variable $z = -ik$ to the above formula, we will obtain the following expression:

$$\Re\{e^{i\omega_c + k\cos(\omega_m)}\} = e^{k\cos(\omega_m)}\cos(\omega_c) \tag{12.20}$$

which, is the basis for the Modified FM synthesis formula; $\omega_c = 2\pi f_c t$ and $\omega_m = 2\pi f_m t$ are the carrier and modulator frequencies, respectively.

One of the most important things about this algorithm is revealed by its expansion:

$$e^{k\cos(\omega_m)}\cos(\omega_c) =$$

$$\frac{1}{e^k}\left(I_0(k)\cos(\omega_c) + \sum_{n=1}^{\infty} I_n(k)\left[\cos(\omega_c + n\omega_m) - \cos(\omega_c - n\omega_m)\right]\right) \qquad (12.21)$$

where $I_n(k)$ are *modified* Bessel functions of the first kind, and constitute the basic (and substantial) difference between FM and ModFM. Their advantage is that (1) they are unipolar and (2) $I_n(k) > I_{n+1}(k)$, which means that spectral evolutions are much more natural here.

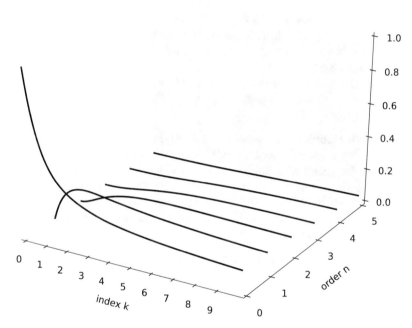

Fig. 12.13 Modified Bessel functions of orders 0 to 5. Unlike the original Bessel functions, these do not oscillate and are unipolar

In particular, the scaled modified Bessels do not exhibit the much-maligned 'wobble' seen in the behaviour of Bessel functions. That very unnatural-sounding characteristic of FM disappears in ModFM. A plot of modified Bessel functions of orders 0 to 5 is shown in Fig. 12.13.

There are several applications of ModFM (as there are of FM) as well as small variations in its design. We will present here first a basic straight implementation of the algorithm. The Csound code uses table lookup to realise the exponential waveshaper in the ModFM formula. Apart from that, all we require is two cosine oscillators, yielding a very compact algorithm (listing 12.18).

Listing 12.18 ModFM UDO

```
/*********************************************
asig ModFM kamp,kfc,kfm,kndx,ifn,imax
kamp - amplitude
kfc - carrier frequency
kfm - modulation frequency
kndx - distortion index
ifn - exp func between 0 and -imax
imax - max  absolute value of exp function
*********************************************/
opcode ModFM,a,kkkkiii
 kamp,kfc,kfm,kndx,iexp,imx xin
 acar oscili kamp,kfc,-1,0.25
 acos oscili 1,kfm,-1,0.25
 amod table -kndx*(acos-1)/imx,iexp,1
          xout acar*amod
endop
```

With ModFM, it is possible to realise typical low-pass filter effects, by varying the index of modulation k. Also, by using the carrier f_c and modulation frequency f_m as the centre of a formant and the fundamental, respectively, it is possible to reproduce the effect of a band-pass filter [70, 73]. In fact, a variant of the ModFM implementation above with phase-synchronous signals can serve as a very efficient alternative to PAF and other formant synthesis techniques (such as FOF [112]). This is show in listing 12.19.

Listing 12.19 ModFM formant synthesis UDO

```
/*********************************************
asig ModForm kamp,kfo,kfc,kbw,ifn,imax
kamp - amplitude
kfo - fundamental frequency
kfc - formant centre frequency
kbw - bandwidth
ifn - exp func between 0 and -imax
imax - max absolute value of exp function
*********************************************/
opcode ModForm,a,kkkkii
 kamp,kfo,kfc,kbw,ifn,itm  xin
 ioff = 0.25
 itab = -1
 icor = 4.*exp(-1)
 ktrig changed kbw
 if ktrig == 1 then
```

```
 k2 = exp(-kfo/(.29*kbw*icor))
 kg2 = 2*sqrt(k2)/(1.-k2)
 kndx = kg2*kg2/2.
endif
kf = kfc/kfo
kfin = int(kf)
ka = kf  - kfin
aph   phasor kfo
acos tablei aph, 1, 1, 0.25, 1
aexp table kndx*(1-acos)/itm,ifn,1
acos1 tablei aph*kfin, itab, 1, ioff, 1
acos2 tablei aph*(kfin+1), itab, 1, ioff, 1
asig = (ka*acos2 + (1-ka)*acos1)*aexp
xout asig*kamp
endop
```

This code synchronises the phase of the carrier and modulation signals, and for this reason, we use a single phase source (aph) for all oscillators (which become just table lookups). There are two sources (carriers) (*acos*1 and *acos*2), whose output is mixed together, interpolated, to set the formant centre frequency more accurately (as is done with PAF).

12.4 Additive Synthesis

Additive methods attempt to implement eq. 12.1 more or less faithfully [90]. In this sense, they are a very powerful, however raw, way of generating new sounds. They can also be quite simple in concept, but often demanding from a computational point of view. The basic form of additive synthesis does not require special unit genera-tors, only table oscillators and readers. In listing 12.20, we show an opcode that represents a single partial. This can be used as a building block for an instrument, by supplying it with an amplitude, frequency and two function tables containing the envelopes for these parameters. A time counter is used to scan through these envelope tables.

Listing 12.20 Additive synthesis partial UDO

```
/***********************************
asig Partial kamp,kfreq,ktime,ifa,iff
kamp - amplitude
kfreq - frequency
ktime - envelope time point (0 - 1)
ifa - amplitude function table
iff - frequency function table
***********************************/
opcode Partial,a,kkkii
```

```
ka,kf,kt,ifa,iff xin
xout(oscili(ka*tablei:k(kt,ifa,1),
          kf*tablei:k(kt,iff,1)))
endop
```

A complete example showing the use of this opcode is shown in listing 12.21. Here we try to create an inharmonic spectrum, with some frequency bending as the sound progresses. A plot of a segment of the output signal produced by this instrument is shown on Fig. 12.14, where it is possible to see that the waveform does not have an obvious periodic pattern. We use two exponential function tables: for amplitude, a simple decaying envelope from 1 to $\frac{1}{1000}$ (-60 dB); for frequency, a downward bend from 1 to 0.9. In the instrument, partials traverse these envelopes at different rates, which are linked to the overall sound duration. Each partial has its own maximum amplitude and frequency, and we can scale all frequencies for each instance with parameter 4. The mix of partials is smoothed by an overall envelope to avoid any clicks at the start or end of a sound. To demonstrate it, the example plays two instrument instances, with different durations and frequency scaling.

Listing 12.21 Additive synthesis example

```
i1 ftgen 1,0,16385,5,1,16384,0.001
i2 ftgen 2,0,16385,-5,1,16384,0.9

instr 1
 ap[] init 8
 ipf[] fillarray 440,480,590,610,700,850,912,990
 ipa[] fillarray 0.8,0.9,0.3,0.7,0.6,0.5,0.1,0.2
 kt = timeinsts()/p3
 ap[0] Partial ipa[0],ipf[0]*p4,kt,1,2
 ap[1] Partial ipa[1],ipf[1]*p4,1.1*kt,1,2
 ap[2] Partial ipa[2],ipf[2]*p4,1.2*kt,1,2
 ap[3] Partial ipa[3],ipf[3]*p4,1.3*kt,1,2
 ap[4] Partial ipa[4],ipf[4]*p4,1.4*kt,1,2
 ap[5] Partial ipa[5],ipf[5]*p4,1.5*kt,1,2
 ap[6] Partial ipa[6],ipf[6]*p4,1.6*kt,1,2
 ap[7] Partial ipa[7],ipf[7]*p4,1.7*kt,1,2
 kcnt = 0
 amix = 0
 while kcnt < 8 do
    amix += ap[kcnt]
    kcnt += 1
  od
  out linen(amix*0dbfs/10,0.01,p3,0.01)
endin
schedule(1,0,20,1.5)
schedule(1,1,19,1)
```

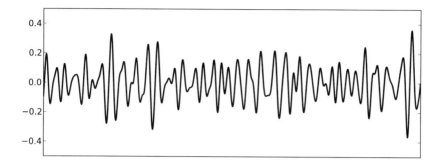

Fig. 12.14 A plot of a short segment of the output from listing 12.21

Implementing simple instruments with a limited number of partials is straightforward[1]. The difficulty starts when the number of components increases and also the amplitude and frequency trajectories have to be individually specified. In these situations, we will need to approach the problem by manipulating partial parameters programmatically, and/or using analytical methods to derive them. In the following sections we will look at two applications of additive synthesis: one that takes a programmatic route, and another that implements synthesis from analysis data.

12.4.1 A Tonewheel Organ Instrument

In a general way, pipe organs have used the principle of additive synthesis for centuries. The idea of combining different registrations, in effect mixing the sound of simple (although not exactly sinusoidal) sources to make a given sound, follows exactly the same principles studied here. So an interesting application of these is to reproduce a classic electronic organ design: the tonewheel instrument. This was pioneered by the Hammond company [62], and it synthesises sound by the addition of simple sources that are very close to sinusoids.

Each component is produced by a rotating wheel, which induces a current in an electric pickup, generating a tone (therefore, it is a *tonewheel*). A note is produced by summing up to nine of these, whose levels are controlled by a mixer (physically, this is a set of sliders called *drawbars*). An organ can have up to 96 of these wheels (arranged in semitones, covering eight octaves). Once the instrument is started, a motor puts the tonewheels into motion. When a key is pressed, nine contacts are made, placing the sound of the corresponding sources in the instrument's nine main audio busses, which are sent to the mixer.

---

[1] Note that in this example we had to call each `Partial` opcode explicitly, outside a loop, as explained in Section 5.2.3. For a variable, dynamic, number of partials, recursion can be used to spawn parallel opcodes as shown in the vocoder example.

We can try to follow this design quite faithfully. The first step is to implement the tonewheel mechanism, which can be modelled by a bank of oscillators. Since the sound of these sources is going to be shared by all the notes in the instrument, it makes sense to place them in a single global location, so we can use an array for that. As already noted in earlier sections, recursion is a useful device to implement banks of unit generators. In listing 12.22 we see the implementation of the tonewheels as a UDO. This is an unusual opcode in that only a single instance of it is required in an orchestra. It places its audio in the global array. The range of tonewheel frequencies is from 32.7 to 7,901.91 Hz.

Listing 12.22 Tonewheel mechanism model UDO

```
giwheels init  96
gawheels[]  init giwheels
/* * * * * * * * * * * * * * * * * * * * * * * * * * *
ToneWheel iwheels
iwheels - number of tonewheels
* * * * * * * * * * * * * * * * * * * * * * * * * * * * * * * */
opcode ToneWheel,0,io
 ibnd,icnt xin
 if icnt < ibnd-1 then
  ToneWheel ibnd, icnt+1
 endif
 gawheels[icnt] oscili 1,cpspch(5+icnt/100)
endop
```

The following step is to implement the note contacts, which are made when a key is pressed. Similarly, the selected tonewheels get added to a bus, which is shared by the whole of the instrument, so we make it a global array of nine audio signals. The important part here is to select the correct tonewheels, which are placed at specific intervals above the fundamental, in semitones: 0, 19, 12, 24, 31, 36, 40, 43 and 48. To simplify, we place these in an array for easy indexing. A UDO which models the organ note is show in listing 12.23. The note to be played is relative to the tonewheel scale (0-95).

Listing 12.23 Tonewheel note UDO

```
gkpart[] fillarray 0,19,12,24,31,36,40,43,48
gabus[] init 9
/* * * * * * * * * * * * * * * * * * * * * * * * * * * * * * * * * * * * *
Note inum
inum - note to be played (relative to tonewheel scale)
* * * * * * * * * * * * * * * * * * * * * * * * * * * * * * * * * * * * */
opcode Note,0,i
inote xin
 kcnt init 0
 knote init 0
 kcnt = 0
```

```
while kcnt < 9 do
   knote = inote+gkpart[kcnt]
   if knote < giwheels  && knote >= 0 then
    gabus[kcnt] = gabus[kcnt] + gawheels[knote]
   endif
   kcnt += 1
 od
endop
```

These two opcodes are responsible for modelling the complete tonewheel and note mechanism. To complement the organ, we need to implement the mixer that takes the drawbar controls and mixes up the nine busses accordingly. This can be accomplished by another UDO, which will also apply some compression to the signal, depending on the configuration of the drawbars and keys used. We assume that instrument 1 will be used to trigger the notes, and we check with the active opcode how many instances of it are currently playing. This UDO is shown in listing 12.24.

Listing 12.24 Tonewheel mixer UDO

```
/********************
asig Mixer kbars[]
kbars[] - drawbar levels(9)
********************/
opcode Mixer,a,k[]
 kbar[] xin
 asig = 0
 kcnt = 0
 kscl = 0
 while kcnt < 9 do
  asig += gabus[kcnt]*kbar[kcnt]
  gabus[kcnt] = 0
  kscl += kbar[kcnt]
  kcnt += 1
 od
 xout asig*0dbfs/(21+kscl*active:k(1))
endop
```

Complementing the model, we will add the typical vibrato/chorus effect, which can be used to modify the organ sound. This is not strictly a part of the additive synthesis mechanism, but a global processor that acts on the mixer output. The Hammond chorus/vibrato is constructed using a very short (1.1 ms) modulated delay line. The modulation wave is triangular, set at a fixed rate (6.87 Hz), and there are three depth settings (45%, 66% and 100%). We implement this in digital form, trying to model these features faithfully. Vibrato and chorus modes are offered; the difference between the two is whether the original signal is mixed to the effect. The basis for this effect will be discussed in more detail in Chapter 13. The effect UDO is shown in listing 12.25.

Listing 12.25 Delay-line vibrato/chorus UDO

```
ifn ftgen 2,0,16384,20,3,1
/*************************
asig VChorus ain,kdepth,ichorus
asig - input signal
kdepth - depth settings (0-3)
ichorus - 1 for chorus, 0 for vibrato.
*************************/
opcode VChorus,a,akp
  asig,ks,ichorus xin
  kdep[] fillarray 0,0.45,0.66,1
  kset = (ks < 0 ? 0 : (ks > 3 ? 3 : ks))
  adel oscili 1.1*kdep[ks],6.87,2
  amix vdelay asig,adel,2
 xout amix + asig*ichorus
endop
```

The functionality in these UDOs can be implemented in instruments in a variety
of ways. The typical use will be to have one single-instance instrument running the
tonewheel, mixer and vibrato/chorus elements, and a minimal instrument (instr
1) calling the note mechanism. We could make it a MIDI-controlled instrument,
since this can be very useful in live performance. An example of this is shown in
listing 12.26. Here, we set the note range from 24 to 120, and use controllers 11
to 19 for the drawbar levels, 20 for effect depth, and 21 to select vibrato or chorus.
These can easily be adjusted for specific set-ups.

Listing 12.26 MIDI-controlled tonewheel organ example

```
instr 1
  Note(notnum()-24)
endin

instr 100
 ToneWheel giwheels
 kbar[]  init 9
 kbar[0] ctrl7 1,11,0,1
 kbar[1] ctrl7 1,12,0,1
 kbar[2] ctrl7 1,13,0,1
 kbar[3] ctrl7 1,14,0,1
 kbar[4] ctrl7 1,15,0,1
 kbar[5] ctrl7 1,16,0,1
 kbar[6] ctrl7 1,17,0,1
 kbar[7] ctrl7 1,18,0,1
 kbar[8] ctrl7 1,19,0,1
 kvib    ctrl7 1,20,0,3
 ksel    ctrl7 1,21,0,1
 asig Mixer kbar
```

```
out    clip(VChorus(asig,kvib,ksel),0,0dbfs)
endin
schedule 100,0,-1
```

12.4.2 Synthesis by Analysis

In many applications of additive synthesis, the number of partials is too large for us to manage manually as part of our program code. In these cases, we will need some external help to automatically detect the frequencies and amplitudes of the target spectrum, and then pass these to a bank of multiple oscillators. This detection, also known as *spectral analysis*, is the centrepiece of a series of techniques that operate in the *frequency domain* (i.e. where we manipulate sinusoidal components, rather than the waveform itself). Additive synthesis is in part a spectral method, and this aspect will be explored in Chapter 14.

The object of the analysis, as we have already noted, is the frequencies and amplitudes of partials. We can make 'one-shot' analyses, where we target a specific point in time of the target sound, or we can take frames at regular intervals to reveal the evolution of its spectrum. In the first case, we will have a 'frozen' moment in time, which can be useful for certain applications. More generally, we will be interested in the time-varying nature of sounds, so a series of analyses spaced in time are best suited for this. Managing the huge amounts of data produced by such processes is often a problem. We can take two approaches:

1. **offline analysis**: in this case we record the waveform of the target sound to a file, and apply the analysis to it, storing the results in another file. The synthesis process will then read the data from the file (Fig. 12.15).
2. **streaming**: here we buffer an input audio stream, analyse it and generate a sequence of analysis frames containing its amplitudes and frequencies. This data can optionally be saved to a file for later use. This is more demanding computationally, as both the analysis and synthesis are happening at the same time. Also, due to the nature of Fourier-based methods, a small amount of inherent latency will exist between input and output because of the buffering involved (Fig. 12.16).

Csound supports both approaches. Traditionally the first kind of analysis-synthesis was the only form available, but with the development of new techniques and more powerful computing platforms, streaming methods were made possible. For this reason, we will concentrate our discussion on examining this approach.

The first issue to be resolved when dealing with spectral streams is where to place the analysis data for easy access by oscillators. As in the tonewheel case, we can use arrays to hold frequencies and amplitudes. Alternatively, we can employ function tables for this purpose. Since both are more or less equivalent, we will use the latter method, for the sake of variety. Before we look into how to obtain the

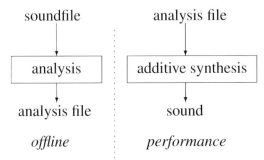

Fig. 12.15 Offline analysis-synthesis

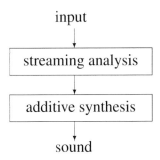

Fig. 12.16 Streaming analysis-synthesis

analysis data, we can design the synthesis process. As with any additive synthesis method, the sources are sinusoidal oscillators. We can build a bank of these, again using recursion. Each partial will take its parameter data from a specific position in two tables, which store amplitude and frequency values, respectively. The code for an oscillator bank UDO is shown in listing 12.27. The opcode has a pitch control, which can be used to transpose the output sound.

Listing 12.27 An oscillator bank UDO

```
/*****************************
asig OscBnk kpitch,ifa,ifn,isize
kpitch - pitch transposition factor
ifa - amplitude function table
iff - frequency function table
isize - oscillator bank size
*****************************/
opcode OscBnk,a,kiiio
 kp,ifa,iff,is,icnt xin
 if icnt < is-1 then
   asig OscBnk kp,ifa,iff,is,icnt+1
```

```
endif
xout asig +
        oscili(port(table(k(icnt),ifa),0.01),
                 kp*table(k(icnt),iff))
endop
```

The oscillator bank design is general enough to work in different scenarios. All it assumes is that a certain number of oscillators are required, and that amplitude and frequency values for each oscillator will be found sequentially in two tables, from 0 to the size of the bank -1. We can use different methods to supply this data to it. A simple way that involves no analysis is to draw a given shape for the spectral envelope (the shape formed by the amplitude of the partials across the frequency bands), and decide what specific partial frequencies we want. This of course will be difficult to manage for time-varying spectra, and so we can get the data from an analysis process instead.

A common way of finding amplitudes and frequencies for a given input is through the phase vocoder. This is a variation on the original channel vocoder discussed in Section 12.2, which follows the same overall principle of breaking the spectrum into a number of bands. Here, however, the analysis is more precise in that it provides both the energy at that band (amplitude) and the frequency for a sinusoidal wave to reproduce it. While we will leave the detailed discussion of the phase vocoder until Chapter 14, we will outline here its relevant parameters for additive synthesis applications.

The analysis breaks the total spectrum into equal-bandwidth portions (as opposed to equal-Q bands seen in the channel vocoder implementation). The width of each 'channel' is therefore dependent on the number of these, which is determined by the size of the analysis frame. The phase vocoder takes a number of waveform samples (the frame), and outputs amplitudes and frequencies representing these. The number of bands is equivalent to half the analysis size (plus one extra, which we can ignore for the moment). For example, if we analyse 1024 waveform samples, we get 512 bands.

For efficiency reasons, using power-of-two size is recommended (a fast algorithm, optimised for these lengths is then used). The width of the channels is simply calculated by dividing the total frequency span (0 Hz to Nyquist frequency) by the number of bands. For instance, at 44,100 Hz sampling rate, with a frame of $N = 1,024$ samples we have 512 channels of $\frac{22050}{512} = 43.0664$ Hz. The analysis is performed at regular periods, which can vary from 1 to $\frac{N}{4}$ samples. This is called the analysis *hopsize*; more frequent analyses require more computation, but can improve quality.

In Csound, the opcode used for streaming phase vocoder analysis is `pvsanal`. The process it employs will be discussed in detail in Chapter 14, but we will discuss its key principles here. It takes an input signal, the analysis size, the hopsize, the analysis window (generally the same as the analysis size) and a window shape (an envelope shape used to smooth the analysis process). It produces a frequency-domain output using the f-sig data type containing the analysis frame consisting of amplitudes and frequencies for each band. These can be placed into tables using the `pvsftw` opcode, which returns a flag indicating whether new analysis data was written to the tables. This is useful because the analysis will not necessarily happen at every k-cycle. With these opcodes, we can construct an analysis UDO to work with our oscillator bank. This is shown in listing 12.28.

Listing 12.28 A streaming frequency analysis UDO

```
/****************************
kflag StreamAnal asig,ifa,iff,isize
asig - input signal
ifa - amplitude function table
iff - frequency function table
isize - number of amp/freq pairs
****************************/
opcode StreamAnal,k,aiii
 asig,ifa,iff,is xin
 fsig pvsanal asig,is*2,is/4,is*2,1
 xout pvsftw(fsig,ifa,iff)
endop
```

An example instrument applying these principles is shown in listing 12.29. Here, we obtain a target spectrum from a soundfile, and use two oscillator banks to resynthesise it with different pitch transpositions.

Listing 12.29 A streaming frequency analysis UDO

```
gioscs init 512
gifn1 ftgen 1,0,gioscs,7,0,gioscs,0
gifn2 ftgen 2,0,gioscs,7,0,gioscs,0
instr 1
 a1 diskin2 "fox.wav",1,0,1
 kfl StreamAnal a1,1,2,gioscs
 a1 OscBnk p4,1,2,gioscs
 a2 OscBnk p5,1,2,gioscs
 out (a1+a2)/2
endin
schedule(1,0,10,1,.75)
```

Various other types of transformations can be applied to the spectral data before resynthesis. For instance, we could shape the spectral envelope in different ways, making filtering effects. The frequency data can be modified to create inharmonic spectra. Two different sounds can have their amplitude and frequency data exchanged or interpolated to create crossed and morphed timbres. In all of these cases, we are venturing into the terrain of spectral processing, which will be followed up in Chapter 14.

12.5 Conclusions

In this chapter, we explored the three classic types of sound synthesis: source-modifier (also known as subtractive), distortion and additive methods. We started by discussing some fundamentals of the spectral representation of audio, introducing the concept of sinusoidal components, harmonics and inharmonic partials. These form an important part of the theory of acoustics that underpins all classic sound synthesis approaches.

Source-modifier synthesis was discussed from the perspective of its elements: the different possibilities of sound generators, and the various types of filters that can be applied to them. The section was completed by two design studies: a virtual analogue model of a synthesiser, and the channel vocoder. In the distortion synthesis section, we explored in detail a variety of synthesis algorithms, their mathematical expressions and corresponding Csound code.

Completing the chapter, an overview of additive synthesis was provided, followed by a complete tonewheel organ model, and an introduction to analysis-synthesis techniques. This provides a point of contact with a number of advanced signal processing methods based on frequency-domain manipulation, which can provide a rich source of interesting instrument designs. We will be complementing this with Chapter 14 dedicated to spectral processing.

Chapter 13
Time-Domain Processing

Abstract This chapter will discuss processing of audio signals through the use of time-domain techniques. These act on the samples of a waveform to deliver a variety of effects, from echoes and reverberation to pitch shifting, timbral modification and sound localisation. The chapter is divided into four sections dealing with the basic methods of fixed and variable delays, filtering and spatial audio. Code examples are provided implementing many of the techniques from first principles, to provide the reader with an insight into the details of their operation.

13.1 Introduction

The manipulation of audio signals can be done in two fundamental ways: by processing the samples of a sound waveform, or by acting on its spectral representation [137]. Time-domain techniques are so named because they implement the former, working on audio as a function of time. The latter methods, which will be explored in Chapter 14, on the other hand process sound in its frequency-domain form.

The techniques explored in this chapter can be divided into three big groups, which are closely related. The first one of these is based on the use of medium to long, *fixed* signal delays, with which we can implement various types of echo and reverberation, as well as timbral, effects. Then we have the short- and medium-length *variable* delay processing that allows a series of pitch and colour transformations. Finally, we have filters, which are designed to do various types of spectral modifications that can be used both in timbre modification and in spatial audio effects.

13.2 Delay Lines

Digital delay lines are simple processors that hold the samples of an audio signal for a specified amount of time [67, 34], releasing them afterwards. Their basic func-

© Springer International Publishing Switzerland 2016
V. Lazzarini et al., *Csound*, DOI 10.1007/978-3-319-45370-5_13

tion is to introduce a time delay in the signal, which is used in a variety of audio processing applications. They can be expressed by a simple equation:

$$y(t) = x(t - d) \tag{13.1}$$

where t is the time, $x(t)$ is the input signal, $y(t)$ is the delay line output, and d the amount of time the signal is delayed.

Conceptually, delay lines are first-in first-out (FIFO) queues, whose length determines the amount of delay imposed on the signal (see Fig. 13.1). Since at each sample period, one sample is placed into the processor and another comes out, the time it takes for a sample to traverse the delay will be

$$d = L \times \frac{1}{sr} \tag{13.2}$$

where L is the length of the delay in samples, and sr, the sampling rate.

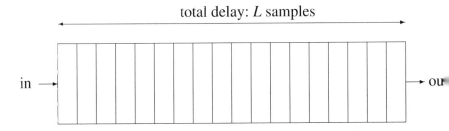

Fig. 13.1 Conceptual representation of a delay line as a FIFO queue, where a sample takes L sampling periods to exit it

The conceptual idea of a sample moving from one position in the FIFO queue to another at each sampling period is useful for us to understand the principles of a delay line. However, it is rarely the case that any delay of more than a few samples is actually implemented in this way. This is because it involves moving L samples around at every sampling period, which can be very inefficient.

Instead, a circular buffer algorithm is used. The idea is to keep the samples in place, and just move the reading position along the delay line, which involves a fraction of the computational cost. Once we read a sample, we can write to the buffer, and move the reading position one slot ahead, and repeat the whole operation. When we reach the end of the buffer, we wrap around to the beginning again. This means that a sample will be held for exactly the time it takes for the reading position to return to where the sample was written, L sampling periods later (Fig. 13.2)

To illustrate this fundamental algorithm, we can implement it as a UDO (although it already exists as internal opcodes in the system). We will use an a-rate array as a circular buffer and work on a sample-by-sample basis (ksmps=1, kr=sr). This is required because we need to access the delay line one sample at a time. The code for this is shown in listing 13.1. The delay length is set according to equation 13.2, making it sure it is rounded to the nearest integral value (array sizes need to be whole numbers), checking that the minimum size is 1. The algorithm follows: read at kpos, write to the same spot, update kpos circularly.

Listing 13.1 Circular buffer delay

```
opcode Delay,a,ai
setksmps 1
asig,idel xin
 kpos init 0
 isize = idel > 1/sr ? round(idel*sr) : 1
 adelay[] init isize
 xout adelay[kpos]
 adelay[kpos] = asig
 kpos = kpos == isize-1 ? 0 : kpos+1
endop
```

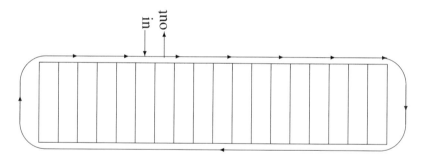

Fig. 13.2 Circular buffer delay line: samples are kept in their stored position, and the read/write position moves along the delay line circularly (in the direction shown by the line arrows). The current read point is L samples behind the last write

This UDO is a version of the delay opcode in Csound, which also implements a circular buffer using a similar algorithm:

```
asig delay ain, idel
```

Csound also allows a delay line to be set with a pair of separate opcodes for reading and writing to it:

```
asig delayr idel
      delayw ain
```

which also implements a circular buffer as in listing 13.1. The minimum delay for `delayr` and `delayw` is one control period (ksmps samples), whereas `delay` does not have this limitation. The basic effect that can be implemented using these opcodes as they are is a single echo. This is done by adding the delayed and original signal together. Further echoes will require other delays to be used, or the use of feedback, which will be discussed in the next section.

13.2.1 Feedback

A fundamental concept in signal processing is *feedback*,which we have already encountered in Chapter 12. As we recall, this involves mixing the output of the process back into its input. Within the present scenario, this allows us to create repeated delays, that will be spaced regularly by the length of the delay line. The feedback signal needs to be scaled by a gain to prevent the system from becoming unstable; without this the output would keep growing as the audio is inserted back into the delay (Fig. 13.3).The expression for this process is

$$y(t) = x(t) + gy(t-d)$$
$$w(t) = y(t-d) \tag{13.3}$$

where $x(t)$ is the input signal, g is the feedback gain, and $y(t-d)$ $(w(t))$ is the output after delay time d. As we can see in Fig. 13.3, there is no direct signal in the output of the feedback delay.

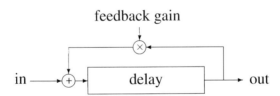

Fig. 13.3 Delay line with feedback. Depending on the value of *feedback gain*, various repeated delays will be produced

The feedback gain will need to be less than 1 to ensure stability. Depending on its value, many repeated delays will be heard. Each time the sound is recirculated through the delay, it will be scaled by a certain amount, so after time T, a short sound going through a delay of length D, and fed back with gain g, will be attenuated by

$$A = g^{\frac{T}{D}} \tag{13.4}$$

For instance, with a delay of 1 s, and a gain of 0.5, the repeats will die off to about $\frac{1}{1000}$th of the original amplitude after 10 seconds. This is an important threshold, equivalent to -60dB, which is often used to measure the *reverberation time* of a system. We can also determine the feedback gain based on a desired reverberation time T_r and D:

$$ g = \left(\frac{1}{1000} \right)^{\frac{D}{T_r}} \tag{13.5} $$

The feedback delay processor is used in many applications. It is a fairly common effect, which can be easily created by mixing the input signal with the feedback path:

```
ay = asig + adelay[kpos]*kg
```

The complete UDO will have an extra parameter that can be used to set the feedback gain. We also add a safety check to make sure the gain does not make the system unstable (by muting the feedback completely).

Listing 13.2 Feedback delay

```
opcode FDelay,a,aki
setksmps 1
asig,kg,idel xin
 kpos init 0
 isize = idel > 1/sr ? round(idel*sr) : 1
 adelay[] init isize
 kg = abs(kg) < 1 ? kg : 0
 ay = asig + adelay[kpos]*kg
 xout adelay[kpos]
 adelay[kpos] = ay
 kpos = kpos == isize-1 ? 0 : kpos+1
endop
```

This processor is also called a *comb filter*. Csound implements this process in the comb opcode:

```
asig comb ain,krvt,idel
```

where the feedback gain is controlled indirectly via a reverberation time parameter krvt, using eq. 13.5. We can also create this feedback delay using delayr and delayw, with a very simple code:

```
asig delayw
delayr ain + asig*kg
```

However in this case, the minimum delay is limited to one control cycle (and so is the feedback loop), whereas the comb filter has no such limit. The name *comb* filter comes from the shape of this unit's amplitude response (Fig. 13.4) whose shape displays a series of regular peaks. These are spaced by the filter fundamental frequency, which is equivalent to the inverse of its delay time, $1/D$. The impulse response is a series of decaying impulses, whose amplitude falls exponentially according to eq. 13.4. The height of the peaks is determined by the gain.

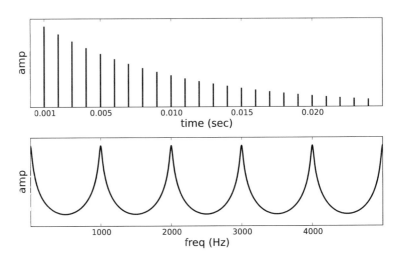

Fig. 13.4 Comb filter impulse (top) and amplitude response (bottom) for $D = 0.001s$ and $g = 0.9$. The amplitude response peaks are spaced by 1,000 Hz

A comb filter can be used for echo effects, or as a way of colouring the spectrum of an input sound. For the latter application, shorter delay times are required, so that the filter peaks are more closely spaced, and its fundamental frequency is in the audio range ($>$20 Hz). In general, comb filters will impart some colour to the input sound, unless the delay time is significantly large for the amplitude response peaks to be bunched together (which is the case in echo applications).

13.2.2 All-Pass Filters

It is possible to create a delay line processor that has a flat amplitude response. This is done by combining a feedforward and a feedback path for the signal, using the same absolute gain value, but with opposite signs. A diagram for this arrangement is shown in Fig. 13.5. This is called an *all-pass filter*, as it passes all frequencies with no attenuation. The expression for this process is

$$y(t) = x(t) + gy(t - d)$$
$$w(t) = y(t - d) - gy(t)$$

$$(13.6)$$

where $x(n)$ is the input, $y(t-d)$ is the delay line output with delay d, and $w(t)$ is the all-pass output. The impulse and amplitude responses for this filter are shown in Fig. 13.6.

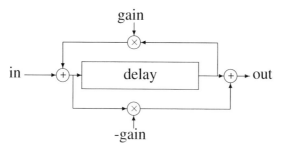

Fig. 13.5 All-pass filter, using a combination of feedback and feedforward delays, with gains of opposite signs, but the same absolute value

From this, we can see how the impulse response decay differs from the comb filter. There is an abrupt drop from the direct sound (whose polarity is reversed at the filter output) to the first delay; the following repetitions are much lower in amplitude than in the comb filter case. So, for an all-pass filter such as the one discussed here, after time T, an impulse going through a delay D and with a feedback gain g will be attenuated by

$$A = (1 - g^2) g^{\frac{T}{D} - 1} \tag{13.7}$$

In order to implement the all-pass filter, we can make some modifications to our comb code, so that it matches the diagram in Fig. 13.5. We take the mix of the feedback signal and input and put this into the feedforward path. The output is a mix of this signal and the delay line output.

Listing 13.3 All-pass filter implementation

```
opcode allpass,a,aki
setksmps 1
asig,kg,idel xin
 isize = idel > 1/sr ? int(idel*sr) : 1
 adelay[] init isize
 kpos init 0
 ay = asig + adelay[kpos]*kg
 xout adelay[kpos] - kg*ay
 adelay[kpos] = ay
 kpos = kpos == isize-1 ? 0 : kpos+1
endop
```

In Csound, an all-pass filter is provided by the `alpass` opcode, which has the same parameters as the comb filter, reverb and delay time (`krvt` and `idel`):

```
asig alpass ain,krvt,idel
```

Like the comb filter, it is possible to implement this with `delayr` and `delayw`:

```
adel delayr idel
amx = ain + adel*kg
asig = adel - kg*amx
delayw amx
```

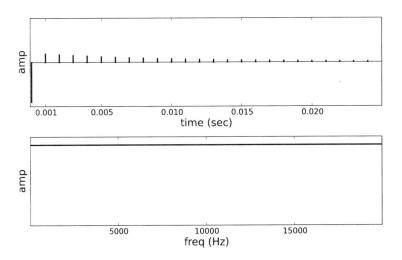

Fig. 13.6 All-pass filter impulse (top) and amplitude response (bottom) for $D = 0.001$ s and $g = 0.9$. The amplitude response is flat throughout the spectrum

A characteristic of the all-pass filter is that, although it does not colour the sound in its steady state, it might 'ring' in response to a transient in the input (e.g. a sudden change in amplitude). This is due to the decay characteristics of its impulse response. The ringing will be less prominent with lower values of its gain g, when the filter tends to decay more quickly.

13.2.3 Reverb

A fundamental application of delay lines is the implementation of reverb effects [30, 34, 67]. These try to add an ambience to the sound by simulating reflections in a given space. There are various ways of implementing reverb. One of them is to use comb and all-pass filters, which are also known as *component reverberators* to create different types of reverb effects. The classic arrangement, also known as *Schroeder* reverb [115], is a number of comb filters in parallel, whose output feeds into a series of all-pass filters. In Csound, this is used in the opcodes `reverb` (four comb + two all-pass), `nreverb` (six comb + five all-pass, also customisable) and `freeverb` (eight comb + three all-pass for each channel in stereo).

To illustrate this design, we will develop a simple UDO that uses four comb and two all-pass filters in the usual arrangement. The main reason for having the comb filters in parallel is that their role is to create the overall diffuse reverb. For this we should try not to make their delay times coincide, so that the four can create different decaying impulse trains to make the reverb more even frequency-wise. As we have seen, comb filters will colour the sound quite a bit, and if we make their spectral peaks non-coincident, this can be minimised. We can achieve this by selecting delay times that are prime numbers. They should lie within a range between 10 and 50 ms for best effect, although we could spread them a bit more to create more artificial-sounding results.

Another issue with standard comb filters is that they sound very bright, unlike most natural reflections, which have less energy at high frequencies. This is because while the comb filter reverb time is the same for the whole spectrum, the most common scenario is that reverb time becomes considerably shorter as frequencies increase. To model this, we can insert a gentle low-pass filter in the feedback signal, which will remove the high end more quickly than the lower [93]. In order to do this, we have to implement our own comb filter with this modification:

```
opcode CombF,a,akki
 asig,kg,kf,idel xin
 kg = 0.001^(idel/kg)
 adel delayr idel
 delayw asig + tone(adel*kg,kf)
 xout adel
endop
```

The all-pass filters, in series, have a different function: they are there to thicken each impulse from the comb filters, so that the reverberation model has enough reflections. For this, they need to have a short delay time, below 10 ms and a small reverb time, of the order of about ten times their delay. We will let the user decide the range of comb filter delays (min, max) and then select the correct times from a prime number list. This is done with an i-time opcode that returns an array with four values. We also report these to the console:

```
opcode DelayTimes,i[],ii
 imin, imax xin
 ipr[] fillarray 11,13,17,19,23,
          29,31,37,41,43,47,53,59,
          61,67,71,73,79
 idel[] init 4
 imin1 = imin > imax ? imax : imin
 imax1 = imax < imin ? imin : imax
 imin = imin1 < 0.011 ? 11 : imin1*1000
 imax = imax1 > 0.079 ? 79 : imax1*1000
 idel[0] = ipr[0]
```

```
icnt = lenarray(ipr)-1
idel[3] = ipr[icnt]
while (idel[3] > imax) do
 idel[3] = ipr[icnt]
 imxcnt = icnt
 icnt -= 1
od
icnt = 0
while (idel[0] <= imin) do
 idel[0] = ipr[icnt]
 imncnt = icnt
 icnt += 1
od
isp = (imxcnt - imncnt)/3
idel[1] = ipr[round(imncnt + isp)]
idel[2] = ipr[round(imncnt + 2*isp)]
printf_i "Comb delays: %d %d %d %d (ms)\n",
         1, idel[0],idel[1],idel[2],idel[3]
 xout idel/1000
endop
```

The reverb UDO sets the four combs in parallel and the two all-pass in series. It uses the above opcodes to implement the comb filter and delay time calculations, as shown in listing 13.4.

Listing 13.4 The implementation of a standard reverb effect, using a combination of four comb and two all-pass filters

```
/****************************************************
asig Reverb ain,krvt,kf,imin,imax
ain - input audio
krvt - reverb time
kf - low-pass frequency cutoff
imin - min comb delay time
imax - max comb delay time
****************************************************/
opcode Reverb,a,akkii
 asig,krvt,kf,imin,imax xin
 idel[] DelayTimes imin,imax
 ac1 CombF  asig,krvt,kf,idel[0]
 ac2 CombF  asig,krvt,kf,idel[1]
 ac3 CombF  asig,krvt,kf,idel[2]
 ac4 CombF  asig,krvt,kf,idel[3]
 ap1 alpass ac1+ac2+ac3+ac4,0.07,0.007
 ap2 alpass ap1,0.05,0.005
 xout ap2
endop
```

Feedback delay networks

Another approach to create reverb effects is to use a *feedback delay network*, or FDN [120, 118]. In this, we have a set of delay lines that are cross fed according to a given feedback matrix. This determines which signals will feed each delay line input. For instance, consider the matrix M in eq. 13.8, and a column vector D of four delay outputs, eq. 13.9:

$$M = \begin{pmatrix} 0 & 1 & 1 & 0 \\ -1 & 0 & 0 & -1 \\ -1 & 0 & 0 & 1 \\ 0 & 1 & -1 & 0 \end{pmatrix} \tag{13.8}$$

$$D = \begin{pmatrix} y_0(t - d_0) \\ y_1(t - d_1) \\ y_2(t - d_2) \\ y_3(t - d_3) \end{pmatrix} \tag{13.9}$$

where d_n is the delay time for line n.

A four-delay FDN can then be expressed as a matrix multiplication (eq. 13.10). It combines the delay outputs, a scalar feedback gain g and the input signal $x(t)$:

$$Y = x(t) + g\{M \times D\} \tag{13.10}$$

The vector Y holds the inputs of the four delay lines:

$$Y = \begin{pmatrix} y_0(t) \\ y_1(t) \\ y_2(t) \\ y_3(t) \end{pmatrix} \tag{13.11}$$

As with the comb filter, the FDN output is taken from the delay line outputs. If we want a stereo output, we can route each of the four outputs to different channels in any combination:

$$FDN = O \times D \tag{13.12}$$

where O is a mix matrix for two channels. For instance, we can have

$$O = \begin{pmatrix} 0.75 & 0.5 & 0.5 & 0.25 \\ 0.25 & 0.5 & 0.5 & 0.75 \end{pmatrix} \tag{13.13}$$

for a basic stereo spread.

A UDO demonstrating these ideas is shown in listing 13.5, and is illustrated in Fig. 13.7. As with the previous reverb implementation, we need to make sure reverb times are not equal across the spectrum, so we add a filter in the feedback path of each delay line. A delay time factor is used to make the FDN longer or shorter, keeping all the individual delay times relative to this. The overall gain is also scaled by $\frac{1}{\sqrt{2}}$, to keep the feedback under control.

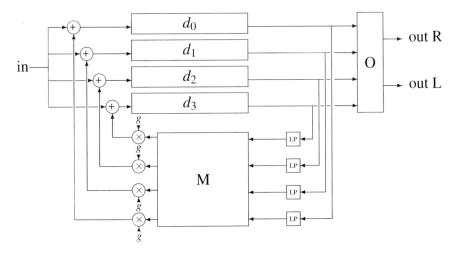

Fig. 13.7 A feedback delay network consisting of four delay lines, and including lowpass filters in the feedback path

Listing 13.5 A feedback delay network UDO

```
/************************************************
al,ar FDN asig,kg,kf,idel
al,ar - left and right outs
asig - input
kg - gain
kf - low-pass cutoff freq
idel - delay size factor
*************************************************/
opcode FDN,aa,akki
 aflt[] init 4
 amix[] init 4
 adel[] init 4
 il[] fillarray .75,.5,.5,.25
 ir[] fillarray .25,.5,.5,.75
 idel[] fillarray 0.023,0.031,0.041,0.047
 imatrix[][] init 4,4
 imatrix[0][0] = 0
 imatrix[0][1] = 1
 imatrix[0][2] = 1
 imatrix[0][3] = 0
 imatrix[1][0] = -1
 imatrix[1][1] = 0
 imatrix[1][2] = 0
```

```
imatrix[1][3] = -1
imatrix[2][0] = -1
imatrix[2][1] = 0
imatrix[2][2] = 0
imatrix[2][3] = 1
imatrix[3][0] = 0
imatrix[3][1] = 1
imatrix[3][2] = -1
imatrix[3][3] = 0

asig,kg,kf,id  xin
idel = idel*id
kg *= $M_SQRT1_2
ki = 0
while ki < 4 do
 kj = 0
 amix[ki] = asig
 while kj < 4 do
  amix[ki] = amix[ki]+aflt[kj]*imatrix[ki][kj]*kg
  kj += 1
 od
 ki += 1
od
adel[0] delay amix[0],idel[0]
aflt[0] tone adel[0],kf
adel[1] delay amix[1],idel[1]
aflt[1] tone adel[1],kf
adel[2] delay amix[2],idel[2]
aflt[2] tone adel[2],kf
adel[3] delay amix[3],idel[3]
aflt[3] tone adel[3],kf
al = 0
ar = 0
kj = 0
while kj < 4 do
 al += aflt[kj]*il[kj]
 ar += aflt[kj]*ir[kj]
 kj += 1
od
xout al, ar
endop
```

FDNs can be combined with other elements to make more sophisticated reverbs. They can provide very good quality effects. In Csound, these structures are used in the `reverbsc` and `hrtfreverb` opcodes.

13.2.4 Convolution

Another way of simulating reflections in a space can be realised through *multitap* delays [67]. In this scenario, we have a single delay line with *taps*, or outputs at different positions along its length, providing smaller delay times. This can be realised in Csound using a tap opcode such as deltap in between delayr and delayw pairs:

```
asig delayr  idel
atap deltap  kdeltap
delayw ain
```

Any number of taps can be placed in a delay line such as this one. Each tap can be scaled by a certain gain and added to the overall mix, to model the contribution of that particular reflection. The extreme case is when we get an output at *every* sample, and scale it by a given amount. This operation is called *convolution* [97], as illustrated by Fig. 13.8. The scaling values for each tap are crucial here: they will define the character of the space we want to reproduce.

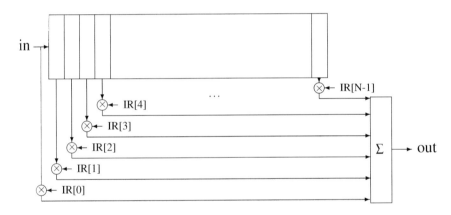

Fig. 13.8 Direct convolution: taps placed at each sample, scaled by each value of the impulse response. The output is the sum of all the scaled taps

In order to model a given system, we can use these values: they are its response to a single impulse, i.e. its impulse response (IR). This is a record of the intensity of all the reflections produced by the system. We have already seen this with respect to all-pass and comb filters, and we can use the same approach to look at real spaces. An IR can be obtained from a recording of an impulse played in a room, and with this, we can implement a convolution-based reverb.

Mathematically stated, this type of convolution can be written as

$$x(t) * h(t) = \sum_{n=0}^{N-1} h(n)x(t-n) \qquad (13.14)$$

where N is the length of the delay line, $h(n)$ are the gain scalers, or the IR used, and $x(t)$ is the input to the delay line.

Convolution is an expensive operation, because it uses N multiplications and additions for each output sample, where N is the IR length in samples. For this reason, direct calculation of convolution using delay lines is only used for short IRs. A faster implementation using spectral processing will be discussed in the next chapter. As an illustration of the process, we present a convolution UDO in listing 13.6. This code is for demonstration purposes only, as a faster internal opcode exists (dconv), which is more practical to use.

Listing 13.6 UDO demonstrating the convolution operation

```
opcode Convolution,a,ai
 setksmps 1
 ain,irt xin
 ilen = ftlen(irt)
 acnv = 0
 kk = ksmps
 a1 delayr (ilen-1)/sr
 while kk < ilen do
   acnv += deltapn(kk)*table(kk,1)
   kk += 1
 od
 delayw ain
 xout acnv + ain*table(0,1)
endop
```

The UDO takes an input signal and the number of a table containing the IR. We need to run this opcode at ksmps = 1, because the minimum delay allowed in delayr and delayw is equivalent to one k-period ($\frac{1}{kr}$ secs, or *ksmps* samples). If we do not mind missing the first *ksmps* samples of the IR, then we could run it at lower control rates. As we can see, the process is very straightforward; we just accumulate the output of each tap (with deltapn, which takes a delay time in samples), and multiply by the IR read from the function table.

It is possible to design a hybrid reverb, whose first hundred or so milliseconds are implemented through convolution, and the rest is based on a standard algorithm or FDN. This will use the IR for the *early reflections*, and the generic reverb for the diffuse part. Often it is the character of the early reflections that is the most significant aspect of a given space, whereas the reverb tail is less distinct. This approach has the advantage, on one hand, of being computationally more efficient than using convolution for the whole duration of the reverberation, and on the other, of providing a more natural feel to the effect (see listing 13.7).

Listing 13.7 Hybrid reverb combining convolution for early reflections and a standard algorithm for diffuse reverberation

```
/**********************************************
asig HybridVerb ain,krvt,kf,kearl,kdiff,irt
ain - input signal
krvt - reverb time
kf - lowpass cutoff factor (0-1)
kearl - level of early reflections
kdiff - level of diffuse reverb
irt - table containing an IR
***********************************************/
opcode HybridVerb,a,akkkki
 asig,krvt,kf,
  kearl,kdiff,irf xin
 kscal = 1/(kearl+kdif)
 ilen = ftlen(irf)
 iert = ilen/sr
 arev nreverb asig,krvt,khf
 acnv dconv asig,ilen,irf
 afad expseg 0.001,iert,1,1,1
     xout (acnv*kearl +
           arev*afad*kdif)*kscal
endop
```

IRs for a variety of rooms, halls etc. are available for download from a number of internet sites. These can be loaded into a function table using GEN 1. The UDO will use the function table size to determine when the transition between convolution and standard reverb happens. For most applications, the function table should ideally have a length equivalent to about 100 ms.

13.3 Variable Delays

A complete class of processes can be implemented by varying the delay time over time [31, 67]. Not only do we have a lengthening or shortening of the time between the direct sound and the delay line output, but some important side effects occur. The various algorithms explored in this section take advantage of these to modify input signals in different ways.

With variable delay times we have to be careful about how the delay line is read. The concerns here are similar to the ones in oscillator table lookup. When the delay is fixed, it is possible to round a non-integral delay time to the nearest number of samples without major consequences. However, when we vary the delay time over time, this is going to be problematic, esp. if the range of delay times is only a few samples. This is because instead of a smooth change of delay time, we get a stepped one. The quality of the output can be significantly degraded.

In order to avoid this problem, for variable-delay effects, we will always employ interpolation when reading a delay line. The simplest, and least costly, method is the linear case, which is a weighted average of the two samples around the desired fractional delay time position. Next, there is cubic interpolation, which uses four points. Higher-order methods are also possible for better precision, but these are more costly. In Csound, we should also try to use audio-rate modulation sources to vary the delay in most cases, to ensure a smooth result.

The following tap opcodes can be used with `delayr` and `delayw` for interpolated reading of the delay line:

- `deltapi`: linear interpolation.
- `deltap3`: cubic interpolation.
- `deltapx`: user-defined higher-order interpolation (up to 1,024 points).

They are generally interchangeable, so any instrument design using one of them can be modified to suit the user's needs in terms of quality/computation load.

13.3.1 Flanger

Flanger is a classic audio effect [6], whose digital implementation employs a modulated feedback delay line [67]. The basic principle of operation is that of a comb filter whose frequency is swept across the spectrum. From Fig. 13.4, we see that the spacing of the peaks in the amplitude response is given by the inverse of the delay time. A one-sample delay will work as a low-pass filter, as the peak spacing is equivalent to sr Hz, which is beyond the frequency range ($0 - \frac{sr}{2}$). A two-sample delay will have peaks at 0 and the Nyquist frequency. As the delays increase, the peaks get closer together, and the result is a sweeping of the filter frequencies over the spectrum.

As the spacing becomes small, the effect is diminished. For this reason, the effect is more pronounced with delays of a few milliseconds. The feedback gain determines how sharp the peaks are, making the effect more present. As its value gets closer to 1, the narrow resonances will create a pitched effect, which might dominate the output sound. In this case, instead of a filter sweep, the result might be closer to a glissando, as the peaks become more like the harmonics of a pulse wave.

An implementation of a flanger UDO is shown in listing 13.8. It uses a sine wave oscillator as a low-frequency oscillator (LFO), modulating the delay time between a minimum and a maximum value. The minimum is set at $\frac{2}{sr}$ ($\frac{1}{kr}$, with ksmps = 2) to allow two samples as a minimum for the cubic interpolation to work properly, and the maximum at 10 ms, which is equivalent to a 100 Hz spacing. The ksmps is set at 2 to allow the delay to go down to the minimum value (remembering that `delayr`/`delayw` have a minimum delay time of one k-period). As the LFO produces a bipolar waveform (which ranges from −kwdth to +kwdth, we need to offset and scale it, so that it is both fully positive and peaks at `kwdth`. Delay times

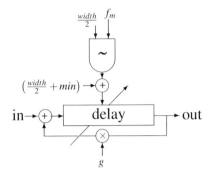

Fig. 13.9 A flanger design using a sine wave LFO. Note that as the LFO produces a bipolar wave-form from $\frac{-width}{2}$ to $\frac{width}{2}$, we need to offset it to be fully positive, with the minimum at *min*

can never be negative. The UDO checks for the input values, making sure they are in the correct range.

Listing 13.8 A flanger UDO using a sinewave delay time modulation

```
/***************************************************
asig Flanger ain,kf,kmin,kmax,kg
ain - input signal
kf - LFO modulation frequency.
kmin - min delay
kmax - max delay
kg - feedback gain
***************************************************/
opcode Flanger,a,akkkk
 setksmps 2
 asig,kf,kmin,kmax,kg xin
 idel = 0.01
 im = 1/kr
 km =  kmin < kmax ? kmin : kmax
 kmx = kmax > kmin ? kmax : kmin
 kmx = (kmx < idel ? kmx : idel)
 km  = (km >  im ? km : im)
 kwdth = kmx - km
 amod oscili kwdth,kf
 amod = (amod + kwdth)/2
 admp delayr idel
 afln deltap3  amod+km
 delayw asig + afln*kg
 xout afln
endop
```

13.3.2 Chorus

The chorus effect works by trying to create an asynchrony between two signals, the original and a delayed copy. To implement this, we set up a delay line whose delay time is modulated by either a period source like the LFO or a noise generator, and combine the delayed and original signals together (Fig. 13.10). The delay times are higher than in the flanger example, over 10 ms, but should use a modulation width of a few milliseconds. A secondary effect is a slight period change in pitch caused by the change in delay time (this will be explored fully in the vibrato effect). The chorus effect does not normally include a feedback path.

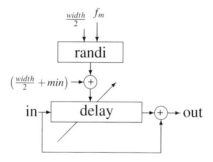

Fig. 13.10 A chorus UDO using a noise generator as a modulation source. As in the flanger case, we need to make sure the modulator output is in the correct delay range

Listing 13.9 A chorus UDO using a noise generator to modulate the delay time

```
/ * * * * * * * * * * * * * * * * * * * * * * * * * * * * * * * * * * * * * * * * * * * * * * * * *
asig Chorus ain,kf,kmin,kmax
ain - input signal
kf - noise generator frequency
kmin - min delay
kmax - max delay
* * * * * * * * * * * * * * * * * * * * * * * * * * * * * * * * * * * * * * * * * * * * * * * * * * /
opcode Chorus,a,akkk
 asig,kf,kmin,kmax xin
 idel = 0.1
 im = 2/sr
 km  =  kmin < kmax ? kmin : kmax
 kmx = kmax > kmin ? kmax : kmin
 kmx = (kmx < idel ? kmx : idel)
 km  = (km >  im ? km : im)
```

```
kwdth = kmx - km
amod randi kwdth,kf,2,1
amod = (amod + kwdth)/2
admp delayr idel
adel deltap3   amod+km
delayw asig
xout adel + asig
endop
```

The chorus implementation in listing 13.9 employs a bandlimited noise generator to create the delay time modulation. This is a good choice for vocal applications,
whereas in other uses, such as the thickening of instrumental sounds, we can use
a sine wave LFO instead. In this code, the two are interchangeable. Minimum and
maximum delays should be set at around 10 to 30 ms, with around 2 to 8 milliseconds difference between them. The larger this difference, the more pronounced the
pitch modulation effect will be.

13.3.3 Vibrato

The vibrato effect (Fig. 13.11) uses an important side-effect of varying the delay
time: frequency modulation (FM). FM happens because in order to create different
delay times, we need to read from the delay memory at a different pace than we
write to it. The difference in speed will cause the pitch to go up or down, depending on whether we are shortening the delay or lengthening it. We can think of this
as recording a segment of audio and reading it back at a different rate, a kind of
dynamic wavetable.

The writing speed to the delay line is always constant: it proceeds at intervals
of $\frac{1}{sr}$ s. In order to shorten the delay, the reading will have to be faster than this,
and so the pitch of the playback will rise. Conversely, to lengthen the delay, we will
need to read at a slower rate, lowering the pitch. If the *rate of change* of delay time
is constant, i.e. the delay gets shorter or longer at a constant pace, the speed of the
reading will be constant (higher or lower than the writing), and the pitch will be
fixed higher or lower than the original. This is the case when we modulate the delay
line with a triangle or a ramp (sawtooth) waveform.

However, if the rate of change of delay line varies over time, the pitch will vary
constantly. If we modulate using a sine wave, then the read-out speed will be variable, resulting in a continuous pitch oscillation. The key here is whether the first-
order difference between the samples of a modulating waveform is constant or varying. A choice of a triangle waveform will make the pitch jump between two values,
above and below the original, as the difference is positive as the waveform rises, and
negative as it drops. With a non-linear curve such as the sine, the pitch will glide
between its maximum and minimum values.

Another important consequence of this is that if we keep the amount of modulation constant, but modify its rate, the pitch change will be different. This is because,

although we are covering the same range of delay times, the reading will proceed at a different pace as the rate changes. So the width of vibrato will be dependent on both the amount and rate of delay modulation.

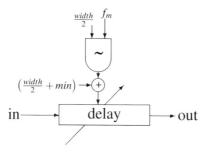

Fig. 13.11 A vibrato UDO using a sine wave LFO. The output consists only of the variable delay signal

It is interesting to consider how the vibrato width relates to the modulation rate and amount. This will depend on the LFO waveform. The resulting frequency modulation is going to be the derivative of the delay time modulation. Let's consider two common cases:

1. triangle: the derivative of a triangular wave is a square wave. So for a triangular modulator with depth Δ_d seconds, and frequency f_m Hz, $\Delta_d \text{Tri}(f_m t)$, we have

$$\Delta_d \frac{\partial}{\partial t} \text{Tri}(f_m t) = \Delta_d f_m \text{Sq}(f_m t) \tag{13.15}$$

where $\text{Sq}(ft)$ is a square wave with frequency f. Thus, the frequency will jump between two values, $f_0(1 \pm \Delta_d f_m)$, where f_0 is the original input sound frequency. For instance, if we have an LFO running with $f_m = 2$ Hz, by modulating the delay between 0.1 and 0.35 s ($\Delta_d = 0.25$), the pitch will alternate between two values, $0.5 f_0$ and $1.5 f_0$.
2. cosine: the derivative of a cosine is a sine. So, if we consider the modulation by $0.5 \cos(2\pi f_m t)$ (remembering that we have scaled it to fit the correct delay range), we have

$$\Delta_d \frac{\partial}{\partial t} 0.5 \cos(2\pi f_m t) = -\Delta_d \pi f_m \sin(2\pi f_m t) \tag{13.16}$$

In this case the frequency will vary in the range $f_0(1 \pm \Delta_d \pi f_m)$.

With these principles in mind, it is possible to consider the frequency modulation effects of a delay line. Note that it is also possible to set f_m to the audio range, for a variety of interesting effects on arbitrary input sounds. This technique is called *Adaptive FM* [76].

Listing 13.10 A vibrato UDO with an optional choice of function table for LFO modulation

```
/******************************************************
asig Vibrato   ain,kf,kmin,kmax[,ifn]
ain - input signal
kf - LFO modulation frequency.
kmin - min delay
kmax - max delay
ifn - LFO function table, defaults to sine
******************************************************/
opcode Vibrato,a,akkkj
 asig,kf,kmin,kmax,ifn xin
 idel = 0.1
 im = 2/sr
 km =  kmin < kmax ? kmin : kmax
 kmx = kmax > kmin ? kmax : kmin
 kmx = (kmx < idel ? kmx : idel)
 km  = (km >  im ? km : im)
 kwdth = kmx - km
 amod oscili kwdth,kf,ifn
 amod = (amod + kwdth)/2
 admp delayr idel
 adel deltap3  amod+km
 delayw asig
 xout adel
endop
```

The vibrato UDO in listing 13.10 implements an optional use of various function tables for its LFO. Vibrato and chorus are very similar in implementation. In Chapter 12, we have implemented these effects as part of the tonewheel organ implementation, and there the differences were only that the chorus effect mixed the original signal, while the vibrato did not. It is important to note that there is some variation in the interpretation of what these effects are supposed to be across different effects implementations. Nevertheless, the ideas discussed here are generally accepted as the basic principles for these processes.

13.3.4 Doppler

The Doppler shift is the perceived change in frequency of a waveform due to a moving source. As it moves towards or away from the listener, the wavefront reaching her is either squeezed or stretched, resulting in a pitch modification. The perceived frequency f_p will be related to the original frequency f_o by the following relationship:

$$f_p = f_o \times \frac{c}{c - v} \tag{13.17}$$

where c is the speed of sound in the air ($\cong 344$ ms^{-1}) and v is the velocity of the sound source. If the source is moving towards the listener, the velocity is positive, and the frequency will rise. In the other direction, the frequency will drop.

The effect of a variable delay line is similar to the Doppler effect. When the delay time is decreased, the effect is similar to making the source move closer to the listener, since the time delay between emission and reception is reduced.

A digital waveform can be modelled as travelling $\frac{c}{sr}$ meters every sample. In this case, a delay of D samples will represent a position at a distance of

$$p = D \times \frac{c}{sr} \, \text{m} \tag{13.18}$$

Thus, if we vary D, the effect can be used to model a sound source moving with velocity V,

$$V = \frac{D}{t} \times \frac{c}{sr} \, \text{m s}^{-1} \tag{13.19}$$

As with the vibrato effect discussed above, Doppler shift is proportional to the rate of change of the delay over time. So a faster moving object will imply a faster change in the delay time. For varying speeds, then the delay time has to change at different rates. If the effect of a sound passing the listener is desired, then a positive shift followed by a negative shift is necessary. This can be easily achieved by decreasing and then increasing the delay times. The rate of change will model the speed of the source. A change in amplitude can also reinforce the effect.

For example, if we want the source to move from a position at maximum distance p_{max} to a minimum distance position p_{min}, we can do this:

1. Set p_{max} and p_{min} in metres.
2. Set the equivalent delay line 'distances' (in seconds), using the expression $d = \frac{p}{c}$.
3. Put the sound through the delay line and vary the delay from the maximum delay to the minimum delay for an approaching source, and vice versa for the opposite direction.
4. Couple the change in delay with change in intensity (e.g. with an envelope).
5. For a variable-speed source, use a non-linear function to change the delay time; for fixed speed, use a linear envelope.

The code in listing 13.11 implements a Doppler shift UDO using the ideas discussed above. It takes a distance in meters, and models the movement of the source according to this parameter, via a delay line change in position and amplitude attenuation.

Listing 13.11 A Doppler shift UDO, which models the movement of a source according to its distance in meters

```
/********************************************************
asig Doppler ain,kpos,imax
ain - input signal
kpos - absolute source distance in meters
imax - maximum source distance
********************************************************/
opcode Doppler,a,aki
 asig,kp,imax xin
 ic = 344
 admp delayr  imax/ic
 adop deltap3 a(kp)/ic
 kscal = kp > 1 ? 1/kp : 1
 delayw asig*kscal
 xout adop
endop
```

13.3.5 Pitch Shifter

The triangle-wave vibrato effect discussed in Section 13.3.3, which creates alternating steady pitch jumps, and indeed the constant-speed doppler effect, suggest that we might be able to implement a pitch shift effect using similar principles. The problem to solve here is how to create a continuously increasing or decreasing delay beyond the range of the delay times available to us. If we could somehow jump smoothly from a maximum delay to a minimum delay (and vice versa), then we could just circularly move around the delay line, as if it were a function table.

The problem with this jump is that there is an abrupt discontinuity in the waveform as we move from one end of the delay to the other. We can use an envelope to hide this transition, but, alone, that would create an amplitude modulation effect. To avoid this, we can use two delay line taps, spaced by $\frac{1}{2}$ delay length, so that when one tap is at the zero point of the envelope, the other is at its maximum. Some modulation artefacts might still arise from the phase differences of the two taps, but we can try to minimise these later.

So, for the pitch shifter effect to work, we need to modulate the delay line with a sawtooth wave, rather than a triangle (we want the pitch to change in just one direction). The straight line of the sawtooth wave will have a constant derivative (except at the transition), which will give us a constant pitch change. As we saw earlier, the amount of pitch change is dependent on the modulation width and frequency. If we want to effect a pitch shift p in a signal with frequency f_0, we can find the sawtooth wave modulation frequency f_m using what we learned from eq. 13.15. According to that, the upwards pitch change pf_0 is equivalent to

$$pf_0 = f_0(1 + \Delta_d f_m) \tag{13.20}$$

We can then determine the f_m required for a given pitch shift factor p and delay modulation width Δ_d as

$$f_m = \frac{p-1}{\Delta_d} \tag{13.21}$$

The pitch shifter effect then is just a matter of using two phase-offset sawtooth waves with frequency set to f_m, which modulate two taps of a delay line. In synchrony with this modulation, we need to use an envelope that will cut out the wrap-around jumps. This can be done with two triangle waves that are run in phase with each modulator.

A Csound UDO implementing these ideas is shown in listing 13.12. As modulators, it employs two phasor opcodes, offset by 0.5 period. Since these produce an up-ramp (inverted) sawtooth, we ran them with negative frequency to produce a down-ramp signal, which will scale the input signal frequency upwards (as the delay time gets shorter). If the pitch shift requested is below 1, then the sawtooth frequency will be positive and the shift will be downwards.

Listing 13.12 A pitch shifter UDO using two sawtooth modulators controlling two taps that will be offset by $\frac{1}{2}$ delay length

```
/ * * * * * * * * * * * * * * * * * * * * * * * * * * * * * * * * * * * * * * * * * * * * * * * * * *
asig PitchShifter ain,kp,kdel,ifn[,imax]
ain - input signal
kp - pitch shift factor (interval ratio)
kdel - delay mod width
ifn - window (envelope) to cut discontinuities
imax - optional max delay (defaults to 1 sec)
* * * * * * * * * * * * * * * * * * * * * * * * * * * * * * * * * * * * * * * * * * * * * * * * * */
opcode PitchShifter,a,akkip
 asig,kp,kdel,ifn,imax xin
 kfm = (kp-1)/kdel
 amd1 phasor -kfm
 amd2 phasor -kfm,0.5
 admp delayr imax
 atp1 deltap3 amd1*kdel
 atp2 deltap3 amd2*kdel
 delayw asig
 xout atp1*tablei:a(amd1,ifn,1) +
     atp2*tablei:a(amd2,ifn,1)
endop
```

The table for the pitch shifter window can be built with a triangle window (using GEN 20):

```
ifn ftgen 1,0,16384,20,3
```

Finally, this pitch-shifting algorithm is prone to amplitude modulation artefacts due to the combination of the two taps, which in certain situations can be very noticeably out of phase. To minimise this, it is possible to pitch track the signal and use this to control the delay modulation width. If we set this to twice the fundamental period of the input sound, then geerally the taps will be phase aligned. Of course, if the input sound does not have a pitch that can be tracked, this will not work. However, in this case, phase misalignment will not play a significant part in the process. The following instrument example shows how this can be set up to create a vocal/instrument harmoniser, which tracks fundamentals in the range of 100 to 600 Hz to control the delay width. The pitch is pegged at this range and we use a `port` opcode to smooth the fluctuations, avoiding any undue modulation.

```
instr 1
 ain inch 1
 kf, ka pitchamdf ain,100,600
 kf = kf < 100 ? 100 : (kf > 600 ? 600 : kf)
 kdel = 2/kf
 kdel port kdel, 0.01, 0.1
 asig PitchShifter ain,1.5,kdel,1
 outs asig+ain,asig+ain
endin
schedule(1,0,-1)
```

13.4 Filters

We have already introduced the main characteristics and applications of filters in source-modifier synthesis (Chapter 12). In this section, we will explore some internal aspects of filter implementation, and the use of these processors in sound transformation.

Filters also depend on delay lines for their operation. In fact, we can describe all of the operations in Sections 13.2 and 13.3 in terms of some sort of filtering. The comb and all-pass delay processors, for instance, are infinite impulse response (IIR) filters, whereas the convolution reverb is a finite impulse response (FIR) filter. These are all high-order filters, employing long delays. In this section, we will start by looking at filters using one or two sample delays, with feedback, and then consider feedforward filters based on longer impulse responses.

13.4.1 Design Example: a Second-Order All-Pass Filter

As an example of how we can implement a filter in Csound, we will look at a second-order all-pass [90]. This design is used in the construction of a phase shifter, which will be our final destination. This effect works by combining a signal with another whose phase has been modified non-linearly across the spectrum. The result is a cancellation of certain frequencies that are out of phase.

An all-pass filter, as seen before, has the characteristic of having a flat amplitude response, i.e. passing all frequencies with no modification. Some of these filters also have the effect of changing the phase of a signal non-linearly across the spectrum. This is the case of the second-order IIR all-pass, which we will implement here. In its output, some frequencies can have up to half a cycle (180°, or π radians) phase shift in relation to the input sound.

As we have seen before, a second-order filter uses up to two samples of delay. Also, as in the case of the high-order all-pass, we will combine a feedback signal path with a feedforward one. The expression is very similar in form:

$$w(t) = x(t) + a_1 w(t-1) - a_2 w(t-2)$$
$$y(t) = w(t) - b_1 w(t-1) + b_2 w(t-2)$$

(13.22)

where a_1 and a_2 are the gains associated with the feedback path of the one- and two-sample delays, and b_1 and b_2 are feedforward gains. Another term used for these is *filter coefficients*. Equation 13.22 is called a *filter equation*.

To make the filter all-pass, we have to balance these gain values to make the feedforward section cancel out the feedback one. For instance, we set b_2 as the reciprocal (inverse) of a_2. The a_1 and b_1 coefficients will be used to *tune* the filter to a certain frequency. We can set all of these according to two common parameters: bandwidth and centre frequency. The former will determine how wide the phase shift region is, and the latter determines its centre. For bandwidth B and frequency f_c, the four coefficients of the all-pass filter will be

$$R = \exp\left(\frac{-\pi B}{sr}\right)$$
$$a_1 = 2R\cos\left(\frac{2\pi f_c}{sr}\right)$$
$$a_2 = R^2$$
$$b_1 = \frac{R}{2}\cos\left(\frac{2\pi f_c}{sr}\right)$$
$$b_2 = \frac{1}{a_2}$$

(13.23)

A plot of the phase response of this filter is shown in Fig. 13.12, using bandwidth $B = 500$ Hz and centre frequency $f_c = 5,000$ Hz. It is possible to see that at the centre frequency the phase is shifted by π radians, or half a cycle.

We can implement this filter as a UDO using the expressions in eqs.13.22 and 13.23. The opcode will take the centre frequency and bandwidth as control-rate

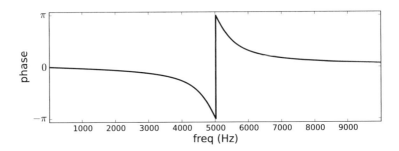

Fig. 13.12 second-order all-pass filter phase response, with bandwidth $B = 500$ Hz and centre frequency $f_c = 5,000$ Hz

arguments. The processing has to be done on a sample-by-sample basis because of the minimum one-sample delay requirement.

Listing 13.13 2nd-order all-pass filter UDO with variable centre frequency and bandwidth

```
/ * * * * * * * * * * * * * * * * * * * * * * * * * * * * * * * * * * * * * * * * * * * * * * * *
asig AP ain,kfr,kbw
ain - input signal
kfr - centre frequency
kbw - bandwitdth
* * * * * * * * * * * * * * * * * * * * * * * * * * * * * * * * * * * * * * * * * * * * * * * * * /
opcode AP,a,akk
 setksmps 1
 asig,kfr,kbw   xin
 ad[] init 2
 kR = exp(-$M_PI*kbw/sr)
 kw = 2*cos(kfr*2*$M_PI/sr)
 kR2 = kR*kR
 aw = asig + kR*kw*ad[0] - kR2*ad[1]
 ay = aw - (kw/kR)*ad[0] + (1/kR2)*ad[1]
 ad[1] = ad[0]
 ad[0] = aw
 xout ay
endop
```

   With this all-pass filter we can build a phase shifter. The idea is to combine the output of the filter with the original signal so that phase differences will create a dip in the spectrum, then modulate the centre frequency to produce a sweeping effect. One second-order all-pass can create a single band-rejecting region. For further ones, we can use more all-pass filters in series, tuning each one to a different frequency. In the design here, we will use three filters, so creating a sixth-order all-pass, with three dips in the spectrum. We will space the filters so that the second and third filters are centred at twice and four times the first frequency. When we combine these and the original signal, the overall effect is not all-pass anymore. In Fig. 13.13, we see a plot of the resulting phase and amplitude responses of the phase shifter, with frequencies centred at 1,000, 2,000 and 4,000 Hz, and bandwidths of 100, 200 and 400 Hz, respectively.

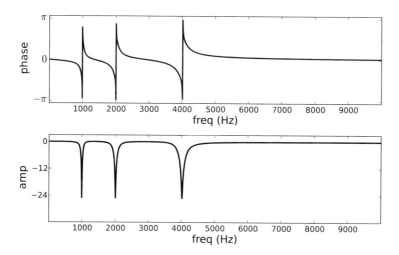

Fig. 13.13 sixth-order phase shifter phase (top) and amplitude (bottom) response, with frequencies centred at 1,000, 2,000 and 4,000 Hz, and bandwidths of 100, 200 and 400 Hz, respectively

A UDO implementing these ideas is shown in listing 13.14. It uses an LFO to modulate the filter centre frequencies, whose range is set between a minimum and a maximum (pegged at 0 and sr/2, respectively). By changing these, it is possible to control modulation depth. The phaser also features a Q control that makes the bandwidths relative to the centre frequencies.

Listing 13.14 Sixth-order phase shifter with LFO modulation, and user-defined frequency ranges and Q

```
/ * * * * * * * * * * * * * * * * * * * * * * * * * * * * * * * * * * * * * * * * * * * * * * * * * *
asig Phaser ain,kfr,kmin,kmax,kQ
ain - input signal
kfr - LFO frequency
kmin - minimum centre frequency
kmax - maximum centre frequency
kQ - filter Q (cf/bw)
* * * * * * * * * * * * * * * * * * * * * * * * * * * * * * * * * * * * * * * * * * * * * * * * * * * /
opcode Phaser,a,akkkk
 as,kfr,kmin,kmax,kQ xin
 km =  kmin < kmax ? kmin : kmax
```

```
kmx = kmax > kmin ? kmax : kmin
km = km > 0 ? km : 0
kmx = kmx < sr/2 ? kmx : sr/2
kwdth = kmax/4 - km
kmd oscili kwdth,kfr
kmd = km + (kmd + kwdth)/2
as1 AP as,kmd,kmd/kQ
as2 AP as1,kmd*2,kmd*2/kQ
as3 AP as2,kmd*4,kmd*4/kQ
xout as3+as
endop
```

A number of variations are possible, by setting a different centre frequency spacing, using more all-pass filters in the cascade connection, and decoupling the bandwidths and frequencies. As with the other variable delay line algorithms discussed earlier, considerable variation exists between different implementations offered by effects processors.

13.4.2 Equalisation

Equalisation is another typical application of filters for sound processing. Equaliser filters tend to be slightly different from the standard types we have seen. They are often designed to boost or cut one particular band without modifying others, whereas an ordinary band-pass filter generally has an effect across all of the spectrum. In Csound, a good equaliser is found in the `eqfil` opcodes, which is based on a well-known design by Regalia and Mitra [106]. This filter has a characteristic response that can be shaped to have a peak or a notch at its centre:

```
asig eqfil ain, kcf, kbw, kgain
```

Its arguments are self-explanatory: `kcf` is the centre frequency, `kbw`, the bandwidth. The gain parameter `kgain` makes the filter boost or cut a given band of frequencies. Its amplitude response will be flat for `kgain` = 1. If `kgain` ¿ 1, a peak will appear at the centre frequency, with bandwitdh set by `kbw`. Outside this band, the response will be flat. A notch can be created with a gain smaller than one.

Listing 13.15 shows a graphic equaliser UDO based on a number of `eqfil` filters arranged in series. It uses recursion to determine the number of bands dynamically. The number of these is taken from the size of a function table containing the gain values for each band. Bands are exponentially spaced in the spectrum, between a minimum and a maximum frequency. This will separate the filters by an equal musical interval. A Q value is also provided as an argument.

Listing 13.15 A graphic equaliser with a user-defined number of bands. The number of gain values in a function table determines the number of filters

```
/******************************************************
```

```
asig Equaliser ain,kmin,kmax,kQ,ifn
ain - input signal
kmin - minimum filter frequency
kmax - maximum filter frequency
kQ - filter Q (cf/bw)
ifn - function table containing the filter gains
*********************************************** /
opcode Equaliser,a,akkkio
 asig,kmin,kmax,
     kQ,ifn,icnt xin
 iend = ftlen(ifn)
 if icnt < iend-1 then
  asig Equaliser asig,kmin,kmax,
                 kQ,ifn,icnt+1
 endif
 print icnt
 kf = kmin*(kmax/kmin)^(icnt/(iend-1))
 xout eqfil(asig,kf,
        kf/kQ,table:k(icnt,ifn))
endop
```

13.4.3 FIR Filters

So far we have not discussed FIR filters in detail, as these are not as widely employed in music synthesis and processing as IIR ones. However, there are some interesting applications for these filters. For instance, we can use them to block out some parts of the spectrum very effectively. In this section, we will look at how we can build FIR filters from a given amplitude response curve. Generally speaking, these filters need to be much longer (i.e. higher order) than feedback types to have a decisive effect on the input sound. Another difficulty is that feedforward filters are not easily transformed into time-varying forms, because of the way they are designed.

FIR filters can be described by the following equation:

$$y(t) = a_0x(t) + a_1x(t-1) + \ldots + a_{N-1}x(t-(N-1))$$
$$= \sum_{n=0}^{N-1} a_nx(t-n) \tag{13.24}$$

where, as before, $y(t)$ is the filter output, $x(t)$ is the input and $x(t-n)$ is the input delayed by n samples. Each delay coefficient a_n can be considered a sample of an impulse response that describes the filter (compare, for instance, eq. 13.24 with eq. 13.14):

$$h(n) = \{a_0, a_1, ..., a_{N-1}\} \tag{13.25}$$

So, effectively, an FIR is a delay line tapped at each sample, with the coefficients associated with each delay point equivalent to its IR. Thus we can implement a feed-forward filter as the convolution of an input signal with a certain specially created IR.

Designing FIR filters can be a complex art, especially if we are looking to minimise the IR length, while approaching a certain amplitude and phase response [88]. However, it is possible to create filters with more generic forms by following some simple steps. The fundamental principle behind FIR design is that we can obtain its coefficients (or its IR) from a given amplitude response via an inverse discrete Fourier transform (IDFT). Thus, we can draw a certain desired spectral curve, and from this we make a frequency response, which can be transformed into an IR for a convolution operation. The theory behind this is explored in more detail in section 14.2.

The filter design process can be outlined as follows:

1. We create a function table with N points holding a desired amplitude curve. Position 0 will be equivalent to 0 Hz and position N to $\frac{sr}{2}$, the Nyquist frequency. The size should be large enough to allow for the transition between passband and stopband[1] to be as steep as we like. The width of this transition region can be as small as $\frac{sr}{2N}$. For $N = 1,024$ and $sr = 44,100$, this is about 21.5 Hz. However, with such small transitions, some ripples will occur at the edges of the transition, which might cause some ringing artefacts. Nevertheless, steep transitions can still be constructed spanning a few table positions, if N is large enough. A table implementing this can be created with GEN7, for instance. For a brickwall (i.e. with a steep transition) low-pass curve, we would have

   ```
   ifn ftgen  1,0,1024,7,1,102,1,10,0,912,0
   ```

   In this case, the passband will be 10% of the spectrum, about 2,205 Hz, and the transition will be 10 table points, or 1%.

2. We use this curve as a set of magnitudes for a 0-phase spectrum, and take its inverse DFT. In Csound, this amounts to copying the table into an array, making its elements complex-valued (with r2c) and taking the transform (rifft):

   ```
   copyf2array iSpec,ifn
   iIR[] rifft r2c(iSpec)
   ```

3. The result is an IR for our filter, but it is still not quite ready to be used. To avoid rounding errors, we need to swap the first and second halves of it, which will not affect the amplitude response, but will make the filter work properly.

   ```
   while icnt < iflen2 do
       itmp = iIR[icnt]
   ```

---

[1] The passband is the region of the spectrum whose amplitude is not affected by the filter. The stopband is the region whose amplitude gets reduced by the filter.

```
        iIR[icnt] = iIR[icnt + iflen2]
        iIR[icnt + iflen2] = itmp
        icnt +=1
    od
```

This process is illustrated by Fig. 13.14. From top to bottom, we have the original amplitude curve, the resulting IR, and the amplitude response of the resulting filter.

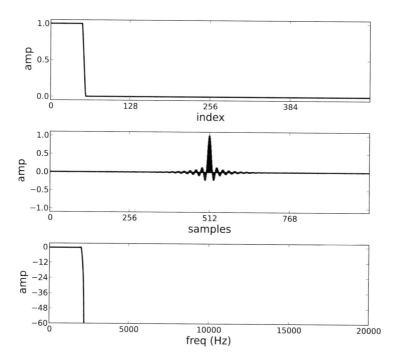

Fig. 13.14 Finite impulse response design. From top to bottom: the original amplitude curve, the resulting IR, and the amplitude response of the resulting filter

Listing 13.16 shows the complete code. It takes a signal and a function table with a given amplitude response, calculates the IR at i-time using a dedicated UDO and then uses it with a direct convolution opcode to produce the output sound.

Listing 13.16 A pair of FIR filter design and performance opcodes, taking an amplitude response from an input function table

```
/ * * * * * * * * * * * * * * * * * * * * * * * * * * * * * * * * * * * * * * * * * * * * * * * * *
irn IR ifn
irn - impulse response output function table number
ifn - amplitude response function table number
* * * * * * * * * * * * * * * * * * * * * * * * * * * * * * * * * * * * * * * * * * * * * * * * * * /
```

```
opcode IR,i,i
 ifn xin
 iflen2 = ftlen(ifn)
 iflen = 2*iflen2
 iSpec[] init iflen2
 icnt init 0
 copyf2array iSpec,ifn
 iIR[] rifft r2c(iSpec)
 irn ftgen 0,0,iflen,7,0,iflen,0
 while icnt < iflen2 do
    itmp = iIR[icnt]
    iIR[icnt] = iIR[icnt + iflen2]
    iIR[icnt + iflen2] = itmp
    icnt +=1
 od
 copya2ftab iIR,irn
 xout irn
endop

/*********************************************
asig FIR ain,ifn
ain - input audio
ifn - amplitude response function table number
**********************************************/
opcode FIR,a,ai
 asig,ifn xin
 irn IR ifn
 xout dconv(asig,ftlen(irn),irn)
endop
```

Filters such as this one can work very well to block a part of the spectrum. Figure 13.15 shows the result of putting a full-spectrum pulse waveform through the filter designed as shown in Fig. 13.14. As can be seen from the plot, the filter is very selective, cutting high frequencies in an effective way.

FIR filters of this kind are fixed, i.e. they cannot be made time-varying as we have done with the more musical IIR designs. It is possible however to have more than one filter running at the same time and cross-fade the output from one to the other. Another aspect to note is that direct convolution can be computationally expensive. This can be replaced with fast convolution, which will be discussed in the next chapter.

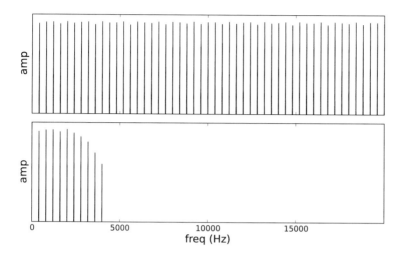

Fig. 13.15 The spectra of a pulse waveform (top) and output of a brickwall FIR filter designed as shown in Fig. 13.14

13.4.4 Head-Related Transfer Functions

Another application of FIR filters is in the simulation of spatial positions through the use of head-related transfer functions (HRTFs) [10]. These are filter frequency responses that model the way our head and outer ears modify sounds before these are transmitted through the middle and inner ear to the brain. With them, it is possible to precisely place sounds on a sphere around the listener, playing the sound directly via headphones, depending on the accuracy of the transfer functions used.

An HRTF encodes the effect of the head and pinna on the incoming sound. As it comes from different directions, the spectrum of a sound is shaped in a particular way, in time (delays) and amplitude. A pair of these, one for each ear, will match every source direction on a sphere. The combination of filtering, inter-aural time delays (ITDs) and inter-aural intensity differences (IIDs) that is encoded in the HRTFs will give the listener's brain the necessary cues to place a sound at a given location [13].

HRTFs are obtained from head-related impulse responses (HRIRs) that are generally measured at spaced points on this sphere, via recordings using dummy heads or real subjects with special in-ear microphones. Once the HRIRs are obtained, we can use a convolution method to implement the filter. In some cases, IIR designs modelled on HRTFs are used, but it is more common to apply them directly in FIR filters. As audio is delivered directly to the two ears and localisation happens through these functions, this process is called *binaural audio*.

It is very common to use generic measurements of HRIRs for binaural applications, but the quality of the localisation effect will vary from person to person, according to how close the functions are to the listener's own. It is well known that HRTFs can be extremely individual, and ideally these should be designed or measured for each person for an accurate effect. However, it is not easy to obtain individualised HRIRs, so most systems providing binaural audio rely on some general set measured for an average person.

This is the case with Csound, where we find a whole suite of opcodes dedicated to different types of HRTF processing [25]. These rely on a very common set measured at MIT, which has also been used in other systems. Here, however there are some clever algorithms that allow for smooth movement of sound sources in space, which are not found elsewhere, as well as high-quality binaural reverb effects.

The HRTF opcodes in Csound are

`hrtfmove` and `hrtfmove2`: unit generators designed to move sound sources smoothly in 3D binaural space, based on two different algorithms.
`hrtfstat`: static 3D binaural source placement.
`hrtfearly`: high-fidelity early reflections for binaural reverb processing.
`hrtfreverb`: binaural diffuse-field reverb based on an FDN design, which can be used in conjuction with `hrtfearly`, or on its own as a standalone effect.

These opcodes rely on two spectral data files, for left and right channels, which have been constructed from the MIT HRTF database specially for these opcodes. They are available in three different sampling rates: 44.1, 48 and 96 kHz and are labelled accordingly in the Csound software distribution. The correct pair of files is required to match the required sr, otherwise the opcodes will not operate correctly.

13.5 Multichannel Spatial Audio

Complementing the discussion of spatial placement of sounds, we will briefly explore two essential methods of multichannel audio composition and reproduction. The first one of these is based on the principle of encoding the sound field in a single description of a sound source's directional properties. The encoded signals then require a separate decoding stage to be performed to obtain the multichannel signals. The other method provides a generalised panning technique for multiple channels, where the output is made up of the actual feeds for loudspeaker reproduction. Both techniques are fully supported by Csound and allow both horizontal and vertical placement of audio sources (2D or 3D sound).

13.5.1 Ambisonics

A classic method of audio spatialisation is provided by *ambisonics*. It encapsulates many models of auditory localisation, except for pinna and specific high-frequency ITD effects, in one single package [51]. The principle of ambisonics, first developed by Michael Gerzon for periphony (full-sphere reproduction, 3D) [46], is to encode the sound direction on a sphere surrounding the listener, and then provide methods of decoding for various loudspeaker arrangements.

The encoded signals are carried in a multiple channel signal set called *b-format*. This set can be constructed in increasing orders of directivity [46] by the use of more channels. Order-0 systems contain only omni directional information; order-1 encodes the signal in three axes, with two horizontal and one vertical component; order-2 splits these into more directions; and so-on. The number of b-format channels required for a system of order n is $(n+1)^2$. These form a hierarchy, so that we can move from one order to the next up by adding $2n+1$ channels to the set.

Encoding of signals is performed by applying the correct gains for each direction. In the case of first-order ambisonics, for sounds on a sphere with constant distance from the subject, we generally have

$$
\begin{aligned}
W &= \frac{\sqrt{2}}{2} \\
X &= \cos(\theta)\sin(\phi) \\
Y &= \sin(\theta)\sin(\phi) \\
Z &= \cos(\phi)
\end{aligned}
\tag{13.26}
$$

where X and Y are the horizontal, and Z is the vertical component. W is the omni directional signal. The parameters θ and ϕ are the angles in the horizontal and vertical planes (azimuth and elevation), respectively. Higher orders will have more channels, with different gain configurations, than these four. In any order, channels involving the height component might be eliminated if 2D-only (pantophonic) reproduction is used.

Decoding is generally more complex, as a number of physical factors, including the spatial configuration of the loudspeakers, will play a part. In particular, non-regularly spaced speakers, e.g. 5.1 surround sound set-ups, have to be treated with care. Other important considerations are the need for near-field compensation, to avoid certain low-frequency effects in reproduction, and dual-band decoding, to account for the fact that ITD and IID perception dominate different areas of the spectrum [51]. In other words, a straight decoding using an inversion of the encoding process might not work well in practice.

Csound includes support for encoding into first-, second- and third-order ambisonics b-format, via the `bformenc1` opcode. This takes a mono signal, an azimuth, and an elevation, and produces an output signal for the order requested. These b-format signals can be mixed together before decoding. Some manipulations, such as rotations, can also be performed, using the correct expressions in UDOs. The process of using ambisonics is often one of producing an encoded mix of several sources, which allows each one of these a separate spatial placement, as a b-format signal, which can then be decoded for specific loudspeaker layouts.

The ambisonic decoder `bformdec1` can be used to decode signals for a variety of configurations. It takes care of finding the correct gains for each one of these, and includes the important features mentioned above, dual-band decoding and near-field compensation. It is also important to note that specialised decoders can be written as UDOs, if any particular requirements that are not met by `bformdec1` are identified. Finally, a combination of decoding and the HRTF opcodes from Section 13.4.4 can be used to provide headphone listening of ambisonic audio. In listing 13.17, we see an example of this for horizontal-plane decoding of order-2 b-format signals.

Listing 13.17 An example of ambisonic to binaural decoding

```
/******************************************************
al,ar Bf2bi aw,ax,ay,az,ar,as,at,au,av
aw,ax ... = 2nd-order bformat input signal
===================================
adapted from example by Brian Carty
******************************************************/
opcode Bf2bi, aa, aaaaaaaaa
 aw,ax,ay,az,ar,as,at,au,av xin
 if sr == 44100 then
  Shl = "hrtf-44100-left.dat"
  Shr = "hrtf-44100-right.dat"
 elseif sr == 48000 then
  Shl = "hrtf-48000-left.dat"
  Shr = "hrtf-48000-right.dat"
 elseif sr == 96000 then
  Shl = "hrtf-96000-left.dat"
  Shr = "hrtf-96000-right.dat"
 else
  al, ar bformdec1 1,aw,ax,ay,az,ar,as,at,au,av
  goto end
 endif
 a1,a2,a3,a4,a5,a6,a7,a8 bformdec1 4,
          aw, ax, ay, az,
          ar, as, at, au, av
 al1,ar1 hrtfstat a2,22.5,0,Shl,Shr
 al2,ar2 hrtfstat a1,67.5,0,Shl,Shr
 al3,ar3 hrtfstat a8,112.5,0,Shl,Shr
 al4,ar4 hrtfstat a7,157.5,0,Shl,Shr
 al5,ar5 hrtfstat a6,202.5,0,Shl,Shr
 al6,ar6 hrtfstat a5,247.5,0,Shl,Shr
 al7,ar7 hrtfstat a4,292.5,0,Shl,Shr
 al8,ar8 hrtfstat a3,337.5,0,Shl,Shr
 al = (al1+al2+al3+al4+al5+al6+al7+al8)/8
 ar = (ar1+ar2+ar3+ar4+ar5+ar6+ar7+ar8)/8
 end:
        xout al,ar
endop
```

The separation between decoding and encoding can be very useful. For instance, composers can create the spatial design for their pieces independently of the particular details of any performance location. Different decoded versions can then be supplied for the specific conditions of each venue.

13.5.2 Vector Base Amplitude Panning

The simple amplitude panning of sources in stereo space (two channels), as implemented for instance by the `pan2` opcode, is extended for multiple channel setups by the vector base amplitude panning (VBAP) method [104]. In amplitude panning, the gains of each channel are determined so that the intensity difference (IID) causes the sound to be localised somewhere between the loudspeakers (the active region). VBAP uses a vector formulation to determine these gains based on the directions of the speakers and the intended source position. One important aspect is that, for a given source, localised at a given point, only two (horizontal plane only) or three (horizontal and vertical planes) loudspeaker channels will be employed at one time. In the former case (2D), the pair around the point, and in the latter case (3D), the triplet defining the active triangle within which the source is to be located will be used.

In Csound, VBAP is implemented by the `vbap` and `vbaplsinit` opcodes. The latter is used to define a loudspeaker layout containing the dimensions used (two or three), the number of speakers, and their directional location in angles with respect to the listener (or the centre of the circle/sphere):

```
vbaplsinit idim, inum, idir1, idir2[, idir3, ...]
```

The `vbap` opcode can then be used to pan a mono source sound:

```
ar1, ar2, [ar3,...] vbap asig,kazim[,kelev,
                              kspread,ilayout]
```

where `kazim` and `kelev` are the horizontal and vertical angles. An optional parameter `kspread` can be used to spread the source, causing more loudspeakers to be active (the default 0 uses only the pair or triplet relative to the active region). Note that loudspeaker positions can be set arbitrarily, and do not necessarily need to conform to regular geometries.

13.6 Conclusions

In this chapter, we have explored the classic techniques of time-domain processing in detail. We saw how the fundamental structure of the digital delay line is implemented, and the various applications of fixed delays, from echoes to component reverberators, such as comb and all-pass filters. The design of a standard reverb was outlined, and the principles of feedback delay networks and convolution reverb were examined.

The various applications of variable delays were also detailed. The common LFO modulation-based effects such as flanger, chorus, and vibrato were presented together with their reference implementation as UDOs. The principle of the Doppler effect and its modelling as a variable delay process was also explored. The discussion of this class of effects was completed with a look at a standard delay-based pitch-shifting algorithm.

The next section of the chapter was dedicated to the study of some aspects of filtering. We looked at an IIR design example, the second-order all-pass filter, and its application in the phaser effect. This was complemented with an introduction to equalisation, where a variable-band graphic equaliser UDO was discussed. The text concluded with the topic of FIR filters and the principles behind their design, with the application of some tools, such as the Fourier transform, that will be explored further in the next chapter. We also saw how these types of filters can be used to give spatial audio through binaural processing. An overview of basic techniques for multichannel audio completed the chapter.

Time-domain techniques make up an important set of processes for the computer musician's arsenal. They provide a great variety of means to transform sounds, which can be applied very effectively to music composition, sound design and production. Together with the classic synthesis methods, they are one of the fundamental elements of computer music.

Chapter 14
Spectral Processing

Abstract In this chapter, we will explore the fundamentals of the theory of frequency-domain processing of audio. The text begins by exploring the most relevant tools for analysis and synthesis: the Fourier transform, the Fourier series and the discrete Fourier transform. Following this, we look at some applications in filter design and implementation. In particular, we show how the fast partitioned convolution algorithm can take advantage of the theory outlined in the earlier sections. The chapter goes on to discuss the phase vocoder, with analysis and synthesis code examples, and its various transformation techniques. Finally, it is completed by an overview of sinusoidal modelling and ATS, and their implementation in Csound.

14.1 Introduction

Spectral processing of sound is based on the principle of manipulating frequency-domain representations of signals. Here, we are looking at data that is organised primarily as a function of frequency, rather than time, as in the case of the methods in previous chapters. Although the primary concern here is the modification of spectral information, we will also consider the time dimension as we process sounds with parameters that change over time. In many places, because of this combination of the two domains, these techniques are also called *time-frequency* methods.

In order to work in the frequency domain, it is necessary to employ an *analysis* stage that transforms the waveform into its spectrum [8, 90, 34]. Complementary, to play back a processed sound it is necessary to employ the reverse operation, *synthesis*, which converts spectra into waveforms. This chapter will first discuss the basic techniques for performing the analysis and synthesis stages, and the key characteristics of the frequency-domain data that they work with. Following this, we will introduce the application of these and the various methods of modifying and processing spectral data.

© Springer International Publishing Switzerland 2016   295
V. Lazzarini et al., *Csound*, DOI 10.1007/978-3-319-45370-5_14

14.2 Tools for Spectral Analysis and Synthesis

The spectrum of audio signals can be obtained through a number of methods. The most commonly used of these are derived from Fourier's theory [43, 17], which tells us that a function, such as a sound waveform, can be decomposed into separate sinusoidal functions (waves) of different amplitudes, frequencies and phases. Related to this fundamental principle there are a number of mathematical tools for analysis and synthesis called *transforms* that are variations with specific application characteristics.

The foundation for all of these is the original continuous-time, continuous-frequency Fourier transform. Closely related to this, we have the continuous-time, discrete-frequency Fourier series, which can be used to synthesise a periodic waveform from its spectrum. Another related variation is given by the discrete-time, continuous-frequency z-transform, commonly used in the spectral description of filters. Finally, we have the discrete-time, discrete-frequency Fourier transform (DFT), which is widely used in spectral processing, and the basis for other algorithms, such as the phase vocoder and sinusoidal modelling. In this section, we will outline the principles and characteristics of the relevant spectral analysis and synthesis tools.

14.2.1 Fourier Transform

The Fourier transform (FT) is the underlying mathematical method for all the processes discussed in this chapter [121]. It is defined as continuous in time t, and it covers all frequencies f, also continuously, from $-\infty$ to ∞ [90]:

$$X(f) = \int_{-\infty}^{\infty} x(t) \left[\cos(2\pi ft) - j\sin(2\pi ft)\right] dt \qquad (14.1)$$

It gives us a spectrum $X(f)$ from a waveform $x(t)$. It does this by multiplying the input signal by a complex-valued sinusoid at frequency f, and then summing together all the values of the function resulting from this product (this is indicated by the integral symbol ($\int$), which means in this case a sum over a continuous interval, from $-\infty$ to ∞).

The result is a pair of spectral coefficients, telling us the amplitudes of a cosine and a sine wave at that frequency. If there is none to be found, these will be zero. If a component is detected, then we might have a sinusoid in cosine or sine phase, or in between these two (this is why a complex multiplication is used, so that the exact phase can be determined). This operation can be repeated by sliding the sinusoid to another frequency, to obtain its coefficients.

In other words, we use a sinusoid as a detector. Once it is tuned to the same frequency as an existing spectral component, it gives a result. If it is not tuned to it, then the result is zero. This is the fundamental operating principle of the FT. While it is mostly a theoretical tool, there are some practical applications, for instance when we restrict the time variable to cover a single cycle of a periodic waveform.

Another important consideration here is that audio signals are real-valued (a waveform is single dimensional), and this determines the shape of the spectra we get from the FT. This fact means each component in a waveform is detected *both* at its negative and positive frequency values, with the amplitude of the sine coefficient with opposite signs on each side (the cosine coefficient is the same). This allows us to be able to construct the negative side ($f < 0$) from the positive one, thus in most cases we only need to work with non-negative frequencies.

Finally, the form of spectral data in amplitudes of cosines and sines is sometimes unwieldy to manipulate. We can convert these into the actual sinusoid amplitudes and phases with the simple relations

$$A = \sqrt{c^2 + s^2} \quad \text{and} \quad Ph = \arctan\left(\frac{s}{c}\right) \tag{14.2}$$

where A and Ph are the amplitude and phase, respectively, of the detected sinusoid, and c and s are its cosine and sine amplitudes. The collection of all detected amplitudes is often called the *magnitude* spectrum. The phases are similarly named. Note that the expressions in eq. 14.2 mean that an audio waveform will have a magnitude spectrum that is symmetric at 0, and a phase spectrum that is anti-symmetric. Again, this allows us to effectively ignore its negative-frequency side.

14.2.2 Fourier Series

The FT has an equivalent inverse transform, also time- and frequency-continuous, that makes a waveform from a given spectrum [121]. This has a form that is very similar to eq. 14.1, but with $X(f)$ as input and $x(t)$ as output. If, however, we constrain our spectrum to be discrete, then we have a variation that is a sum of (an infinite number of) sinusoids at frequencies $k = \{-\infty, ..., -2, -1, 0, 1, 2, ..., \infty\}$:

$$x(t) = \frac{1}{2\pi} \sum_{k=-\infty}^{\infty} X(k) \left[\cos(2\pi kt) + j\sin(2\pi kt)\right] \tag{14.3}$$

This is called the Fourier series (FS), where the continuous frequency f is restricted to a series of integers k. Here, the sinusoids are complex-valued, but the result is restricted to be real-valued (it is an audio waveform), and thus it can be simplified to a real-valued expression using only the non-negative coefficients:

$$
\begin{aligned}
x(t) &= \frac{1}{\pi} \left[\frac{a_0}{2} + \sum_{k=1}^{\infty} a_k \cos(2\pi kt) - b_k \sin(2\pi kt) \right] \\
&= \frac{1}{\pi} \left[\frac{a_0}{2} + \sum_{k=1}^{\infty} A_k \cos(2\pi kt + Ph_k) \right]
\end{aligned}
\tag{14.4}
$$

where a_k and b_k are the sine and cosine amplitudes of each frequency component f, and in the second form, A_k and Ph_k are the magnitudes and phases. A key characteristic of the FS is that its output is periodic, repeating every time t increases by 1. So eq. 14.4 describes a single cycle of a waveform with a possibly infinite number of harmonics. Its parameters can be determined by analysis, from an existing waveform, using the FT limited to the interval of a single cycle. A simplified version of the Fourier transform is used, for instance, in GEN 10 (eq. 12.2), whereas more complete versions are implemented in GEN 9 and GEN 19.

14.2.3 Discrete Fourier Transform

Both the FT and the FS are defined in continuous time, so for digital audio applications, which are discrete in time, they will remain somewhat theoretical. The continuous-frequency, discrete-time, variation, called the z-transform, is used in the study of digital filters, to determine their frequency response [97]. Although related, its applications are beyond the scope of this chapter.

The next step is to consider a transform that is discrete in time and frequency. This is called the discrete Fourier transform [65, 55, 56] and can be expressed by the following expression:

$$X(k) = \frac{1}{N} \sum_{t=0}^{N-1} x(t) \left[\cos(2\pi k t/N) - j \sin(2\pi k t/N) \right] \qquad k = 0, 1, ..., N-1 \quad (14.5)$$

Comparing it with eq. 14.1, we can see that here both time and frequency are limited to N discrete steps. Sampling in time makes our frequency content limited to $\frac{sr}{2}$. Given that the frequency spectrum is both negative and positive, the DFT will produce N values covering the full range of frequencies, $-\frac{sr}{2}$ to $\frac{sr}{2}$.

The first $\frac{N}{2}$ samples will correspond to frequencies from 0 to the Nyquist. The $\frac{N}{2}$ point refers both to $\frac{sr}{2}$ and to $-\frac{sr}{2}$. The second half of the DFT result will then consist of the negative frequencies continuing up to 0 Hz (but excluding this point). In Fig. 14.1, we see a waveform and its corresponding magnitude spectrum, with the full positive and negative spectrum. As indicated in Section 14.2.1, we can ignore the negative side, and there will be $\frac{N}{2} + 1$ pairs of spectral coefficients. They will correspond to equally spaced frequencies between 0 and $\frac{sr}{2}$ (inclusive).

We can think of each one of these frequency points as a band, channel, or bin. Here, instead of a sliding sinusoid, we have a stepping one, as a component detector. Another way to see this is that we are analysing one period of a waveform whose harmonics are multiples of $\frac{sr}{N}$ Hz, the fundamental analysis frequency. If the input signal is perfectly periodic within $\frac{N}{sr}$ secs, then the analysis will capture these harmonics perfectly (Fig. 14.1).

If we try to analyse a signal that does not have such characteristics, the analysis will be smeared, i.e. the component detection will be spread out through the various

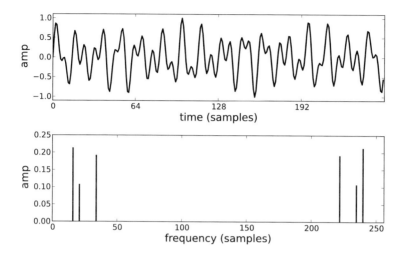

Fig. 14.1 A waveform with three harmonics (top) and its magnitude spectrum (bottom). The wave period is an integer multiple of the DFT size (*N*=256), giving a perfect analysis. Note that the full spectrum shows that the components appear in the positive and negative sides of the spectrum. The first half of the plot refers to positive frequencies (0 to $\frac{sr}{2}$), and the second, to negative ones (from $-\frac{sr}{2}$ to 0 Hz)

frequency points (or bins), as shown in Fig. 14.2. This is because the DFT will always represent the segment analysed as if it were periodic. That is, the input will always be modelled as made of components whose frequencies are integer multiples of the fundamental analysis frequency.

This means that if we apply the inverse DFT, we will always recover the signal correctly:

$$x(t) = \sum_{k=0}^{N-1} X(k) \left[\cos(2\pi kt/N) + j\sin(2\pi kt/N) \right] \qquad t = 0, 1, ..., N-1 \qquad (14.6)$$

Comparing eq. 14.6 with the FS in eq. 14.4, we can see that they are very similar. The IDFT can be likened to a bandlimited, complex-valued, version of the FS.

Windows

The smearing problem is an issue that affects the clarity of the analysis data. By looking at the cases where it occurs, we see that it is related to discontinuities created at the edges of the waveform segment sent to the DFT. This is because to do the analysis we are effectively just extracting the samples from a signal [49]. The process is called *windowing* [66], which is the application of an envelope to the signal.

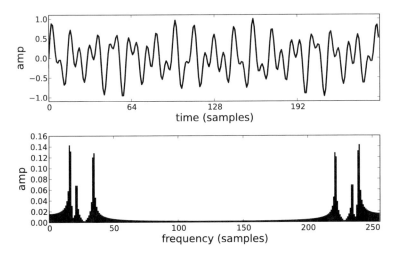

Fig. 14.2 A three-component waveform (top) and its magnitude spectrum (bottom). The wave period is not an integer multiple of the DFT size (*N*=256), resulting in a smeared analysis. The spectral peaks are located at the component frequencies, but there is a considerable spread over all frequency points

In this case, the window shape used is equivalent to a rectangular shape (i.e. zero everywhere, except for the waveform segment, where it is one).

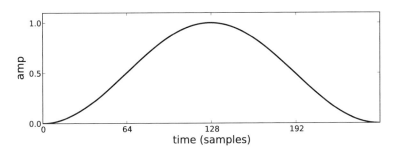

Fig. 14.3 A Hanning window, which is an inverted cosine, raised and scaled to fit in the range of 0 to 1

There are many different types of window shapes that allow us to minimise the smear problem. These will tend to smooth the edges of the analysed segment, so that they are very close to zero. A common type of window used in DFT analysis is the Hanning window. This shape is created by an inverted cosine wave, that is offset and scaled to range between 0 and 1, as seen in Fig. 14.3. Its application to an input waveform is shown in Fig. 14.4, reducing the amount of smearing by concentrating the magnitude spectrum around the frequency of its three components.

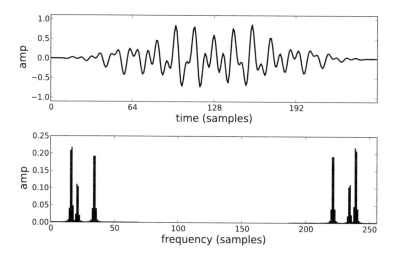

Fig. 14.4 A waveform windowed with a Hanning shape, and its resulting spectrum with a reduced amount of smearing

The fast Fourier transform

The DFT as defined by eqs. 14.5 and 14.6 can be quite heavy to compute with large window (segment) sizes. However, there are fast algorithms to calculate it, which exploit some properties of these expressions. These are called the *fast Fourier transform* (FFT), and they produce outputs that are equivalent to the original DFT equations. The classic FFT algorithm works with specific window lengths that are set to power-of-two sizes.

In addition, with audio signals, it is common to use transforms that are optimised for real-valued inputs. In Csound, these are represented by the following opcodes for forward (wave to spectrum) and inverse real-signal DFTs:

```
xSpec[]   rfft xWave[]
xWave[]   rifft xSpec[]
```

These both work with i-time and k-rate arrays. The xSpec[] will consist of N pairs of numbers containing the cosine and sine amplitudes for one for each non-negative point, except for 0 Hz and $\frac{sr}{2}$, which are amplitude-only (0 phase), and are packed together in array positions 0 and 1.

The full (negative and positive) spectrum DFTs are implemented by

```
xSpec[]   fft xWave[]
xWave[] fftinv xSpec[]
```

With these, both the inputs and outputs are expected to be complex.

Applications

An application of the DFT has already been shown in Section 13.4.3, where the inverse transform was employed to generate the FIR filter coefficients for an amplitude response designed in the spectral domain. In listing 13.16, we employed the `rifft` opcode to do a real-valued inverse discrete Fourier transform, and were able to get the IR that defined the desired filter amplitude response.

This use of the transform is enabled by the fundamental idea that a filter frequency response is the spectrum of its impulse response (Fig. 14.5). So it is also possible to a look at the amplitude curve of a filter by taking the DFT of its impulse response and converting it into a magnitude spectrum. We can do likewise with the phases, and see how it affects the different frequencies across the spectrum. This makes the DFT a very useful tool in the design and analysis of filters.

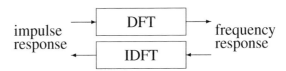

Fig. 14.5 The relationship between impulse response and frequency response of a digital filter

The DFT can also be used to determine the harmonic amplitudes and phases of any periodic waveforms. This allows us to be able to reproduce these by applying the Fourier series, or indeed the IDFT, to generate wavetables for oscillators. More generally, if we use a sequence of DFTs, we will be able to determine time-varying spectral parameters. For instance, a sequence of magnitude spectra taken at regular intervals yields a time-frequency representation called a *spectrogram*. In Fig. 14.6, we see a 2D plot of a sequence of magnitudes over time, where the darker lines indicate the spectral peaks. Another way to show a spectrogram is through a 3D graph called a *waterfall* plot (Fig. 14.7), where the three dimensions of frequency, time and amplitude are placed in separate axes.

14.3 Fast Convolution

Another direct application of the DFT that is related to filters is *fast convolution* [65, 90]. As we have seen in Chapter 13, direct calculation of convolution using

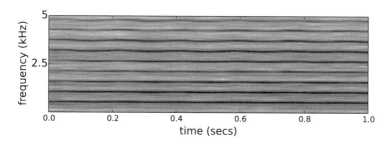

Fig. 14.6 The spectrogram of a C4 viola sound. The dark lines indicate the harmonic partials detected by the analysis

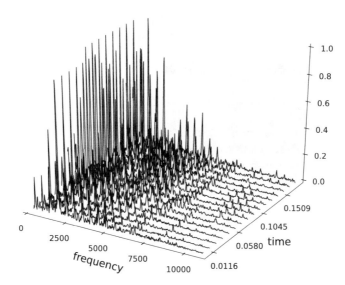

Fig. 14.7 A 3D plot of the spectrogram of a C4 viola sound, showing the sequence of magnitude spectra taken at regular interval.

tapped delay lines, multiplications and additions can be computationally expensive for large IR sizes. The DFT offers a fast method, taking advantage of the following principles:

1. The DFT of an IR corresponds to the FIR filter spectrum.
2. Time-domain convolution corresponds to frequency-domain multiplication.
3. The spectrum of an input signal can also be obtained with the DFT.

So, the operation can be implemented by analysing a signal and an IR using the DFT, multiplying them together, and transforming back the result using the IDFT

(Fig. 14.8). Since the transforms can be implemented with fast algorithms, for large IR sizes it is more efficient to use this method. The major difficulty to overcome is that in order to do this, we need to have the complete input signal before we could take its DFT. This is not practical for streaming and real-time applications, where the signal is available only on a sample-by-sample basis. It means we need to wait for the complete input signal to finish before we are able to hear the convolution, which also means an unacceptable latency between input and output.

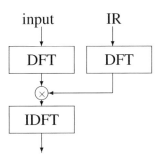

Fig. 14.8 Fast convolution: multiplication in the frequency domain is equivalent to convolution in the time domain

A solution is to slice the input signal to match the size of the IR, and take a sequence of DFTs, then overlap-add the result. This works well for arbitrary input sizes, so the operation can be performed in a streaming fashion. However, depending on the size of the IR, audible latency will remain. We can minimise this by further slicing both the input signal and the IR into a smaller size, applying a slightly different algorithm to reconstruct the output, called *partitioned convolution*.

In this case, we will need to balance the need for lower latency with the computational load, as smaller partitions are slower to calculate. Effectively this works as an in-between solution in that the larger the partitions, the closer we are to a full DFT-based fast convolution, and the smaller, the closer to a direct method (which would be equivalent to a partition size of one sample).

In Csound, the ftconv is a drop-in fast partitioned convolution replacement for dconv, in that, similarly, it reads an IR from a function table. We can set the partition size, which will determine the amount of initial delay. This opcode can deal in real time with much larger IRs than dconv, and it can be used to implement, for instance, convolution reverbs. Another alternative which takes an IR from a soundfile is pconvolve.

In fact, we can do much better by combining these two opcodes, ftconv and dconv. It is possible to completely avoid the DFT latency by using direct convolution to cover the first partition, then do the rest with the fast method [45]. To make this work, we have to split the IR into two segments: one containing the first P samples, where P is the partition size, and another segment with the remaining samples. Once this is done, we use dconv with the first part, and ftconv with the second, and mix the two signals.

This can be further enhanced by having multiple partition sizes, starting with a small partition, whose first block is calculated with direct convolution, and growing towards the tail of the IR, using the DFT method. An example with three partitions is shown in Fig. 14.9, where the N-sample IR is segmented into sections with 128, 896 and $N - 1,024$ samples each. The first section is always calculated with direct convolution (partition size = 1), followed in this case by a segment with seven partitions of 128 samples, and the rest using 1024-sample blocks.

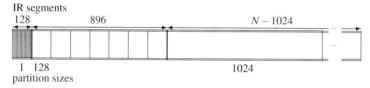

Fig. 14.9 Multiple-partition fast convolution, three partitions with sizes 1, 128, and 1,024. The N-sample IR is segmented into three sections, with 128, 896 and $N - 1,024$ samples each. In this case, the first partition size is 128 and the partition growth ratio is 8

The following example in listing 14.1 implements this scheme. It takes the first partition size `ipart`, a partition growth ratio `irat` and the total number of partition sizes `inp`. Each new partition size will be `irat` times the previous size. The audio signal is then processed first with direct and then with fast convolution to produce a zero-latency output, very efficiently. The size, number and ratio of partitions can be adjusted to obtain the best performance.

Listing 14.1 Convolution UDO using a combination of direct and fast methods, with multiple partition sizes

```
/ * * * * * * * * * * * * * * * * * * * * * * * * * * * * * * * * * * * * * * * * * * * *
asig ZConv ain,ipart,irat,inp,ifn
ain - input signal
ipart - first partition size in samples
irat - partition ratio
inp - total number of partition sizes
ifn - function table number containing the IR
* * * * * * * * * * * * * * * * * * * * * * * * * * * * * * * * * * * * * * * * * * * * * /
opcode ZConv,a,aiiiio
 asig,iprt,irat,inp,ifn,icnt xin
 if icnt < inp-1 then
  acn ZConv asig,iprt,irat,inp,ifn,icnt+1
 endif
 if icnt == 0 then
   a1 dconv asig,iprt,ifn
 elseif icnt < inp-1 then
   ipt = iprt*irat^(icnt-1)
```

```
    isiz = ipt*(irat-1)
    a1 ftconv asig,ifn,ipt,ipt,isiz
 else
    ipt = iprt*irat^(icnt-1)
    a1 ftconv asig,ifn,ipt,ipt
 endif
 xout a1 + acn
endop
```

14.4 The Phase Vocoder

The DFT is a single-shot analysis. If we step it over time, and take a sequence of regular transform frames, we can start to capture aspects of time-varying spectra that cannot be manipulated with just one window. This analysis is also called the short-time Fourier transform (STFT). Furthermore, we will be able to compare analysis data from successive frames to determine the time-varying frequency of each signal component. This is the basis for a practical implementation of the phase vocoder (PV) algorithm [42, 35].

The DFT can provide us with the magnitudes and phases of each analysis point (bin), via the expressions in eq. 14.2. However, this does not tell us directly what the actual frequencies of each component are, except in the trivial case where these are exact multiples of the fundamental analysis frequency. So we need to do more work. The key to this is to know that frequency is the time derivative of the phase. This can be translated to differences between phases from two successive DFT analysis frames.

The PV analysis algorithm transforms an audio signal into amplitude and frequency signals, and these can be converted back into the time domain through PV synthesis (Fig. 14.10). The analysis output consists of $\frac{N}{2} + 1$ bins (bands, where N is the DFT size), each with an amplitude-frequency pair at each time point of the analysis. These can be spaced by one or, more commonly, several samples. The spacing between each pair of time positions is called the analysis *hopsize*. PV bins will have a constant bandwidth, equivalent to $\frac{sr}{N}$ Hz. They will be centred at $0, \frac{sr}{N}, 2\frac{sr}{N}, ..., \frac{sr}{2}$ Hz.

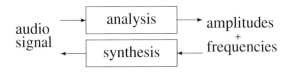

Fig. 14.10 Phase vocoder analysis and synthesis, which transform audio signals into their constituent amplitudes and frequencies and vice versa

The PV analysis algorithm can be outlined as follows [64]:

1. Apply a window of size N to an input signal:

```
kWin[] window kIn,krow*ihop
```

2. Take the DFT of this window, generating data for $2N + 1$ points (non-negative frequencies):

```
kSpec[] rfft kWin
```

3. Convert this data from cosine and sine amplitudes into magnitudes and phases:

```
kMags[] mags kSpec
kPha[] phs kSpec
```

4. Take the phase difference between the current and the previous analysis frame, bin by bin:

```
kDelta[] = kPha - kOlph
kOlph = kPha
```

5. Convert these differences to Hz:

```
kDelta unwrap kDelta
while kk < isize/2 do
    kPha[kk] = (kDelta[kk] + kk*iscal)*ifac
    kk += 1
od
```

6. Hop the window position by $\frac{N}{o}$ samples (hopsize), where o is the number of overlapped frames, and continue from the top.

The analysis output at each time point is a frame of amplitudes and frequencies for each analysis point. The 0 Hz and Nyquist frequency amplitudes are often packed together in the first two positions of the array (as in rfft). This data format is very easy to manipulate, as we will see in later sections. New data frames will be produced every $\frac{hopsize}{sr}$ s, which is the analysis period. In general, the hopsize should be no larger than $\frac{1}{4}$ of the DFT size, but it can be smaller to guarantee a better quality of audio ($\frac{1}{8}$ is a good choice).

The full listing of a PV analysis UDO is shown in listing 14.2. This code requires the hopsize to be an integral multiple of ksmps, to allow the shiftin opcode to correctly copy the input samples into an array.

Listing 14.2 Phase vocoder analysis opcode

```
/********************************************************
kMags[],kFreqs[],kflg PVA asig,isize,ihop
kMags[] - output magnitudes
kFreqs[] - output frequencies
kflg - new frame flag (1=new frame available)
asig - input signal
```

```
isize - DFT size
ihop - hopsize
************************************************/
opcode PVA,k[]k[]k,aii
 asig,isize,ihop xin
 iolaps init isize/ihop
 kcnt init 0
 krow init 1
 kIn[] init isize
 kOlph[] init isize/2 + 1
 ifac = (sr/(ihop*2*$M_PI))
 iscal = (2*$M_PI*ihop/isize)
 kfl = 0
 kIn shiftin asig
 if kcnt == ihop then
   kWin[] window kIn,krow*ihop
   kSpec[] rfft kWin
   kMags[] mags kSpec
   kPha[] phs kSpec
   kDelta[] = kPha - kOlph
   kOlph = kPha
   kk = 0
   kDelta unwrap kDelta
   while kk < isize/2 do
    kPha[kk] = (kDelta[kk] + kk*iscal)*ifac
    kk += 1
   od
   krow = (krow+1)%iolaps
   kcnt = 0
   kfl = 1
 endif
 xout kMags,kPha,kfl
 kcnt += ksmps
endop
```

Analysis data can be resynthesised by applying the reverse process:

1. Convert frequencies in Hz back into phase differences:

```
while kk < isize/2 do
    kFr[kk] = (kFr[kk] - kk*iscal)*ifac
    kk += 1
od
```

2. Integrate (add together) the phase differences to get the current phases:

```
kPhs = kFr + kPhs
```

3. Convert this data from magnitudes and phases into cosine and sine amplitudes:

```
kSpec[] pol2rect kMags,kPhs
```

4. Take the IDFT:

```
kRow[] rifft kSpec
```

5. Window, and overlap-add the data into the output stream:

```
kWin[] window kRow, krow*ihop
  kOut setrow kWin, krow
  kOla = 0
  kk = 0
  until kk == iolaps do
    kRow getrow kOut, kk
    kOla = kOla + kRow
    kk += 1
  od
```

6. Hop the window position by $\frac{N}{o}$ samples, where o is the number of overlapped frames.

The full listing of a PV analysis UDO is shown in listing 14.3. This is designed to work with the data generated by the PVA opcode. This code also requires the hopsize to be an integral multiple of ksmps to allow the shiftout opcode to perform the overlap-add operation correctly.

Listing 14.3 Phase vocoder synthesis opcode

```
/*******************************************************
asig PVS kMags[],kFreqs[],kflg,isize,ihop
kMags[] - input magnitudes
kFreqs[] - input frequencies
kflg - new frame flag (1=process new frame)
isize - DFT size
ihop - hopsize
*******************************************************/
opcode PVS,a,k[]k[]kii
 kMags[],kFr[],kfl,isize,ihop xin
 iolaps init isize/ihop
 ifac = ihop*2*$M_PI/sr;
 iscal = sr/isize
 krow init 0
 kOla[] init isize
 kOut[][] init iolaps,isize
 kPhs[] init isize/2+1
 if kfl == 1 then
  kk = 0
  while kk < isize/2 do
```

```
   kFr[kk] = (kFr[kk] - kk*iscal)*ifac
   kk += 1
  od
  kPhs = kFr + kPhs
  kSpec[] pol2rect kMags,kPhs
  kRow[] rifft kSpec
  kWin[] window kRow, -krow*ihop
  kOut setrow kWin, krow
  kOla = 0
  kk = 0
  until kk == iolaps do
   kRow getrow kOut, kk
   kOla = kOla + kRow
   kk += 1
  od
  krow = (krow+1)%iolaps
 endif
 xout shiftout(kOla)/iolaps
endop
```

These two opcodes implement streaming PV analysis and synthesis, to which modifications can be made on the fly. They demonstrate the process from first principles, for didactical purposes mostly. For practical applications, users should employ the internal opcodes pvsanal and pvsynth, which are equivalent, but are more efficient and convenient:

```
fsig   pvsanal asig,isize,ihop,iwinsize,iwintype
asig   pvsynth fsig
```

The input parameters to the analysis are input signal, DFT size, hop size, window size and window type. Window size can be larger than DFT size, but in most applications it is the same. There are a variety of available window shapes, the most commonly used being the Hanning, which is type 1. The analysis data is carried in a spectral type (f-sig), which conveniently wraps the spectral data, its description (DFT size etc.) and a framecount to allow other opcodes to operate correctly at the PV analysis rate. Streaming spectral signals can then be synthesised with pvsynth. There are no limitations as to the size of the hop with these opcodes. However, if the hopsize is less than ksmps, analysis and resynthesis is done sample by sample using the sliding DFT algorithm (which can be very expensive in computational terms). The PV algorithm is also used in other Csound opcodes, such as temposcal and mincer, which can be used for timescaling and pitch-shifting effects. The Csound utility pvanal also implements PV analysis, producing PVOCEX-format spectral files, which can be used with the streaming opcodes. A great variety of transformation techniques can be applied to PV data [75, 92].

Sliding phase vocoder

In the case of hopsizes that are smaller than the number of samples in a time-domain processing block (ksmps), or if it is, is very small (≤ 10 samples), `pvsanal` will switch to a special algorithm called the *sliding* phase vocoder [19]. This uses an iterative version of the DFT [18, 41], proposed originally by J. A. Moorer [94], which uses the fact that if we use a hopsize-1 transform, there will be a lot of redundancy between two consecutive analysis frames. The downside to this is that calculations have to be made on a sample-by-sample basis, so the process can be quite heavy on ordinary processors (although it is highly parallel, and has been implemented in graphics processing units (GPUs) to take advantage of this.[1]

On the other hand, the sliding algorithm produces a smoother result, and as it runs at the audio rate, it allows some processes to modulate the spectral data with an audio signal. In listing 14.4, we have an example of this, where the input to the sliding phase vocoder has its frequency modulated by an oscillator (using `pvshift`, see Section 14.4.1 below), whose is locked in a 1:2.33 ratio with the pitch detected in the input signal. Note that this code is unlikely to perform in real time (using current CPU technology), but can be rendered to file output.

Listing 14.4 Sliding phase vocoder frequency modulation

```
instr 1
 Sname = p5
 p3 = filelen(Sname)
 asig   diskin Sname
 kcps,kamp ptrack asig, 1024
 kcps port kcps,0.01
 amod oscili p4*kcps,kcps*2.33
 fs1 pvsanal asig,1024,1,1024,1
 fs2 pvshift fs1,amod,0
 ahr pvsynth fs2
 out ahr
endin
schedule(1,0,0,5,"cornetto.wav")
```

14.4.1 Frequency Effects

The phase vocoder allows frequencies to be manipulated in a variety of ways. For instance, we can scale it by a certain amount, which will result in pitch shifting. In this case, all frequencies in a PV analysis frame get multiplied by a scalar value. This is implemented by the `pvscale` opcode, which works with f-sigs:

---

[1] Csound opcodes are available for this, but they require specialist hardware in the form of specific GPU cards [69].

```
fsig  pvsanal fsigin,kscale
```

where `kscale` is the scaling value (pitch shift interval ratio), > 1 for upwards, and < 1 for downwards transposition. This process is based on multiplying the frequency data and then moving it to the correct bin, if necessary, as the scaling operation might place the new frequencies beyond their original bin bandwidth. For wider scaling with large transposition ratios, reducing the hopsize will improve the quality of the effect. The `pvscale` process is alias-free: upwards transpositions do not introduce frequencies beyond the Nyquist.

An alternative to `pvscale` is given by both the `mincer` and `temposcal` opcodes. These opcodes read audio data from a function table and can pitch scale it, producing an audio signal at the output. They employ a variant of the PV algorithm that includes *phase locking*, which can reduce some of the artefacts that may appear as a result of the process. Both opcodes transpose pitch using a different method to pvscale, by resampling in the time domain prior to the PV analysis and synthesis. Because of this, some care needs to be taken to avoid aliasing in upwards shift by employing a filter (such as an FIR designed as per Section 13.4.3), if necessary. These opcodes will be discussed in more detail in Section 14.4.5.

Another interesting PV effect is frequency shifting, which instead of scaling the data, offsets it by a given amount. This causes the spectrum to either stretch or compress, depending on the sign of the shift parameter. For instance, if we shift the frequencies by 150 Hz, and the input signal has a harmonic spectrum with a 440 Hz fundamental, the output partial frequencies will be 590, 1030, 1470,..., which make up an inharmonic spectrum. Frequency shifting is implemented by

```
fsig pvshift fsigin, kshift, klowest
```

where `kshift` is the frequency offset, and `klowest` is the lowest frequency affected by the process.

14.4.2 Formant Extraction

Pitch shifting of vocal sounds can suffer from a spectral distortion effect because all the amplitudes also get shifted along with the frequencies. This is true of all forms of the effect, regardless of how they are implemented. In order to fix this, we need to correct the amplitudes so that they are not distorted. This is done by extracting the *formants* of the input sound. Formants are regions of resonance in the spectrum. In the case of the voice, each different vowel sound will have characteristic formants, which are more or less fixed and do not change with fundamental frequency. When pitch shifting, these are also transposed, causing a noticeable distortion in the spectrum (the 'Donald Duck' effect). We call the overall contour of the amplitude spectrum the *spectral envelope*. Formants are 'bumps' or 'mountains' in the spectral envelope, each one with a specific centre frequency and bandwidth. The channel vocoder discussed in Section 12.2 depends on the presence of clear formant regions for its effect clarity.

Both `pvscale` and `pvshift` have optional working modes where formants are extracted and reapplied to the transposed sound, correcting the distortion caused by the frequency changes. These are particularly important if we want to create realistic harmonisation or pitch correction effects:

```
fsig pvsanal fsigin,kscale,ikeepform
fsig pvshift fsigin, kshift, klowest,ikeepform
```

There are three options for `ikeepform`: 0 for no effect, 1 selects plain cepstrum processing and 2 uses the true envelope method. Mode 1 is the least computationally expensive and generally performs well. Listing 14.5 shows a harmoniser using `pvscale` and formant correction. Note how we need to use a delay line to time align the direct and PV signals, as the PV imposes a small latency of $N + h$ samples, where N is the DFT size and h the hopsize.

Listing 14.5 Harmoniser example, with formant correction

```
instr 1
 isiz = 2048
 Sf = "cornetto.wav"
 p3 = filelen(Sf)
 asig diskin2 Sf,1
 fs1 pvsanal asig,isiz,isiz/8,isiz,1
 fs2 pvscale fs1,p4,1
 ahr pvsynth fs2
 adi delay  asig,isiz*1.125/sr
    out ahr*p6+adi*p5
endin
schedule(1,0,1,1.25,0.75,0.75)
schedule(1,0,1,.75,0,0.75)
```

The *cepstrum* is another important tool in frequency-domain processing. It is defined as the DFT of the log magnitude spectrum. It detects the undulations in the amplitudes of a DFT frame, where the wider ones will correspond to low-index cepstral coefficients and the narrower ones to high-index ones. If we think of the magnitude spectrum as a waveform, then we can visualise how this is the case: long fluctuations appear in the low part of the spectrum, short ones are to do with high frequencies.

Formants are detected by removing the narrow undulations in the amplitude spectrum, and keeping the wider ones that define resonance regions. With the cepstrum we can do this by *liftering* (removing) the high-order coefficients, then taking the inverse cepstrum to recover the spectral envelope with the formant regions. A plot demonstrating this result is shown in Fig. 14.11, where we can see a formant curve that has been obtained from an underlying spectrum using the cepstrum method. Once we have this, we can apply it to shape the amplitudes of the transposed signal.

The spectral envelope can also be manipulated independently of the frequencies, in case we want to deliberately distort it to change the character of an input sound. The `pvswarp` opcode is designed to do this:

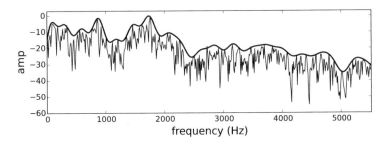

Fig. 14.11 The spectral envelope (thick line), obtained using the cepstrum method, and its underlying amplitude spectrum

```
fsig pvswarp fsigin, kscal, kshift
```

where `kscal` scales the spectral envelope, stretching (> 1) or compressing it (< 1), and `kshift` shifts it linearly by a certain offset (positive or negative).

14.4.3 Spectral Filters

It is possible to manipulate the amplitudes in the PV stream fairly freely. This allows us to construct time-varying filters, which will shape the spectrum in some way. There are a number of opcodes in Csound that allow f-sigs to be filtered. The first of these is a pair of band-pass and band-reject filters:

```
fsig pvsbandp fsigin, xlowcut, xlowfull,
                 xhighfull, xhighcut[,ktype]
fsig pvsbandr fsigin, xlowcut, xlowfull,
                 xhighfull, xhighcut[,ktype]
```

These filters pass or reject a certain trapezoid-shaped band in the spectrum, defined by `xlowcut`, `xlowfull`, `xhighfull` and `xhighcut`. The first two parameters determine the lower transition, and the other two, the higher transition bands, in Hz. These parameters are normally k-rate, but can work at audio rate if hopsize 1 is used. The optional `ktype` parameter can be used to determine the shape of the transition curve (defaults to linear). More generally, we can draw any shape on a function table and use it as the amplitude curve of a filter that is applied with

```
fsig pvsmaska fsrc, ifn, kdepth
```

Here, the function table `ifn` should have at least $\frac{N}{2} + 1$ points (one for each bin), with an arbitrary shape that will be used to filter `fsrc`. The amount of filtering ca be controlled dynamically with `kdepth`.

We can also use an arbitrary PV stream to filter another. This is very similar to the previous case, but now we are using an fsig as the (time-varying) amplitude curve for the filter. This is a generalisation of the idea of filtering, and the result is an emphasis on the common spectral elements between the two PV streams used (input and filter):

```
fsig pvsfilter fsigin, fsigfil, kdepth[, igain]
```

where `fsigfil` is the 'filter' used. For instance, a sine wave would work as a very narrow band-pass filter. Other, arbitrary, time-varying signals will have a variety of effects, making this process a rich field for continuous experimentation.

Another type of filtering that is possible to do with the phase vocoder is *stencilling*. This is done by comparing one signal to a mask, bin-per-bin, and changing its amplitude if it falls below that of the mask:

```
fsig pvstencil fsigin, kgain, klevel, ifn
```

The mask is provided by the function table `ifn`, with at least $\frac{N}{2}$ points. If the amplitude of a bin falls below the mask multiplied by `klevel`, then it is scaled by `kgain`. For noise reduction applications, a noise print function table can be created with GEN43. This reads PV data from a file and creates a single frame of amplitudes containing their average over the duration of the sound. However, the mask function table can be created with any arbitrary means for different types of amplitude transformation. The `kgain` parameter can also be set to > 1 for reverse-masking effects.

14.4.4 Cross-synthesis and Morphing

The discussion of filtering in the previous section touched on an important point for PV processing: the possibility of combining the characteristics of two spectra. This is generally called *cross-synthesis*, and there are a number of ways in which it can be achieved (including some of the filtering operations above). The first one of these is a straight combination of amplitudes from one PV stream with the frequencies of another:

```
fsig pvscross fsrc, fdest, kamp1, kamp2
```

In this case, depending on the values for `kamp1` and `kamp2`, the output signal will have the frequencies of `fsrc` and a mix of the amplitudes of `fsrc` and `fdest`. If `kamp2` = 0, no cross-synthesis will take place, and if `kamp1` = 0, then only the `fdest` amplitudes will be used (provided that `kamp2` is not zero).

A similar operation is provided by `pvsvoc`, but instead of the raw bin amplitudes, we apply the spectral envelope of one stream to the frequencies of another. The difference between the two can be appreciated in Fig. 14.11.

```
fsig pvsvoc famp, fexc, kdepth, kgain
```

Here, with `kdepth` = 1, the spectral envelope of `famp` will be combined with the frequencies of `fexc`. With lower `kdepth` values, less of the `fexc` signal is used. This opcode can produce results that are similar to the channel vocoder, but with a different character.

Finally, the `pvsmorph` opcode implements spectral interpolation of amplitudes and frequencies. Two signals are combined depending on the value of the interpolation parameters:

```
fsig pvsmorph fsig1, fsig2, kampint, kfrqint
```

The `kampint` and `kfrqint` arguments control linear amplitude and frequency interpolation, where 0 corresponds to `fsig1` and 1 to `fsig2`. In between these two values, it is possible to morph the two spectra. The seamlessness of the interpolation will depend on the similarity of the two inputs. If they have spectra whose components are not in overlapping regions, the effect of changing the interpolation parameters over time might sound more like a cross-fade. If, however, they have good correspondences in terms of their partials, the effect can be quite striking. A simple UDO demonstrating this application is shown in listing 14.6, where the spectra of two signals can be interpolated independently in amplitude and frequency.

Listing 14.6 Morphing UDO

```
/***************************************************
asig Morph ain1,ain2,kaint,kfint
ain1 - input signal 1
ain2 - input signal 2
kaint - amplitude interpolation (0 < kaint < 1)
kaint - frequency interpolation (0 < kfint < 1)
***************************************************/
opcode Morph,a,aakk
 as1,as2,ka,kf xin
 isiz = 2048
 ihop = isiz/8
 fs1 pvsanal as1,isiz,ihop,isiz,1
 fs2 pvsanal as2,isiz,ihop,isiz,1
 fsm pvsmorph fs1,fs2,ka,kf
 xout pvsynth(fsm)
endop
```

14.4.5 Timescaling

Timescaling is the effect of changing the duration of an audio signal without affecting its frequencies. As we know from the previous chapters, by reading a sound from a function table or a delay line at a speed that is different than the one we originally used to write the data we can change its pitch. With the phase vocoder, we have already seen that the pitch of the signal can be manipulated independently. It turns out that its duration can also be changed.

Pure timescaling is an inherently non-real-time operation. Although we can play back sounds and modify their durations on the fly, we need to have these stored somewhere previously (as we cannot predict the future or have infinite memory for past events). It is possible to record audio into a memory buffer, and then play it back at a different rate, but eventually we will either run out of input signal (if we compress the durations) or run out of memory (if we stretch them).

The PV process can stretch or compress audio data by playing it back at a different rate than it was written. Since what it uses is a series of amplitude-frequency frames, when the data is synthesised, it will not have its frequencies altered (unless we do so explicitly). Time and frequency are bundled together in a waveform. If we change one, we also modify the other. In the spectral domain, however, we can break the ties that link them together. A PV frame is a full description of a segment of a waveform. The playback rate of PV frames does not change this fact, and so, in general, a sound can have its duration modified without affecting its frequencies (within certain bounds, as some audible artefacts might appear in extreme stretching).

Timescaling is done by reading through a sequence of audio frames at different rates. This can mean repeating, or interpolating between, frames for stretching, and skipping frames for compression. The two function-table-reading opcodes `mincer` and `temposcal` both implement this process internally, doing a full analysis-synthesis cycle to produce a timescaled (and independently pitch-transposed) output. They implement a technique called *phase locking* that can reduce the artefacts found in extreme time stretching:

```
asig mincer atimpt, kamp, kpitch, ktab,
        klock[,ifftsize,idecim]
```

This opcode takes an audio-rate time position in secs (`atimpnt`) and plays the audio from function table `ktab` at that point. To make the sound play back at the original rate, we need to increment the time position linearly without scaling it. It is possible to go backwards and to scratch back and forth, or to freeze the sound at given points. There is total time flexibility with this opcode, and the pitch can be transposed independently. The parameter `klock` switches phase locking on or off, and we can control the DFT and hop size by changing the default parameters (2048 and 256, respectively):

```
asig temposcal ktimescal, kamp, kpitch, ktab,
        klock [,ifftsize, idecim, ithresh]
```

The `temposcal` opcode is a variation, which plays back the audio according to a timescaling parameter, 1 for no change, below 1 for stretching and above 1 for compression (like a speed control). In addition, it attempts to find attack transients in the audio so that these are not stretched (the time stretching is momentarily suppressed). This allows a better result in scaling the tempo of recordings, so that hits and note onsets are not smeared (Fig. 14.12). This is done by comparing the power of the audio in subsequent frames, looking for abrupt changes that might indicate attacks. The optional `ithresh` parameter can be used to change the detection threshold (1 dB).

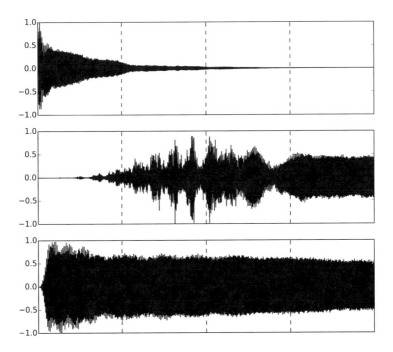

Fig. 14.12 A xylophone strike (top) and two time-stretched versions (10 x original duration). The middle one shows a clear smearing of the note onset, whereas the bottom one preserves the onset (produced by `temposcal`)

It is also possible to timescale streaming PV signals. We can do this by either reading from a table or a file, or by writing and reading from a memory buffer. In the first case, the opcode `pvstanal` is a variant on `temposcal` that produces fsigs as output. It has similar parameters and behaviour, with the exception that it does not perform phase locking, as this is not available for PV streams.

Phase vocoder data files can be generated with the utility `pvsanal`. This can be accessed via a dedicated menu option in some frontends (e.g. CsoundQt), and is available through the `-U pvsanal` command option:

```
csound -U pvsanal <input> <output.pvx> <options>
```

where <input> is the name of an input file in any of the formats accepted by Csound, and <output.pvx> is a file containing the PV data, which uses the special PVOCEX format. This is the PV file format used throughout Csound. A number of optional parameters (<options>) can be used to control the analysis. Details on these can be found in the Reference Manual. Such PV analysis files can be read using the opcodes

```
fsig pvsfread ktimpt, Sname [, ichan]
fsig pvsdiskin Sname,ktscal,kgain[,ioffset, ichan]
```

The difference between them is that the first one loads the whole file into memory and reads from there. The second one reads directly from the disk. Their parameters are also slightly different, with `pvsfread` using a time position to read from the file (`ktimpt`), and `pvsdiskin` using a timescaling factor (`ktscal`). In both opcodes, `Sname` is a string containing the name of the PVOCEX analysis file. They will output a PV stream that can be used with other opcodes, and synthesised with `pvsynth`. These opcodes use interpolation for timescaling, so the resulting output is somewhat different from `mincer` and `temposcal`. In addition, there is no phase locking of PV streams, so artefacts might result in very long stretching. However, these can also be explored creatively in the design of new sounds.

Finally, there are a pair of opcodes that set up a memory buffer write/read scheme for PV streams. These are

```
ihandle, ktime  pvsbuffer fsig, ilen
fsig pvsbufread  ktime, khandle
```

The first opcode sets up a circular buffer `ilen` seconds long and writes `fsig` to it. Its outputs are a reference to the circular buffer (`ihandle`) and the current write time position into the buffer. The reading opcode reads from the buffer referred to by `khandle`, at `ktime`, producing `fsig`. Note that its time position is completely independent from the write time, and that it can be decremented (for backwards playback). Any number of readers can refer to one single circular buffer (single writer, many readers). The writing `ktime` can be used as an indicator of the current time for the readers.

It is important to point out that as the buffer memory is finite, time compressing will eventually make the reader go beyond the range of the buffer, and also overtake the writing position. Time stretching can cause the writing to overtake the reading position. In both cases, the results will depend on the input sound. In any case, reading wraps around the ends of the buffer, in both directions. These opcodes can be used for a range of different real-time PV stream time manipulations.

14.4.6 Spectral Delays

The `pvsbuffer` opcode can also be thought of as a spectral delay line, with any number of taps implemented by `pvsbufread`. In addition, the reading can be done in specific frequency bands, so different delay times can be set for different areas of the spectrum. This is enabled by two optional arguments, `ilo` and `ihi`, which define the frequency band to be read:

```
fsig pvsbufread  ktime, khandle[, ilo, ihi]
```

This characteristic is further enhanced by `pvsbufread2`, which allows the user to set a per-bin delay time for amplitudes and frequencies, independently. Thus with a single opcode, multiple delay times can be achieved for different spectral bands, down to a bin bandwidth resolution:

```
fsig pvsbufread2  ktime, khandle, ift1, ift2
```

The function tables `ift1` and `ift2` contain the delay times in seconds for each one of the bins, for amplitude and frequency, respectively. They should be at least $\frac{N}{2}+1$ positions long. These opcodes can be used to create many effects where delays are attached to different areas of the spectrum, de-synchronising the PV stream. Listing 14.7 shows the effect applied to the viola sound of Fig. 14.6. Its spectrogram is shown in Fig. 14.13. Note how the function table is used to delay frequencies in ascending order from bin 0 to 128 (5,512.5 Hz).

Listing 14.7 Spectral delay example

```
ifn ftgen 1,0,514,7,0,128,1,256,1,128,1
instr 1
 Sf = "violac3.wav"
 p3 = filelen(Sf)
 a1 diskin2 Sf,1
 fs1 pvsanal a1,1024,128,1024,1
 ih,kt pvsbuffer fs1, 2
 fs2 pvsbufread2 kt,ih,1,1
 a2 pvsynth fs2
    out a2
endin
schedule(1,0,1)
```

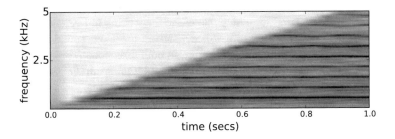

Fig. 14.13 The C4 viola sound (Fig. 14.6) played through a spectral delay with delay times that increase with frequency. It is possible to see how it makes the harmonics arpeggiate as their onset is spread out in time

14.4.7 Miscellaneous Effects

A number of non-standard frequency data manipulation effects can be applied to PV streams:

- Spectral arpeggios: the `pvsarp` opcode transforms the amplitudes of an input PV stream by boosting one bin and attenuating all the others around it. It can be used with an LFO to create partial arpeggiation effects.
- Blurring: a PV stream can be blurred by averaging out its amplitude and frequency over time using `pvsblur`.
- Demixing: `pvsdemix` takes a stereo signal and attempts to extract mixed sources by searching positions using a reverse pan pot.
- Freezing: it is possible to freeze an input signal at a given point using `pvsfreeze`.
- Mixing: the `pvsmix` opcode can be used to do an ultra-seamless mix by combining the loudest bins from two different inputs.
- Smoothing: the frequency and amplitude parts of a PV stream can be smoothed with a first-order low-pass filter through the use of `pvsmooth`.
- Signal generation: `pvsosc` produces a variety of audio waveforms directly in the spectral domain.

In addition to these, PV streams can be written and read to/from tables via the `pvsftw` and `pvsftr` opcodes, and to/from arrays with `pvs2array` and `pvsfromarray`. The pitch of a signal can be tracked with `pvspitch`, as well as its centroid, with `pvscent`. We can read individual bins with `pvsbin`, and display the data using `pvsdisp` (this will also depend on frontend implementation). The streaming phase vocoder subsystem in Csound is a rich source of opportunities for the exploration of spectral manipulation of audio.

14.5 Sinusoidal Modelling

Another approach to spectral processing is to model a sound as a sum of time-varying sinusoidal tracks [74, 9].This, in contrast to the DFT-frame approach of the phase vocoder, will identify peaks in the spectrum, over a certain time, and link them to make continuous lines. Each one of these will be modelling a sinusoid, with a variable amplitude, frequency and phase [85]. So, for instance, while in PV analysis a single-component glissando would be detected at various bins in successive frames, here it will create a single track.

The spectral data, in this case, is a collection of tracks. These can be provided as part of a streaming process in the same way as in the phase vocoder. However, the number of tracks will be variable, as some can die off, and new ones can appear as the sound changes over time. The most common way of reconstructing the time-domain audio signal from these is to use additive synthesis. One of the advantages of the sinusoidal method is that it is possible to keep the phase information intact while resynthesising, which is not the case with the phase vocoder, where phases are discarded.

The central component of sinusoidal modelling is partial tracking. This is done by first searching for peaks in the spectrum, determining their position and then matching them with previously detected ones in an earlier time point. The process is done successively in time, spaced by a hopsize. So a track will emerge from a peak, and if a continuation is found at the following time point, it will be kept. This is done until no connection is found, and the track is killed. Tracks are linked by frequency and amplitude proximity. If too much of a change is found in either, two peaks will not be connected as a track. The analysis can require a number of points to exist before a track is allowed to exist, as well as small gaps in its continuation. A plot of a series of frequency tracks of a piano sound is shown in Fig. 14.14.

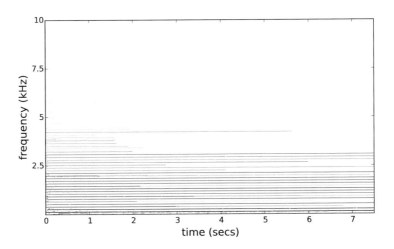

Fig. 14.14 Partial frequencies of a piano note recording as tracked by a sinusoidal model. Amplitudes are shown in grey scale (set according to the maximum amplitude of each track)

Csound provides a suite of streaming sinusoidal modelling opcodes, for analysis transformation, and synthesis. The process of creating the sinusoidal tracks can be outlined as follows:

1. An initial analysis step produces frames of amplitudes, frequencies, and phases from an input signal. This is done in Csound by applying the Instantaneous Frequency Distribution (IFD) [1, 44] to provide a frame of frequencies. This is also based on the DFT, but uses a different method to the phase vocoder. The process also yields amplitudes and phases for the same analysis frame. The opcodes pvsifd or tabifd are used for this:

```
ffr,fph pvsifd  asig, isize, ihop, iwin
ffr,fph tabifd  ktimpt,kamp,kpitch,isize,ihop,iwin,ifn
```

Each one produces a pair of f-sig outputs, one containing amplitudes and frequencies (PVS_AMP_FREQ format), and another containing amplitudes and phases (PVS_AMP_PHASE). The main difference between them is that `pvsifd` takes an audio input signal, whereas `tabifd` reads from a function table.

2. With these, we can run the tracking of sinusoidal partials that makes up the model:

```
ftrks partials   ffr,fphs,kthresh,imin,igap
```

The opcode takes two f-sigs containing amplitude, frequency and phase, as produced by `pvsifd`, and outputs an f-sig containing sinusoidal track data (PVS_TRACKS format). This is a different format to a normal PV stream and will only work with track-manipulating opcodes. The input parameters are as follows: `kthresh`, an amplitude threshold used to control the partial tracking, where partials below this will not be considered. It is defined as a fraction of the loudest partial. The `imin` determines the minimum number of detected peaks at successive time points that will make up a track. For instance, `imin = 1` will make every peak make a track, higher values will imply waiting to see whether a track emerges from a sequence of peaks. The `igap` parameter allows for a number of gaps to exist in tracks before it is defined as dead.

Streaming sinusoidal modelling consists of connecting these two opcodes to obtain the track data. This can be further transformed, and then resynthesised using additive synthesis.

14.5.1 Additive Synthesis

The additive resynthesis of sinusoidal tracks can be performed in various ways. Csound implements three opcodes for this, which have slightly different characteristics:

1. `sinsyn`: this synthesises tracks very faithfully, employing the phases in a cubic interpolation algorithm to provide very accurate reconstruction. However, it is not possible to scale frequencies with it. It is also slower than the other opcodes.
2. `resyn`: this opcode uses phases from track starting points, and cubic interpolation, allowing frequency scaling.
3. `tradsyn`: an amplitude-frequency-only, linear-interpolation, additive synthesiser. It does not use phases, and it is more efficient than its counterparts, although not as accurate.

The first opcode should be used whenever a precise resynthesis of the partial tracks is required. The other two can be used with transformed data, but `tradsyn` is the more flexible of these, as it does not require correct phases to be preserved anywhere in the stream.

In addition to these opcodes, it is also possible to convert the partial track stream into a PV signal using the `binit` opcode:

```
fsig binit fin, isize
```

This converts the tracks into an equal-bandwidth bin-frame signal
(PVS_AMP_FREQ) with DFT size `isize`, which can be reconstructed with overlap-
add synthesis (`pvsynth`). It can also be further processed using streaming PV opcodes. The conversion works by looking for suitable tracks to fill each frequency
bin. If more than one track fits a certain bin, the one with highest amplitude will be
used, and the other(s) discarded. PV synthesis can be more efficient than the additive
methods, depending on the number of partials required.

14.5.2 Residual Extraction

Modelling in terms of sinusoids tends to limit the spectrum to stable components.
More transient and noisy parts are somewhat suppressed. This allows us to separate
the sinusoidal tracks from these other components, which are called the analysis
residual. The separation is done by employing `sinsyn` to reproduce the track data
accurately, and then performing a time-domain subtraction from the original signal
of its resynthesis. The result is what was not captured by the sinusoidal modelling.
This is implemented in listing 14.8.

Listing 14.8 Residual extraction using sinusoidal modelling of an input signal

```
/*****************************************************
ares,asin Residual ain,kthresh,isize,ifcos
ares - residual output
asin - sinusoidal output
kthr - analysis threshold
isize - DFT size
ihop - hopsize
ifcos - function table containing a cosine wave
*****************************************************/
opcode Residual, aa, akiii
 ain,kthr,isiz,ihsiz,ifcos  xin
 idel = isiz-ihsiz*(isiz/(2*ihsiz)-1)
 ffr,fphs pvsifd    ain, isiz, ihsiz, 1
 ftrk partials ffr, fphs,kthr, 1, 1, 500
 aout sinsyn    ftrk, 2, 500, ifcos
 asd  delayr   idel/sr
 asig deltapn  idel
      delayw   ain
 aenv linsegr  0,idel/sr,0,1/sr,1,1,1
 xout aout*aenv-asig,aout
endop
```

This code provides both the residual and the sinusoidal signals as output. The
sinusoidal modelling process places a small latency equivalent to $N - h\left(\frac{N}{2h} - 1\right)$,

where N is the DFT size, and h, the hopsize. So in order to align the signals (and more importantly, their phases) correctly, we need to delay the input signal by this amount. The `sinsyn` opcode also adds a small onset to the signal, starting before the actual tracked sinusoids start playing, and so we apply an envelope to avoid any bleed into the residual output. It is also important to supply the opcode with a cosine wavetable, instead of a sine, otherwise the phases will be shifted by a quarter of a cycle. Such a table can be created with GEN 9:

```
ifn ftgen 1,0,16384,9,1,1,90
```

The `kthresh` parameter can be used to adjust the tracking (a value of the order of 0.003 is a good start). The computational load will depend on both the input signal and the threshold value. An input with many components will increase the resource requirements, and reducing the `kthresh` parameter will also produce more partials. In Fig. 14.15, we can see the waveform plot of a piano note sinusoidal model and its residual. This latter captures very well the moment the hammer strikes the string at the note onset.

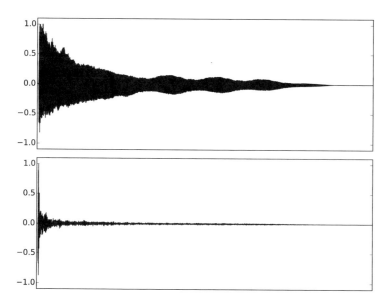

Fig. 14.15 The normalised plots of the sinusoidal model (top) and residual (bottom) of a piano sound. Note how the energy of the residual part is mostly concentrated at the note onset

14.5.3 Transformation

Various transformation opcodes are present in Csound to manipulate sinusoidal tracks. They will take and produce f-sig variables in the PVS_TRACKS format, and will work in a similar fashion to the streaming PV opcodes.

trcross: this opcode takes two track signals and performs cross-synthesis of amplitudes and frequencies. Note that this works in a different way to PV cross-synthesis, as a search for matching tracks is used (since there are no bins to use for matching):
trfilter: this implements track filtering using an amplitude response taken from a function table.
trhighest: the highest-frequency track is extracted from a streaming track input signal.
trlowest: similarly, the lowest-frequency track is extracted.
trmix: this opcode mixes two partial track streams.
trscale: frequency scaling (transposition) of tracks.
trshift: frequency shifting (offsetting) of partials.
trsplit: this splits tracks into two streams depending on a k-rate frequency threshold.

Transformations using partial track data can be very different from other spectral processes. For certain processes, such as filtering and selecting sub-sets of partials, this data format is very flexible. Also, due to the way the data is stored, synthesis with limited numbers of tracks will provide a non-linear filtering effect, resulting in suppression of the shortest-living tracks.

14.6 Analysis Transformation and Synthesis

Each sound can be located on a continuous scale between harmonic and noisy. The discrete Fourier transform is based upon the paradigm of harmonicity, so consequently it has weaknesses in analysing and resynthesising noisy sounds or noisy parts of natural sounds. In the previous section, sinusoidal modelling has been explained as one method to overcome this restriction. Analysis Transformation and Synthesis (ATS) is another well-established method for the same goal [99]. It separates the sinusoidal (deterministic) part of a sound from its noisy (stochastic or residual) part and offers different methods of resynthesis and transformation.

14.6.1 The ATS Analysis

Like in the other techniques which are described in this chapter, the analysis part of ATS starts with performing a short-time Fourier transform (STFT). As real-time analysis is beyond the scope of ATS, there is no particular need for the efficiency of the power-of-two FFT algorithm. This means that frames can be of any size, which can be useful for some cases.[2] Following the STFT, spectral peaks are detected for each frame. Very similarly to the partial-tracking technique, these peaks are now compared to each other, to extract spectral trajectories. The main difference can be seen in the application of some psychoacoustic aspects, mainly the Signal-to-Mask Ratio (SMR). Spectral peaks are compared from frame to frame. For creating a track, they need to fall into a given maximum frequency deviation, and also accomplish other requirements, like not to be masked. The minimal number of frames required to form a track can be set as an analysis parameter (default = 3), as can the number of frames which will be used to "look back" for a possible peak candidate. Tracks will disappear if they cannot find appropriate peaks any more, and will be created if new peaks arise.

These tracks are now resynthesised by additive synthesis, taking into also phase information. This is important because the main goal of this step is to subtract this deterministic part in the time domain from the signal as a whole (a process similar to the one in listing 14.8 above). The residual, as the difference of the original samples and its sinusoidal part, represents the noisy part of the signal. It is now analysed using the Bark scale, which divides (again because of psychoacoustic reasons) the range of human hearing into 25 critical bands.[3] Noise analysis is performed, to get the time-varying energy in each of the 25 bands of a Bark scale. This is done by STFT again, using the same frame size as in the analysis of the deterministic part. The full analysis process is outlined in Fig. 14.16. So, for the ATS resynthesis model, sinusoidal and noise components can be blended into a single representation as sinusoids plus noise-modulated sinusoids [99]:

$$P_{synth} = P_{amp}P_{sin} + P_{noi}P_{sin} \tag{14.7}$$

where P_{synth} is a synthesised partial, P_{sin} its sinusoidal component, P_{amp} its amplitude and P_{noi} its noise component.

In MUSIC N, P_{noi} is defined as `randi` because it can be represented by a linear-interpolation random generator. As it generates random numbers at a lower rate, it can produce band-limited noise. The opcode `randi` produces a similar output, so this formula is nearly literally Csound code:

$$P_{noi} = randi(P_E, P_{bw}) \tag{14.8}$$

where P_E is the time-varying energy of the partial, and P_{bw} represents the spectral bandwidth, used as the rate of the `randi` generator.

---

[2] For instance, if we know the frequency of a harmonic sound, we can choose a window size which is an integer multiple of the samples needed for the base frequency.

[3] See the example in listing 14.12 for the Bark frequencies.

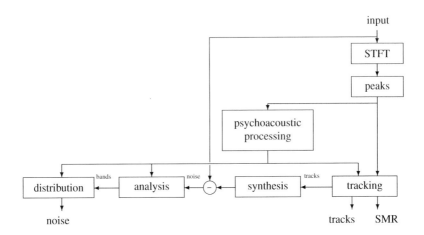

Fig. 14.16 ATS analysis [99], taking in an input audio signal and producing noise in 25 perceptual bands, sinusoid tracks containing amplitude, frequency and phase, and the signal-to-mask ratio (SMR)

14.6.2 The ATS Analysis File Format

Both parts of the analysis are stored in the analysis file. The sinusoidal trajectories are stored in the usual way, specifying the amplitude, frequency and phase of each partial. In the corresponding frame, the residual energy is analysed in 25 critical bands which represent the way noisy sounds are recognised in pitch relations. So one frame stores this information for N partials:

```
amp 1 freq 1 phase 1
amp 2 freq 2 phase 2
. . .
amp N freq N phase N
residual energy for critical band 1
residual energy for critical band 2
. . .
residual energy for critical band 25
```

The number of partials very much depends on the input sound. A sine wave will produce only one partial, a piano tone some dozen, and complex sounds some hundreds, depending on the analysis parameters.

In Csound, ATS analysis is performed with the utility ATSA. The various analysis parameters differ from the usual STFT settings in many ways and should be tweaked by the analysis to really "meet" an individual sound. They have been described implicitly above and are in detail described in the manual. Not setting these sixteen parameters but using the defaults instead, the command line for analysing the "fox.wav" looks like this:

```
$ csound -U atsa fox.wav fox.ats
```

The content of the "fox.ats" analysis file can be queried by the `ATSinfo` opcode, which returns:

```
Sample rate = 44100 Hz
Frame Size = 2205 samples
Window Size = 8821 samples
Number of Partials = 209
Number of Frames = 58
Maximum Amplitude = 0.151677
Maximum Frequency = 15884.540555 Hz
Duration = 2.756667 seconds
ATS file Type = 4 (footnote)
```

The default value of 20 Hz for the minimum frequency leads to the huge frame size of 2,205 samples, which in the case of "fox.ats" results in smeared transitions. Setting the minimum frequency to 100 Hz and the maximum frequency to 10,000 Hz has this effect:

```
$ csound -U atsa -l 100 -H 10000 fox.wav fox.ats
```

```
Frame Size =     441 samples
Window Size =   1765 samples
Number of Partials = 142
Number of Frames = 278
Maximum Amplitude = 0.407017
Maximum Frequency = 9974.502118 Hz
```

The time resolution is better now and will offer better results for the resynthesis. We will use this file ("fox.ats") in the following examples as basic input to show some transformations.

14.6.3 Resynthesis of the Sinusoidal Part

There are basically two options to resynthesise the deterministic part of the sound: `ATSread` and `ATSadd`. The opcode `ATSread` returns the amplitude and frequency of one single partial. This is quite similar to using the `pvread` opcode (or for streaming use, the `pvsbin` opcode), which returns a single FFT bin. The difference between the two can be seen clearly here: the frequency of the bin moves rapidly, whilst a partial in ATS stays at a certain pitch, just fading in or out.[4]

---

[4] The maximum frequency deviation is $\frac{1}{10}$ of the previous frequency as default.

```
kfreq, kamp ATSread ktimepnt, iatsfile, ipartial
```

The following example uses the frequency and amplitude output of ATSread to create an arpeggio-like structure. The frequencies of the tones are those from partials 5, 7, 9, ..., 89. The start of one note depends on the maximum of a single partial track. This maximum amplitude is queried first by the check_max instrument and written to the appropriate index of the gkPartMaxAmps array. A note is then triggered by p_trigger when this maximum minus idBdiff (here 2 dB) is crossed. In case this threshold is passed again from below later, a new tone is triggered. The result will very much depend on the threshold: for idBdiff = 1 we will only get one note per partial, for idBdiff = 6 already several. Both check_max and p_trigger are called by partial_arp in as many instances as there are partials to be created. The percussive bit, played by the pling instrument, is nothing more than one sample which is fed into a mode filter. The amplitude of the sample is the amplitude returned by ATSread at this point. The frequency is used as the resonant frequency of the filter, and the quality of the filter is inversely proportional to the amplitude, thus giving the soft tones more resonance than the stronger ones.

Listing 14.9 ATS partials triggering an arpeggio-like structure

```
gS_ats = "fox.ats"
giDur ATSinfo gS_ats, 7 ;duration
giLowPart = 5 ;lowest partial used
giHighPart = 90 ;highest partial used
giOffset = 2 ;increment
gkPartMaxAmps[] init giHighPart+1

instr partial_arp
 p3 = giDur*5 ;time stretch
 iCnt = giLowPart
 while iCnt <= giHighPart do
  ;check maximum in each partial and write in array
  schedule "check_max", 0, 1, iCnt
  ;call instr to play a note if threshold is crossed
  schedule "p_trigger", 0, p3, iCnt
  iCnt += giOffset
 od
 ;create time pointer for all instances of p_trigger
 gkTime line 0, p3, giDur
endin

instr check_max
 kMaxAmp, kTim init 0
 while kTim < giDur do
   kFq,kAmp ATSread kTim, gS_ats, p4
   kMaxAmp = kAmp > kMaxAmp ? kAmp : kMaxAmp
   kTim += ksmps/sr
```

```
  od
  gkPartMaxAmps[p4] = dbamp(kMaxAmp)
  turnoff
endin

instr p_trigger
 idBdiff = 2
 kMax = gkPartMaxAmps[p4] ;get max of this partial (dB)
 kPrevAmp init 0
 kState init 0 ;0=below, 1=above thresh
 kFq,kAmp ATSread gkTime, gS_ats, p4
 if kAmp > kPrevAmp &&
    dbamp(kAmp) > kMax-idBdiff &&
    kState == 0 then
  event "i", "pling", 0, p3*2, kFq, kAmp
  kState = 1
 elseif kAmp < kPrevAmp &&
         dbamp(kAmp) < kMax-idBdiff &&
         kState == 1 then
  kState = 0
 endif
 kPrevAmp = kAmp
endin

instr pling
 aImp mpulse p5*3, p3
 aFilt mode aImp, p4, 1/p5*50
 out aFilt
endin
schedule("partial_arp",0,1)
```

Rather than creating a number of sound generators on our own, we can use the opcode ATSadd to perform additive resynthesis by driving a number of interpolating oscillators. We can choose how many partials to use, starting at which offset and with which increment, and a gate function can be applied to modify the amplitudes of the partials.

```
ar ATSadd ktimepnt, kfmod, iatsfile, ifn, ipartials
              [,ipartialoffset, ipartialincr, igatefn]
```

Like ATSread and similar opcodes, ktimepnt requires a time pointer (in seconds) and iatsfile the ATS analysis file. The kfmod parameter offers a transposition factor (1 = no transposition), and ipartials is the total number of partials which will be resynthesised by a bank of interpolating oscillators. Setting ipartialoffset to 10 will resynthesise starting at partials 11, and ipartialincr = 2 will skip partial 12, 14, ... The optional gate function igatefn scales the amplitudes in the way that the x-axis of the table represents analysed amplitude values (normalised 0 to 1), and the y-axis sets a multiplier which is applied. So this table will leave the analysis results untouched:

```
giGateFun ftgen  0, 0, 1024, 7, 1, 1024, 1
```

The next table will only leave partials with full amplitude unchanged. It will reduce lower amplitudes progressively, and eliminate partials with amplitudes below 0.5 (-6 dB relative to the maximum amplitude):

```
giGateFun ftgen  0, 0, 1024, 7, 0, 512, 0, 512, 1
```

The next table will boost amplitudes from 0 to 0.125 by multiplying by 8, and then attenuate partials more and more from 0.125 to 1:

```
giGateFun ftgen  0, 0, 1024, -5, 8, 128, 8, 896, 0.Q01
```

So, similarly to the pvstencil opcode, a mask can be laid on a sound by the shape of this table. But unlike pvstencil, this will only affect the sinusoidal part.

14.6.4 Resynthesis of the Residual Part

The basic opcodes to resynthesise the residual part of the sound are ATSreadnz and ATSaddnz. Similarly to the ATSread/ATSadd pair, ATSreadnz returns the information about one single element of the analysis, whilst ATSaddnz offers access to either the whole noisy part, or a selection of it. The main difference to ATSread is that we now deal with the 25 noise bands, instead of partials.

The ATSreadnz opcode returns the energy of a Bark band at a certain time:

```
kenergy ATSreadnz ktimepnt, iatsfile, iband
```

The kenergy output represents the intensity; so following the relation $I = A^2$ following $A = \sqrt{I}$ we will use the square root of the kenergy output to obtain proper amplitude values. This code resynthesises the noise band using randi.

Listing 14.10 ATS resynthesis of one noise band

```
gS_ats = "fox.ats"
giDur ATSinfo gS_ats, 7 ;duration

instr noise_band
 iBand = 5 ;400..510 Hz
 p3 = giDur
 ktime line 0, giDur, giDur
 kEnergy ATSreadnz ktime, gS_ats, iBand
 aNoise randi sqrt(kEnergy), 55
 aSine poscil .25, 455
 out aNoise*aSine
endin
schedule("noise_band",0,1)
```

And this resynthesises all noise bands together:

Listing 14.11 ATS resynthesis of all noise bands in standard manner

```
gS_ats = "fox.ats"
giDur ATSinfo gS_ats, 7
giBark[] fillarray 0,100,200,300,400,510,630,770,920,
                1080,1270,1480,1720,2000,2320,2700,
                3150,3700,4400,5300,6400,
                7700,9500,12000,15500,20000

instr noise_bands
 p3 = giDur
 gktime line 0, giDur, giDur
 iBand = 1
 until iBand > 25 do
  iBw = giBark[iBand] - giBark[iBand-1]
  iCfq = (giBark[iBand] + giBark[iBand-1]) / 2
  schedule "noise_band", 0, giDur, iBand, iBw, iCfq
  iBand += 1
 od
endin

instr noise_band
 kEnergy ATSreadnz gktime, gS_ats, p4
 aNoise randi sqrt(kEnergy), p5
 out aNoise * poscil:a(.2, p6)
endin
schedule("noise_bands",0,1)
```

The ATS analysis data can be used for other noise generators, and further trans-formations can be applied. In the following example, we change the way the time pointer moves in the noise_bands master instrument. Instead of a linear pro-gression processed by the line opcode, we use transeg to get a concave shape. This leads to a natural ritardando, in musical terms. The noise_band_gauss in-strument, one instance of which is called per noise band, uses a gaussian random distribution to generate noise. This is then filtered by a mode filter, which simulates a mass-spring-damper system. To smooth the transitions in proportion to the tempo of the time pointer, the portk opcode is applied to the kEnergy variable, thus creating a reverb-like effect at the end of the sound.

Listing 14.12 ATS resynthesis noise bands with modifications

```
gS_ats = "fox.ats"
giDur ATSinfo gS_ats, 7 ;duration
giBark[] fillarray 0,100,200,300,400,510,630,770,920,
                1080,1270,1480,1720,2000,2320,2700,
                3150,3700,4400,5300,6400,7700,9500,
```

```
                    12000,15500,20000

instr noise_bands
 p3 = giDur*5
 gkTime transeg 0, p3, -3, giDur
 iBand = 1
 until iBand > 23 do ;limit because of mode max freq
  iBw = giBark[iBand] - giBark[iBand-1]
  iCfq = (giBark[iBand] + giBark[iBand-1]) / 2
  schedule "noise_band_gauss", 0, p3, iBand, iBw, iCfq
  iBand += 1
 od
endin

instr noise_band_gauss
 xtratim 2
 kEnergy ATSreadnz gkTime, gS_ats, p4
 aNoise gauss sqrt(portk(kEnergy,gkTime/20))
 aFilt mode aNoise, p6, p6/p5
 out aFilt/12
endin
schedule("noise_bands",0,1)
```

So, like ATSread, ATSreadnz has its own qualities as it gives us actual access to the analysis data, leaving the application to our musical ideas.

ATSaddnz is designed to resynthesise the noise amount of a sound, or a selection of it. The usage is very similar to the ATSadd opcode. The total number of noise bands can be given (ibands), as well as an offset (ibandoffset) and an increment (ibandincr):

```
ar ATSaddnz ktimepnt, iatsfile, ibands[, ibandoffset,
                      ibandincr]
```

The resynthesis works internally with the randi facility as described above, so the main influence on the resulting sound is the number and position of noise bands we select.

14.6.5 Transformations

We have already seen several transformations based on an ATS analysis file. The time pointer can be used to perform time compression/expansion or any irregular movement. We can select partials and noise bands by setting their number, offset and increment, and we can transpose the frequencies of the deterministic part by a multiplier. The ATSsinnoi opcode combines all these possibilities in a very compact way, as it adds the noise amount of a specific Bark region to the partials which fall in this range. Consequently, we do not specify any noise band, but only the partials; but the sinus/noise mix can be controlled via the time-varying variables `ksinlev` and `knzlev`:

```
ar ATSsinnoi ktimepnt, ksinlev, knzlev, kfmod,
    iatsfile, ipartials[, ipartialoffset, ipartialincr]
```

Cross-synthesis is a favourite form of spectral processing. ATS offers cross-synthesis only for the deterministic part of two sounds. The first sound has to be given by an `ATSbufread` unit. So the syntax is:

```
ATSbufread ktimepnt1, kfmod1, iatsfile1,
       ipartials1 [, ipartialoffset1, ipartialincr1]
  ar ATScross ktimepnt2, kfmod2, iatsfile2, ifn,
       klev2, klev1,
       ipartials2 [, ipartialoffset2, ipartialincr2]
```

`ATSbufread` has the same qualities as `ATSadd`, except that it reads an .ats analysis file in a buffer which is then available to be used as first sound by `ATScross`. In the example below, its time pointer `ktimepnt1` reads the `iatsfile1` "fox.ats" from 1.67 seconds to the end: "over the lazy dog" are the words here. Extreme time stretching is applied (about 100:1) and the pitch (`kfmod1`) moves slowly between factor 0.9 and 1.1, with linear interpolation and a new value every five seconds. This part of the sound is now crossed with a slightly time-shifted variant of itself, as the `ATScross` opcode adds a random deviation between 0.01 and 0.2 seconds to the same time pointer. This deviation is also slowly moving by linear interpolation. Both sounds use a quarter of the overall partials; the first with a partial offset of 1 and an increment of 3, the second sound without offset and an increment of 2. The level of the first (`kLev1`) and the second (`kLev2`) sound are also continuously moving, between 0.2 and 0.8 as maxima. The result is an always changing structure of strange accords, sometimes forming quasi-harmonic phases, sometimes nearly breaking into pieces.

Listing 14.13 ATS cross-synthesis with a time-shifted version of the same sound

```
gS_file = "fox.ats"
giPartials ATSinfo gS_file, 3
giFilDur ATSinfo gS_file, 7
giSine ftgen 0, 0, 65536, 10, 1

instr lazy_dog
  kLev1 randomi .2, .8, .2
  kLev2 = 1 - kLev1
  kTimPnt linseg 1.67, p3, giFilDur-0.2
  ATSbufread kTimPnt, randi:k(.1,.2,0,1),
                   gS_file, giPartials/4, 1, 3
  aCross ATScross kTimPnt+randomi:k(0.01,0.2,0.1), 1,
                   gS_file, giSine, kLev2,
                   kLev1, giPartials/4, 0, 2
```

```
  outs aCross*2
endin

schedule("lazy_dog",0,100)
```

14.7 Conclusions

In this chapter, we have explored the main elements of spectral processing supported by Csound. Starting with an overview of the principles of Fourier analysis, and a tour of the most relevant tools for frequency-domain analysis and synthesis, we were able to explore the fundamental bases on which the area is grounded. This was followed by a look at how filters can take advantage of Fourier transform theory, for their design and fast implementation. The technique of partitioned convolution was also introduced, with an example showing the zero-latency use of multiple partitions.

The text dedicated significant space to the phase vocoder and its streaming implementation in Csound. We have detailed how the algorithm is constructed by implementing the analysis and synthesis stages from first principles. The various different types of transformations that can be applied to PV data were presented, with examples of Csound opcodes that implement these.

A discussion of the principles of sinusoidal modelling completed the chapter. We introduced the frequency analysis and partial-tracking operations involved, as well as the variants of additive synthesis available to reconstruct a time-domain signal. Residual extraction was demonstrated with an example, and a suite of track manipulation opcodes was presented. In addition to this, ATS, a specialised type of spectral modelling using sinusoidal tracking, was introduced and explored in some detail. Spectral processing is a very rich area for sonic exploration. It provides a significant ground for experimentation in the design of novel sounds and processes. We hope that this chapter provides an entry point for readers to sample the elements of frequency-domain sound transformation.

Chapter 15
Granular Synthesis

Abstract In this chapter, we will look at granular synthesis and granular effects processing. The basic types of granular synthesis are discussed, and the influence of parameter variations on the resulting sound is shown with examples. For granular effects processing on a live input stream, we write the input sound to a buffer used as a source waveform for synthesising grains, enabling granular delays and granular reverb designs. We proceed to look at manipulations of single grains by means of grain masking, and look at aspects of gradual synchronisation between grain-scheduling clocks. The relationship between regular amplitude modulation and granular synthesis is studied, and we use pitch synchronous granular synthesis to manipulate formants of a recorded sound. Several classic types of granular synthesis are known from the literature, originally requiring separate granular synthesis engines for each type. We show how to implement all granular synthesis types with a single generator (the `partikkel` opcode), and a parametric automation to do morphing between them.

15.1 Introduction

Granular synthesis (or particle synthesis as it is also called) is a very flexible form of audio synthesis and processing, allowing for a wide range of sounds. The technique works by generating small snippets (grains) of sound, typically less than 300~400 milliseconds each. The grains may have an envelope to fade each snippet in and out smoothly, and the assembly of a large number of such grains makes up the resulting sound. An interesting aspect of the technique is that it lends itself well to gradual change between a vast range of potential sounds. The selection and modification of the sound fragments can yield highly distinctive types of sounds, and the large number of synthesis parameters allows very flexible control over the sonic output. The underlying process of assembling small chunks of sound stays basically the same, but the shape, size, periodicity and audio content of grains may be changed dynamically. Even if the potential sounds can be transformed and gradually morphed from

© Springer International Publishing Switzerland 2016   337
V. Lazzarini et al., *Csound*, DOI 10.1007/978-3-319-45370-5_15

one type to another, it can be useful to describe some of the clearly defined types of sound we can attain. These types represent specific combinations of parameter settings, and as such represent some clearly defined locations in the multidimensional sonic transformation space. Curtis Roads made a thorough classification of different granular techniques in his seminal book Microsound [109]. Roads' book inspired the design of the Csound opcode `partikkel`, as a unified generator for all forms of time-based granular sound [22]. We will come back to Roads' classification later, but will start with a more general division of the timbral potential of granular techniques. The one aspect of the techniques that has perhaps the largest impact on the resulting sound is the density and regularity of the grains. Because of this, we will do a coarse classification of the types of sounds based on this criteria. This coarse division is based on the perceptual threshold between rhythm and tone: below approximately 20 Hz, we tend to hear separate events and rhythm; when the repetition rate is higher the events blend together and create a continuous tone. The boundary is not sharp, and there are ways of controlling the sound production so as to avoid static pitches even with higher grain rates. Still it serves us as a general indication of the point where the sound changes function.

15.1.1 Low Grain Rates, Long Grains

When we have relatively few grains per second (< 50) and each grain is relatively long (> 50 milliseconds), then we can clearly hear the timbre and pitch of the sound in each grain. In this case the output sound is largely determined by the waveform inside each grain. This amounts to quickly cross-fading between segments of recorded sound, and even though the resulting texture may be new (according to the selection and combination of sounds), the original sounds used as source material for the grains are clearly audible. Csound has a large selection of granular synthesis opcodes. The `partikkel` opcode has by far the highest flexibility, as it was designed to enable all forms of time-based granular synthesis. There are also simpler opcodes available that might be easier to use in specific use cases. The following two examples accomplish near identical results, showing granular synthesis first with `syncgrain`, then with `partikkel`.

Listing 15.1 Example using syncgrain

```
giSoundfile ftgen 0,0,0,1,"fox.wav",0,0,0
giSigmoWin ftgen 0,0,8193,19,1,0.5,270,0.5

instr 1
kprate linseg 1,2.3,1,0,-0.5,2,-0.5,0,1,1,1
kGrainRate = 25.0
kGrainDur = 2.0
kgdur = kGrainDur/kGrainRate
kPitch = 1
```

```
a1 syncgrain ampdbfs(-8),kGrainRate,kPitch,kgdur,
    kprate/kGrainDur,giSoundfile,giSigmoWin,100
 out a1
endin

schedule(1,0,5.75)
```

Listing 15.2 Example using `partikkel`

```
giSoundfile ftgen 0,0,0,1,"fox.wav",0,0,0
giSine ftgen 0,0,65536,10,1
giCosine ftgen 0,0,8193,9,1,1,90
giSigmoRise ftgen 0,0,8193,19,0.5,1,270,1
giSigmoFall ftgen 0,0,8193,19,0.5,1,90,1

instr 1
asamplepos1 linseg 0,2.3,0.84,2,0.478,1.47,1.0
kGrainRate = 25.0
async = 0.0 ; (disable external sync)
kGrainDur = 2.0
kgdur = (kGrainDur*1000)/kGrainRate
kwavfreq = 1
kwavekey1 = 1/(tableng(giSoundfile)/sr)
awavfm = 0 ; (FM disabled)
a1 partikkel kGrainRate,0,-1,async,0,-1,
    giSigmoRise,giSigmoFall,0,0.5,kgdur,
    ampdbfs(-13),-1,kwavfreq,0.5,-1,-1,awavfm,
    -1,-1,giCosine,1,1,1,-1,0,\
    giSoundfile,giSoundfile,giSoundfile,giSoundfile,-1,
    asamplepos1,asamplepos1,asamplepos1,asamplepos1,
    kwavekey1,kwavekey1,kwavekey1,kwavekey1,100
 out a1
endin

schedule(1,0,5.75)
```

As can be seen from the two above examples, `syncgrain` requires less code and may be preferable for simple cases. The `partikkel` opcode provides more flexibility and thus also requires a bit more code. One notable difference between the two opcodes for this simple case is that the time pointer into the source waveform is handled differently. `Syncgrain` uses a *rate of time pointer movement* specified in relation to grain duration. `Partikkel` uses a *time pointer value as a fraction of the source waveform duration*. `Syncgrain`'s method can be more convenient if the time pointer is to move at a determined rate through the sound. `Partikkel`'s time pointer is more convenient if one needs random access to the time point. For the remaining examples in this chapter, partikkel will be used due to its flexibility to do all different kinds of granular synthesis.

The next example shows a transition from a time pointer moving at a constant rate, then freezing in one spot, then with increasingly random deviation from that spot. The global tables from listing 15.2 are used.

Listing 15.3 Example with random access time pointer

```
instr 1
asamplepos1 linseg 0,1.2,0.46,1,0.46
adeviation rnd31 linseg(0,1.3,0,2,0.001,2,
            0.03,1,0.2,1,0.2),1
asamplepos1 = asamplepos1 + adeviation
kGrainRate = 30.0
async = 0.0 ; (disable external sync)
kGrainDur = 3.0
kgdur = (kGrainDur*1000)/kGrainRate
kwavfreq = 1
kwavekey1 = 1/(tableng(giSoundfile)/sr)
awavfm = 0 ; (FM disabled)
a1 partikkel kGrainRate,0,-1,async,0,-1,
    giSigmoRise,giSigmoFall,0,0.5,kgdur,
    ampdbfs(-13),-1,kwavfreq,0.5,-1,-1,awavfm,
    -1,-1,giCosine,1,1,1,-1,0,
    giSoundfile,giSoundfile,giSoundfile,giSoundfile,-1,
    asamplepos1,asamplepos1,asamplepos1,asamplepos1,
    kwavekey1,kwavekey1,kwavekey1,kwavekey1,100
 out a1
endin
```

15.1.2 High Grain Rates, Periodic Grain Clock

When the grain rate is high ($> 30{\sim}50$ Hz) and strictly periodic, we will in most cases hear a pitch with fundamental frequency equal to the grain rate. Any sound that is periodic (the exact same thing happening over and over again at regular intervals) will constitute a clearly defined pitch, and the pitch is defined by the rate of repetition. If the contents of our individual grains are identical, the output waveform will have a repeating pattern over time (as illustrated in Fig 15.1), and so also constitute a clearly perceptible pitch. It is noteworthy that very precise exact repetition is a special case that sounds very different from cases where repetitions are not exact. The constitution of pitch is quite fragile, and deviations from the periodicity will create an unclear or noisy pitch (which of course might be exactly what we want sometimes). In the case of high grain rate with short grains, the audio content of each grain is not perceivable in and as itself. Then the waveform content of the grain will affect the timbre (harmonic structure), but not the perceived pitch since pitch will be determined by the rate of repetition (i.e. the grain rate). More details on this are given in Section 15.5. The following example shows the use of a high grain rate to constitute pitch.

Listing 15.4 Example with high grain rate, pitch constituted by grain rate

```
instr 1
 kamp adsr 0.0001, 0.3, 0.5, 0.5
 kamp = kamp*ampdbfs(-6)
 asamplepos1 = 0
 kGrainRate = cpsmidinn(p4)
 async = 0.0 ; (disable external sync)
 kGrainDur = 1.0
 kgdur = (kGrainDur*1000)/kGrainRate
 ka_d_ratio = p5
 kwavfreq line 200, p3, 500
 kwavekey1 = 1
 awavfm = 0 ; (FM disabled)
 a1 partikkel kGrainRate, 0, -1, async, 0, -1,
     giSigmoRise, giSigmoFall, 0, ka_d_ratio, kgdur,
     kamp, -1, kwavfreq, 0.5, -1, -1, awavfm,
     -1, -1, giCosine, 1, 1, 1, -1, 0,
     giSine, giSine, giSine, giSine, -1,
     asamplepos1,asamplepos1,asamplepos1,asamplepos1,
     kwavekey1, kwavekey1, kwavekey1, kwavekey1, 100
  out    a1
endin

schedule(1,0,1,48,0.5)
schedule(1,1,1,51,0.5)
schedule(1,2,1,53,0.5)
schedule(1,3,1,55,0.5)
schedule(1,4,3,58,0.5)

schedule(1,8,1,48,0.1)
schedule(1,9,1,51,0.1)
schedule(1,10,1,53,0.1)
schedule(1,11,1,55,0.1)
schedule(1,12,3,58,0.1)
```

Fig. 15.1 Close-up of a grain waveform, repeating grains constitute a stable pitch

15.1.3 Grain Clouds, Irregular Grain Clock

When the grain generation is very irregular, usually also combined with fluctuations in the other parameters (e.g. pitch, phase, duration) the resulting sound will have turbulent characteristics and is often described as a cloud of sound. This is a huge class of granular sound with a large scope for variation. Still, the irregular modulation creates a perceptually grouped collection of sounds.

Listing 15.5 Example of a sparse grain cloud

```
instr 1
kamp  adsr 2, 1, 0.5, 2
kamp  = kamp*ampdbfs(-10)
asamplepos1 = 0
kGrainRate = randh(30,30)+32
async = 0.0 ; (disable external sync)
kGrainDur = randh(0.5,30)+0.7
kgdur = (kGrainDur*1000)/kGrainRate
ka_d_ratio = 0.2
kwavfreq = randh(300,30)+400
kwavekey1 = 1
awavfm = 0 ; (FM disabled)
a1 partikkel kGrainRate, 0, -1, async, 0, -1,
    giSigmoRise, giSigmoFall, 0, ka_d_ratio, kgdur,
    kamp, -1, kwavfreq, 0.5, -1, -1, awavfm,
    -1, -1, giCosine, 1, 1, 1, -1, 0,
    giSine, giSine, giSine, giSine, -1,
    asamplepos1,asamplepos1,asamplepos1,asamplepos1,
    kwavekey1, kwavekey1, kwavekey1, kwavekey1, 100
out a1
endin

schedule(1,0,6)
```

Listing 15.6 If we change these lines of code, we get a somewhat denser grain cloud

```
kGrainRate  = randh(80,80)+100
```

```
kgdur = 30 ; use static grain size (in millisecs)...
;...so we comment out the relative
;    grain dur calculation
; kgdur = (kGrainDur*1000)/kGrainRate;
kwavfreq = randh(100,80)+300
```

15.2 Granular Synthesis Versus Granular Effects Processing

By granular synthesis, we usually mean a granular technique applied to synthesised or pre-recorded source sounds. If we use a live audio stream as the source of our grains, we use the term *granular effects processing*. Technically, the creation of audio grains is the same in both cases, only the source material for grains differs. Applying granular techniques to a live audio stream allows us the rich transformational potential of granular synthesis while retaining the immediacy and interactivity of a live audio effect. To record live audio into a table for use as a source waveform, we use the `tablewa` opcode, as outlined in listing 15.7. This is a circular buffer, and to control the delay time incurred by the difference between record (write) and playback (read) positions in the table, we write the record pointer to a global k-rate variable (gkstartFollow). We will reference this value when calculating the read position for creating grains. Also note that the global variable 0dbfs *must* be set to 1, otherwise the system will blow up when using feedback in the granular processing.

Listing 15.7 Record live audio to table, for use as a source waveform for granular processing

```
0dbfs = 1

; audio buffer table for granular effects processing
giLiveFeedLen = 524288 ; 11.8 seconds buffer at 44.1
giLiveFeedLenSec = giLiveFeedLen/sr
giLiveFeed ftgen 0, 0, giLiveFeedLen+1, 2, 0

instr 1
a1 inch 1
aFeed chnget "partikkelFeedback"
kFeed = 0.4
a1 = a1 + (aFeed*kFeed)
iLength = ftlen(giLiveFeed)
gkstartFollow tablewa giLiveFeed, a1, 0
; reset kstart when table is full
gkstartFollow = (gkstartFollow > (giLiveFeedLen-1) ?
     0 : gkstartFollow)
; update table guard point (for interpolation)
tablegpw giLiveFeed
endin
```

15.2.1 Grain Delay

For all the granular-processing effects, the live stream is recorded into a circular buffer (listing 15.7), from which the individual grains are read. The difference between the record and playback positions within this buffer will affect the time it takes between sound being recorded and played back, and as such allows us to control the delay time. With `partikkel` the read position is set with the `samplepos` parameter. This is a value in range 0.0 to 1.0, referring to the relative position within the source audio waveform. A `samplepos` value of 0.5 thus means that grains will be read starting from the middle of the source sound. Delay time in seconds can be calculated as `samplepos` × buffer length in seconds. We must make sure not to read before the record pointer (which could easily happen if we use random deviations from zero delay time with the circular buffer). Otherwise we will create clicks and also play back audio that is several seconds (buffer length) delayed in relation to what we expected. Crossing the record pointer boundary will also happen if we use zero delay time and upwards transposition in the grains. In this situation, we start reading at the same table address as where we write audio, but we read faster than we write, and the read pointer will overtake the write pointer. To avoid this, we need to add a minimum delay time as a factor of pitch × duration for the grains we want to create. When manipulating and modulating the time pointer into this circular buffer, we may want to smooth it out and upsample to a-rate at the final stage before processing. This requires special care, as we do not want to filter the ramping signal when it resets from 1.0 to 0.0. For this purpose we can use a UDO (listing 15.8), doing linear interpolation during upsampling of all sections of the signal except when it abruptly changes from high to low. If we did regular interpolation, the time pointer would sweep fast through the whole buffer on phase reset, and any grains scheduled to start during this sweep would contain unexpected source audio (see Fig 15.2)

Listing 15.8 UDO for interpolated upsampling of the time pointer in the circular buffer

```
gikr = kr
opcode UpsampPhasor, a,k
kval xin
setksmps 1
kold init 0
if (abs(kold-kval)<0.5) && (abs(kold-kval)>0) then
reinit interpolator
elseif abs(kold-kval)>0.5 then; when phasor restarts
kold = kold-1
reinit interpolator
endif
interpolator:
aval linseg i(kold), 1/gikr, i(kval), 1/gikr, i(kval)
rireturn
kold = kval
xout aval
```

```
endop
```

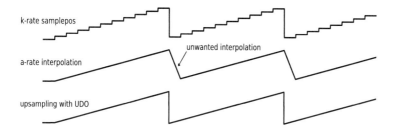

Fig. 15.2 Upsampling a k-rate time pointer (`samplepos`), we need to disable interpolation on phase reset to avoid reading the table quickly backwards when the phase wraps around

For granular effects processing, the grain rate is usually low, as we are interested in hearing the timbral content of the live audio stream. One interesting aspect of granular delays is that a change in the delay time does not induce pitch modulation, as is common with traditional delay techniques. This allows us separate control over delay time and pitch, and also to scatter grains with differing time and pitch relations into a continuous stream of echo droplets. Writing the output of the grain delay process back into the circular buffer (mixed with the live input) allows for grain delay feedback, where the same transformational process is applied repeatedly and iteratively on the same source material. This can create cascading effects.

Listing 15.9 Simple grain delay with feedback and modulated delay time

```
instr 2

; grain clock
 kGrainRate = 35.0
 async   = 0.0

; grain shape
 kGrainDur = 3.0
 kduration = (kGrainDur*1000)/kGrainRate

; grain pitch (transpose, or "playback speed")
 kwavfreq = 1
 kfildur1 = tableng(giLiveFeed) / sr
 kwavekey1 = 1/kfildur1
 awavfm = 0

; automation of the grain delay time
 ksamplepos1 linseg 0, 1, 0, 2, 0.1, 2,
```

```
        0.1, 2, 0.2, 2, 0, 1, 0
kpos1Deviation randh 0.003, kGrainRate
ksamplepos1 = ksamplepos1 + kpos1Deviation

; Avoid crossing the record boundary
ksamplepos1 limit ksamplepos1,
      (kduration*kwavfreq)/(giLiveFeedLenSec*1000),1
; make samplepos follow the record pointer
ksamplepos1  =
      (gkstartFollow/giLiveFeedLen) - ksamplepos1
asamplepos1 UpsampPhasor ksamplepos1
asamplepos1 wrap asamplepos1, 0, 1

a1  partikkel kGrainRate, 0, -1, async, 0, -1,
 giSigmoRise, giSigmoFall, 0, 0.5, kduration, 0.5, -1,
 kwavfreq, 0.5, -1, -1, awavfm,
 -1, -1, giCosine, 1, 1, 1,
 -1, 0, giLiveFeed, giLiveFeed,
 giLiveFeed, giLiveFeed, -1,
 asamplepos1,asamplepos1,asamplepos1,asamplepos1,
 kwavekey1, kwavekey1, kwavekey1, kwavekey1, 100

; audio feedback in granular processing
aFeed dcblock a1
chnset aFeed, "partikkelFeedback"

 out a1*ampdbfs(-6)

endin
```

Listing 15.10 Grain delay with four-voice pitching and scattered time modulations

```
instr 2

; grain clock
 kGrainRate = 35.0
 async  = 0.0

; grain shape
 kGrainDur = 2.0
 kduration = (kGrainDur*1000)/kGrainRate

; different pitch for each source waveform
 kwavfreq = 1
 kfildur1 = tableng(giLiveFeed) / sr
 kwavekey1 = 1/kfildur1
```

```
 kwavekey2 = semitone(-5)/kfildur1
 kwavekey3 = semitone(4)/kfildur1
 kwavekey4 = semitone(9)/kfildur1
 awavfm = 0

 ; grain delay time, more random deviation
 ksamplepos1 = 0
 kpos1Deviation randh 0.03, kGrainRate
 ksamplepos1 = ksamplepos1 + kpos1Deviation
 ; use different delay time for each source waveform
 ; (actually same audio, but read at different pitch)
 ksamplepos2 = ksamplepos1+0.05
 ksamplepos3 = ksamplepos1+0.1
 ksamplepos4 = ksamplepos1+0.2

 ; Avoid crossing the record boundary
#define RecordBound(N)#
 ksamplepos$N. limit ksamplepos$N.,
       (kduration*kwavfreq)/(giLiveFeedLenSec*1000),1
 ; make samplepos follow the record pointer
 ksamplepos$N.  =
       (gkstartFollow/giLiveFeedLen) - ksamplepos$N.
 asamplepos$N. UpsampPhasor ksamplepos$N.
 asamplepos$N. wrap asamplepos$N., 0, 1
#
$RecordBound(1)
$RecordBound(2)
$RecordBound(3)
$RecordBound(4)

 ; activate all 4 source waveforms
 iwaveamptab ftgentmp 0, 0, 32, -2, 0, 0, 1,1,1,1,0

 a1  partikkel kGrainRate, 0, -1, async, 0, -1,
  giSigmoRise, giSigmoFall, 0, 0.5, kduration, 0.5, -1,
  kwavfreq, 0.5, -1, -1, awavfm,
  -1, -1, giCosine, 1, 1, 1,
  -1, 0, giLiveFeed, giLiveFeed, giLiveFeed, giLiveFeed,
  iwaveamptab, asamplepos1, asamplepos2,
  asamplepos3, asamplepos4,
  kwavekey1, kwavekey2, kwavekey3, kwavekey4, 100

 ; audio feedback in granular processing
 aFeed dcblock a1
 chnset aFeed, "partikkelFeedback"
```

```
out a1*ampdbfs(-3)
```

```
endin
```

15.2.2 Granular Reverb

We can use granular techniques to create artificial reverberant effects [37]. These are not necessarily directed towards modelling real acoustic spaces, but provide a very flexible tool for exploring novel and imaginary reverberant environments. As traditional artificial reverb techniques are based on delay techniques, so are granular reverbs based on granular delays. In addition to complex delay patters, we can also exploit the independence of time and pitch in granular synthesis to create forms of time stretching of the live stream. Time stretching is something that would be highly unlikely to occur in an acoustic space, but since reverberation can also be heard as a certain prolongation of the sound, we can associate this type of transformation with a kind of reverb. When time stretching a live signal we quickly run into a practical problem. Time stretching involves gradually increasing the delay time between input and output, and if the sound is to be perceived as a "immediately slowed down", there is a limit to the actual delay we want to have before we can hear the slowed down version of the sound. To alleviate this increasing delay time, we use several overlapping time stretch processes, fading a process out when the delay has become too large for our purposes and simultaneously fading a new process in (resetting the delay to zero for the new process). See Fig. 15.3 and listing 15.11.

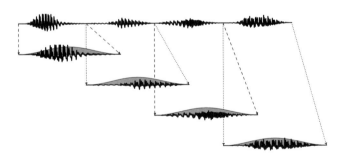

Fig. 15.3 Time pointers and source wave amps for time stretching real-time input. Each of the four stretched segments constitutes a granular process on a source waveform for `partikkel`

Listing 15.11 Granular reverb skeleton: four-voice overlapping time stretch

```
instr 2
```

```
; grain clock
 kGrainRate = 110.0
 async  = 0.0

; grain shape
 kGrainDur = 7.0
 kduration = (kGrainDur*1000)/kGrainRate

; same pitch for all source waveforms
 kwavfreq = 1
 kfildur1 = tableng(giLiveFeed) / sr
 kwavekey1 = 1/kfildur1
 awavfm = 0

 ; grain delay time,
 ; gradually increasing delay time
 ; to create slowdown effect.
 kplaybackspeed = 0.25 ; slow down
 koverlaprate = 0.8 ; overlap rate
 koverlap = 1 ; amount of overlap between layers

 ; four overlapping windows of slowdown effect,
 ; fading in and out,
 ; reset delay time to zero on window boundaries
#define Overlaptime(N'P)#
 koverlaptrig$N. metro koverlaprate, $P.
 if koverlaptrig$N. > 0 then
 reinit timepointer$N.
 endif
timepointer$N.:
 ksamplepos$N. line 0, i(kfildur1),
       1-i(kplaybackspeed)
 itimenv$N. divz i(koverlap), i(koverlaprate), .01
 kampwav$N. oscilli itimenv$N.*0.1, 1,
       itimenv$N., giSigmoWin
 rireturn
#
$Overlaptime(1'0.0)
$Overlaptime(2'0.25)
$Overlaptime(3'0.50)
$Overlaptime(4'0.75)

 ktimedev = 4/(giLiveFeedLenSec*1000)
#define TimeDeviation(N)#
 kdevpos$N. rnd31 ktimedev, 1
```

```
ksamplepos$N. = ksamplepos$N.+kdevpos$N.
#
$TimeDeviation(1)
$TimeDeviation(2)
$TimeDeviation(3)
$TimeDeviation(4)

 ; Avoid crossing the record boundary
#define RecordBound(N)#
 ksamplepos$N. limit ksamplepos$N.,
       (kduration*kwavfreq)/(giLiveFeedLenSec*1000),1
 ; make samplepos follow the record pointer
 ksamplepos$N. =
       (gkstartFollow/giLiveFeedLen) - ksamplepos$N.
 asamplepos$N. UpsampPhasor ksamplepos$N.
 asamplepos$N. wrap asamplepos$N., 0, 1
#
$RecordBound(1)
$RecordBound(2)
$RecordBound(3)
$RecordBound(4)

 ; channel masking table, send grains alternating to
 ; left and right output, for stereo reverb
 ichannelmasks ftgentmp 0, 0, 16, -2, 0, 1, 0, 1

 ; activate all 4 source waveforms
 iwaveamptab ftgentmp 0, 0, 32, -2, 0, 0, 1,1,1,1,0

 ; write amp envelope for overlapping
 ; slowdown windows to wave mix mask table
 tablew kampwav1, 2, iwaveamptab
 tablew kampwav2, 3, iwaveamptab
 tablew kampwav3, 4, iwaveamptab
 tablew kampwav4, 5, iwaveamptab

 a1, a2  partikkel kGrainRate, 0, -1, async, 0, -1,
   giSigmoRise, giSigmoFall, 0, 0.5, kduration, 1, -1,
   kwavfreq, 0.5, -1, -1, awavfm,
   -1, -1, giCosine, 1, 1, 1, ichannelmasks,
   0, giLiveFeed, giLiveFeed, giLiveFeed, giLiveFeed,
   iwaveamptab, asamplepos1, asamplepos2,
   asamplepos3, asamplepos4,
   kwavekey1, kwavekey1, kwavekey1, kwavekey1, 100
```

```
; audio feedback in granular processing
aFeed dcblock a1
; empirical adjustment of feedback
; scaling for stability
aFeed = aFeed*0.86
chnset aFeed, "partikkelFeedback"

outs a1*ampdbfs(-6), a2*ampdbfs(-6)

endin
```

Adding feedback, i.e. routing the output from the granular slowdown process back into the live recording buffer, we get a longer and more diffuse reverb tail. In the following example, we used a somewhat less extreme slowdown factor, and also increased the random deviation to the time pointer. As is common in artificial reverb algorithms, we've also added a low-pass filter in the feedback path, so high frequencies will decay faster than lower spectral components.

Listing 15.12 Granular reverb with feedback and filtering: listing only the differences from listing 15.11

```
kFeed = 0.3
...
kplaybackspeed = 0.35 ; slow down
...
ktimedev = 12/(giLiveFeedLenSec*1000)
...
aFeed butterlp aFeed, 10000
```

We can also add a small amount of pitch modulation to the reverb. This is a common technique borrowed from algorithmic reverb design, to allow for a more rounded timbre in the reverb tail.

Listing 15.13 Granular reverb with pitch modulation: listing only the differences from listing 15.12. Of the partikkel parameters, only the four wavekeys have been changed

```
kFeed = 0.5
...
kpitchmod = 0.005
#define PitchDeviation(N)#
kpitchdev$N. randh kpitchmod, 1, 0.1
kwavekey$N. = 1/kfildur1*(1+kpitchdev$N.)
#
$PitchDeviation(1)
$PitchDeviation(2)
$PitchDeviation(3)
$PitchDeviation(4)
...
a1, a2  partikkel kGrainRate, 0, -1, async, 0, -1,
```

```
giSigmoRise, giSigmoFall, 0, 0.5, kduration, 1, -1,
kwavfreq, 0.5, -1, -1, awavfm,
-1, -1, giCosine, 1, 1, 1, ichannelmasks,
0, giLiveFeed, giLiveFeed, giLiveFeed, giLiveFeed,
iwaveamptab, asamplepos1, asamplepos2,
asamplepos3, asamplepos4,
kwavekey1, kwavekey2, kwavekey3, kwavekey4, 100
```

Exploiting the independence of time and pitch transformations in granular synthesis, we can also create reverb effects with shimmering harmonic tails or we can create tails that have a continually gliding pitch. The design will normally need at least one granular process for each pitch-shifted component. In our example we are still using a single (four-voice) grain generator, creating a simple and raw version of the effect at a very low computational cost. We do pitch shifting on a grain-by-grain basis (with pitch masking, every third grain is transposed, see Section 15.3) instead of using a separate granular voice for each pitch. As a workaround to get a reasonably dense reverb with this simple design, we use very long grains. To get finer control over the different spectral components and also denser reverb, we could use several granular generators, feeding into each other.

Listing 15.14 Granular reverb, some grains are pitch shifted up an octave, creating a shimmering reverb tail: listing only the differences from listing 15.13. Of the `partikkel` parameters, only the `iwavfreqstarttab` and `iwavfreqendtab` values have been changed

```
kFeed = 0.7
...
kGrainDur = 12.0
...
kplaybackspeed = 0.25 ; slow down
...
; pitch masking tables
iwavfreqstarttab ftgentmp 0, 0, 16, -2, 0, 2, 1,1,2
iwavfreqendtab ftgentmp 0, 0, 16, -2, 0, 2, 1,1,2
...
a1, a2  partikkel kGrainRate, 0, -1, async, 0, -1,
  giSigmoRise, giSigmoFall, 0, 0.5, kduration, 1, -1,
  kwavfreq, 0.5, iwavfreqstarttab, iwavfreqendtab,
  awavfm, -1, -1, giCosine, 1, 1, 1, ichannelmasks,
  0, giLiveFeed, giLiveFeed, giLiveFeed, giLiveFeed,
  iwaveamptab, asamplepos1, asamplepos2,
  asamplepos3, asamplepos4,
  kwavekey1, kwavekey2, kwavekey3, kwavekey4, 100
...
aFeed butterlp aFeed, 8000
aFeed = aFeed*0.6
```

The observant reader may have noticed that we have a specification of feedback level in two different places in the code. Also, that these two operations essentially do the same thing (effecting the level of audio feedback in the granular process). The rationale for this is to provide an intuitive control over the feedback level, affecting the reverberation time, much like in FDN reverbs (see Section 13.2.3). We assume that one would expect such a reverb algorithm to have (an approximation of) infinite reverb time when feedback is set to 1.0. This intuitive feedback level control is the first one (`kFeed = 0.7`). The second adjustment (`aFeed = aFeed*0.6`) can be viewed as a system parameter that needs to be recalculated when the reverb design changes. The specifics of the granular processing can significantly change the gain of the signal. No automatic method for adjustment has been provided here, but we can state that the gain structure is mainly dependent on the amount of grain overlap, as well as pitch and time deviations for single grains. The grain overlap is easily explained due to more layers of sound generally creating a higher amplitude. The pitch and time deviations affect the feedback potential negatively, e.g. changing the pitch of the signal in the feedback loop effectively lowers the potential for sharp resonances. For the reverb designs presented here, the feedback scaling was adjusted empirically by setting `kFeed = 1.0` and then adjusting `aFeed = X` until an approximation of infinite reverb time was achieved. Due to the presence of random elements (for time and pitch modulation) the effective feedback gain will fluctuate, especially so in the design we have used here with only one granular generator. Using more generators will make a more complex delay network, and the resulting effect of random fluctuations of parameters will be more even.

15.3 Manipulation of Individual Grains

In a regular granular synthesis situation, we may generate tens or hundreds of grains per second. For precise control over the synthesis result, we may want to modify each of these grains separately. To facilitate this, we can use a technique called grain masking. In Microsound [109] the term is used to describe selective muting of individual grains, implying an amplitude control that may also be gradual. By extending the notion of grain masking to also include pitch trajectories, frequency modulation, output channel and source waveform mixing, we can also extend the range of control possibilities over the individual grains. At high grain rates, any kind of masking will affect the perceived pitch. This is because masking even a single grain affects the periodicity of the signal (see listing 15.15). Masking single grains intermittently will add noise to the timbre; masking every second grain will let the perceived pitch drop by one octave, as the repetition period doubles (listing 15.17). Further subharmonics may be generated by dropping every third, fourth or fifth grain and so on. These pitch effects will occur regardless of which grain parameter is masked, but the timbral effect will differ somewhat with the masking of different parameters (listing 15.19). At lower grain rates, the masking techniques can be used to create elaborate rhythmic, spatial and harmonic patterns. If we use different mask lengths, we can achieve polyrhythmic relationships between the masking patterns of different parameters. This is quite effective to add vividness and complexity to a timbral evolution. In Csound's `partikkel` opcode, grain masking is specified by a masking table (figs. 15.4 and 15.5). The masking table is read incrementally when generating grains, so values next to each other in the table will apply to neighbouring grains successively. As most grain-masking patterns are periodic, the masking table index can be looped at user-specified indices. For non-periodic patterns, we can simply use arbitrarily large masking tables or rewrite the table values continuously.

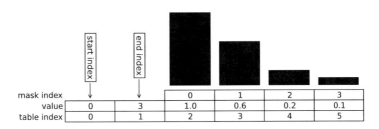

		mask index	0	1	2	3
start index	end index	value	0	3	1.0	0.6	0.2	0.1
		table index	0	1	2	3	4	5

Fig. 15.4 Amplitude masking table, the two first indices control loop start and end, the remaining indices are amplitude values for each successive grain. Table values can be modified in real time to create dynamically changing patterns

Fig. 15.5 Amplitude masking using the mask table shown in Fig. 15.4

Listing 15.15 Synchronous granular synthesis with high grain rate and increasing amount of random masking during each note

```
instr 1
kamp adsr 0.0001, 0.3, 0.5, 0.5
kamp = kamp*ampdbfs(-6)
asamplepos1 = 0
kGrainRate = cpsmidinn(p4)
async  = 0.0 ; (disable external sync)
kGrainDur = 1.0
kgdur  = (kGrainDur*1000)/kGrainRate
ka_d_ratio = 0.5
```

```
kwavfreq = kGrainRate*4
kwavekey1 = 1
awavfm = 0 ; (FM disabled)
krandommask line 0, p3, p5
a1 partikkel kGrainRate, 0, -1, async, 0, -1,
    giSigmoRise, giSigmoFall, 0, ka_d_ratio, kgdur,
    kamp, -1, kwavfreq, 0.5, -1, -1, awavfm,
    -1, -1, giCosine, 1, 1, 1, -1, krandommask,
    giSine, giSine, giSine, giSine, -1,
    asamplepos1,asamplepos1,asamplepos1,asamplepos1,
    kwavekey1, kwavekey1, kwavekey1, kwavekey1, 100
out a1
endin
```

Listing 15.16 Score for the instrument in listing 15.15

```
i1 0 2 48 0.3
i1 2 . 51 .
i1 4 . 53 .
i1 6 4 60 .
s
i1 0 2 48 1.0
i1 2 . 51 .
i1 4 . 53 .
i1 6 4 60 .
```

Listing 15.17 Synchronous granular synthesis with high grain rate, gradually decreasing the amplitude of every second grain during each note, creating an octaviation effect

```
instr 1
kamp adsr 0.0001, 0.3, 0.5, 0.5
kamp = kamp*ampdbfs(-6)
asamplepos1 = 0
kGrainRate = cpsmidinn(p4)
async = 0.0 ; (disable external sync)
kGrainDur = 0.5
kgdur = (kGrainDur*1000)/kGrainRate
ka_d_ratio = 0.5
kwavfreq = kGrainRate*2
kwavekey1 = 1
awavfm = 0 ; (FM disabled)
krandommask = 0
igainmasks ftgen 0, 0, 4, -2, 0, 1, 1, 1
koctaviation linseg 1, 0.5, 1, p3-0.5 , 0
tablew koctaviation, 2, igainmasks

a1 partikkel kGrainRate, 0, -1, async, 0, -1,
```

```
        giSigmoRise, giSigmoFall, 0, ka_d_ratio, kgdur,
        kamp, igainmasks, kwavfreq, 0.5, -1, -1, awavfm,
        -1, -1, giCosine, 1, 1, 1, -1, krandommask,
        giSine, giSine, giSine, giSine, -1,
        asamplepos1,asamplepos1,asamplepos1,asamplepos1,
        kwavekey1, kwavekey1, kwavekey1, kwavekey1, 100
  out a1
  endin
```

Listing 15.18 Score for the instrument in listing 15.17

```
i1 0 2 48
i1 2 . 51
i1 4 . 53
i1 6 4 60
```

Listing 15.19 As above (listing 15.17), but this time using pitch masks to create the octaviation effect. The pitch of every second grain is gradually changed, ending at an octave above. Still, the perceived pitch drops by one octave, since the rate of repetition is doubled when every second grain is different. Use the score in listing 15.18

```
instr 1
kamp adsr 0.0001, 0.3, 0.5, 0.5
kamp = kamp*ampdbfs(-6)
asamplepos1 = 0
kGrainRate = cpsmidinn(p4)
async = 0.0 ; (disable external sync)
kGrainDur = 0.5
kgdur = (kGrainDur*1000)/kGrainRate
ka_d_ratio = 0.5
kwavfreq = kGrainRate*2
kwavekey1 = 1
awavfm = 0 ; (FM disabled)
krandommask = 0

; pitch masking tables
iwavfreqstarttab ftgentmp 0, 0, 16, -2, 0, 1, 1,1
iwavfreqendtab ftgentmp 0, 0, 16, -2, 0, 1, 1,1

koctaviation linseg 1, 0.5, 1, p3-0.5 , 2
tablew koctaviation, 2, iwavfreqstarttab
tablew koctaviation, 2, iwavfreqendtab

a1 partikkel kGrainRate, 0, -1, async, 0, -1,
    giSigmoRise, giSigmoFall, 0, ka_d_ratio, kgdur,
    kamp, -1, kwavfreq, 0.5,
    iwavfreqstarttab, iwavfreqendtab, awavfm,
```

```
    -1, -1, giCosine, 1, 1, 1, -1, krandommask,
    giSine, giSine, giSine, giSine, -1,
    asamplepos1, asamplepos1, asamplepos1, asamplepos1,
    kwavekey1, kwavekey1, kwavekey1, kwavekey1, 100
out a1
endin
```

15.3.1 Channel Masks, Outputs and Spatialisation

Using masking techniques with parameters such as amplitude, pitch and modulation is unambiguous: we can always describe exactly what effect they will have on the resulting sound. With channel masking, the situation is somewhat different. Channel masking describes which output channel (of the grain generator) this grain should be sent to. The `partikkel` opcode in Csound can use up to eight audio outputs, and the masking values can also be fractional, distributing the sound between two outputs. Now, what we do with the sound at each output is arbitrary. We may distribute the audio signals to different locations in the listening space, or we might process each output differently. As an example, we might send every fifth grain to a reverb while we pan every other grain right and left, and create a bank of filters and send grains to each of the filters selectively (listing 15.20). We could also simply route the eight outputs to eight different speakers to create a granular spatial image in the room. If we need more than eight separate outputs, several instances of `partikkel` can be linked and synchronised by means of the `partikkelsync` opcode. See Section 15.4 for more details.

Listing 15.20 Example of channel masking, every second grain routed to stereo left and right, every fifth grain to reverb, and routing of random grains to a high-pass and a low-pass filter

```
nchnls = 2

giSine ftgen 0, 0, 65536, 10, 1
giCosine ftgen 0, 0, 8193, 9, 1, 1, 90
; (additive) saw wave
giSaw ftgen 0, 0, 65536, 10, 1, 1/2, 1/3, 1/4, 1/5,
       1/6, 1/7, 1/8,  1/9, 1/10, 1/11, 1/12, 1/13,
       1/14, 1/15, 1/16, 1/17, 1/18, 1/19, 1/20
giSigmoRise ftgen 0, 0, 8193, 19, 0.5, 1, 270, 1
giSigmoFall ftgen 0, 0, 8193, 19, 0.5, 1, 90, 1

instr 1
kamp = ampdbfs(-12)
asamplepos1 = 0
kGrainRate transeg p4, 0.5, 1, p4, 4, -1, p5, 1, 1, p5
async = 0.0 ; (disable external sync)
```

```
kGrainDur = 0.5
kgdur = (kGrainDur*1000)/kGrainRate
ka_d_ratio = 0.5
kwavfreq = 880
kwavekey1 = 1
awavfm = 0 ;  (FM disabled)
krandommask = 0


; channel masking table,
; output routing for individual grains
; (zero based, a value of 0.0 routes to output 1)
; output 0 and 1 are used for stereo channels L and R,
; output 2 and 3 for reverb left and right
; output 4 and 5 is sent to two different filters
ichannelmasks ftgentmp 0, 0, 16, -2, 0, 9,
      0, 1, 0, 1, 2, 1, 0, 1, 0, 3

; randomly route some grains (mask index 4 or 5)
; to the first filter
kfilt1control randh 1, kGrainRate, 0.1
kfilt1trig = (abs(kfilt1control) > 0.2 ? 4 : 0)
kfilt1trig =
    (abs(kfilt1control) > 0.6 ? 5 : kfilt1trig)
if kfilt1trig > 0 then
tablew 4, kfilt1trig, ichannelmasks ; send to output 4
else
tablew 0, 4, ichannelmasks ; reset to original values
tablew 1, 5, ichannelmasks
endif


; randomly route some grains (mask index 9 or 10)
; to the second filter
kfilt2control randh 1, kGrainRate, 0.2
kfilt2trig = (abs(kfilt2control) > 0.2 ? 9 : 0)
kfilt2trig =
   (abs(kfilt2control) > 0.6 ? 10 : kfilt2trig)
if kfilt2trig > 0 then
tablew 5, kfilt2trig, ichannelmasks ; send to output 5
else
tablew 1, 9, ichannelmasks ; reset to original values
tablew 0, 10, ichannelmasks
endif


a1,a2,a3,a4,a5,a6 partikkel kGrainRate, 0, -1, async,
   0, -1, giSigmoRise, giSigmoFall, 0, ka_d_ratio,
```

```
    kgdur, kamp, -1, kwavfreq, 0.5, -1, -1, awavfm,
    -1, -1, giCosine, 1, 1, 1, ichannelmasks,
    krandommask, giSaw, giSaw, giSaw, giSaw, -1,
    asamplepos1, asamplepos1, asamplepos1,asamplepos1,
     kwavekey1, kwavekey1, kwavekey1, kwavekey1, 100

outs a1,a2
chnset a3, "reverbLeft"
chnset a4, "reverbRight"
chnset a5, "filter1"
chnset a6, "filter2"
endin

instr 11
a1 chnget "reverbLeft"
a2 chnget "reverbRight"
ar1,ar2 freeverb a1,a2, 0.8, 0.3
idry = 0.2
ar1 = ar1+(a1*idry)
ar2 = ar2+(a2*idry)
outs ar1,ar2
a0 = 0
chnset a0, "reverbLeft"
chnset a0, "reverbRight"
endin

instr 12
a1 chnget "filter1"
a2 chnget "filter2"
afilt1 butterlp a1, 800
afilt2 butterhp a2, 2000
asum = afilt1+afilt2
outs asum, asum
a0 = 0
chnset a0, "filter1"
```

Listing 15.21 Score for the instrument in listing 15.20

```
i1 0 20 330 6
```

15.3.2 Waveform Mixing

We can also use grain masking to selectively alter the mix of several source sounds for each grain. With `partikkel`, we can use five source sounds: four sampled waveforms and a synthesised trainlet (see section 15.7.3 for more on trainlets). Each mask in the waveform mix masking table will then have five values, representing the amplitude of each source sound. In addition to the usual types of masking effects, the waveform mixing technique is useful for windowed-overlap-type techniques such as we saw in granular reverb time stretching. In that context we combine waveform mixing with separate time pointers for each source waveform, so we can fade in and out the different layers of time delay (in the source waveform audio buffer) as needed.

Listing 15.22 Example of source waveform mixing, rewriting the amplitude values for each source wave in a single wave mask

```
giSoundfile ftgen 0, 0, 0, 1,"fox.wav",0,0,0
giSine ftgen 0, 0, 65536, 10, 1
giCosine ftgen 0, 0, 8193, 9, 1, 1, 90
; (additive) saw wave
giSaw ftgen 0, 0, 65536, 10, 1, 1/2, 1/3, 1/4, 1/5,
        1/6, 1/7, 1/8,  1/9, 1/10, 1/11, 1/12, 1/13,
        1/14, 1/15, 1/16, 1/17, 1/18, 1/19, 1/20
giNoiseUni ftgen 0, 0, 65536, 21, 1, 1
giNoise ftgen 0, 0, 65536, 24, giNoiseUni, -1, 1
giSigmoRise ftgen 0, 0, 8193, 19, 0.5, 1, 270, 1
giSigmoFall ftgen 0, 0, 8193, 19, 0.5, 1, 90, 1

instr 1
kamp = ampdbfs(-3)
kwaveform1 = giSine
kwaveform2 = giSaw
kwaveform3 = giNoise
kwaveform4 = giSoundfile
asamplepos1 = 0.0 ; phase of single cycle waveform
asamplepos2 = 0.0
asamplepos3 = 0.0
asamplepos4 = 0.27; start read position in sound file

kGrainRate = 8
async = 0.0 ; (disable external sync)
kGrainDur = 1.0
kgdur = (kGrainDur*1000)/kGrainRate
ka_d_ratio = 0.5
kwavfreq = 1 ; set master transposition to 1
; source 1 and 2 are single cycle source waveforms,
```

```
; so the pitch is determined by cycles per second
kwavekey1 = 440 ; (a single cycle sine)
kwavekey2 = 440 ; (a single cycle saw)
; source 3 and 4 are tables with audio sample data,
; so the playback frequency should be relative to
; table length and sample rate
kwavekey3 = 1/(tableng(kwaveform3)/sr); (noise at sr)
kwavekey4 = 1/(tableng(kwaveform4)/sr); (soundfile)
awavfm = 0 ; (FM disabled)
krandommask = 0

; wave mixing by writing to the wave mask table
iwaveamptab ftgentmp 0, 0, 32, -2, 0, 0, 1,0,0,0,0
kamp1 linseg 1, 1, 1, 1, 0, 5, 0, 1, 1, 1, 1
kamp2 linseg 0, 1, 0, 1, 1, 1, 1, 1, 0, 1, 0
kamp3 linseg 0, 3, 0, 1, 1, 1, 1, 1, 0, 1, 0
kamp4 linseg 0, 5, 0, 1, 1, 1, 1, 1, 0, 1, 0
tablew kamp1, 2, iwaveamptab
tablew kamp2, 3, iwaveamptab
; we do additional scaling of source 3 and 4,
; to make them appear more equal in loudness
tablew kamp3*0.7, 4, iwaveamptab
tablew kamp4*1.5, 5, iwaveamptab

a1 partikkel kGrainRate, 0, -1, async, 0, -1,
    giSigmoRise, giSigmoFall, 0, ka_d_ratio,
    kgdur, kamp, -1, kwavfreq, 0.5, -1, -1,
    awavfm, -1, -1, giCosine, 1, 1, 1, -1,
    krandommask, kwaveform1, kwaveform2,
    kwaveform3, kwaveform4, iwaveamptab,
    asamplepos1, asamplepos2, asamplepos3,
    asamplepos4,  kwavekey1, kwavekey2,
    kwavekey3, kwavekey4, 100

out a1
endin

schedule(1,0,10)
```

Listing 15.23 Example of source waveform masking, changing to a new source waveform for each new grain

```
giSoundfile ftgen 0, 0, 0, 1,"fox.wav",0,0,0
giSine ftgen 0, 0, 65536, 10, 1
giCosine ftgen 0, 0, 8193, 9, 1, 1, 90
; (additive) saw wave
```

```
giSaw ftgen 0, 0, 65536, 10, 1, 1/2, 1/3, 1/4, 1/5,
       1/6, 1/7, 1/8,  1/9, 1/10, 1/11, 1/12, 1/13,
       1/14, 1/15, 1/16, 1/17, 1/18, 1/19, 1/20
giNoiseUni ftgen 0, 0, 65536, 21, 1, 1
giNoise ftgen 0, 0, 65536, 24, giNoiseUni, -1, 1
giSigmoRise ftgen 0, 0, 8193, 19, 0.5, 1, 270, 1
giSigmoFall ftgen 0, 0, 8193, 19, 0.5, 1, 90, 1

instr 1
kamp = ampdbfs(-3)
kwaveform1 = giSine
kwaveform2 = giSaw
kwaveform3 = giNoise
kwaveform4 = giSoundfile
asamplepos1 = 0.0 ; phase of single cycle waveform
asamplepos2 = 0.0
asamplepos3 = 0.0
asamplepos4 = 0.27; start read position in sound file

kGrainRate transeg 400, 0.5, 1, 400, 4,
                   -1, 11, 1, 1, 11
async = 0.0 ; (disable external sync)
kGrainDur = 1.0
kgdur = (kGrainDur*1000)/kGrainRate
ka_d_ratio = 0.5
kwavfreq = 1
kwavekey1 = 440
kwavekey2 = 440
kwavekey3 = 1/(tableng(kwaveform3)/sr)
kwavekey4 = 1/(tableng(kwaveform4)/sr)
awavfm = 0 ; (FM disabled)
krandommask = 0

; wave masking, balance of source waveforms
; specified per grain
iwaveamptab ftgentmp 0, 0, 32, -2, 0, 3, 1,0,0,0,0,
       0,1,0,0,0,
       0,0,1,0,0,
       0,0,0,1,0

a1 partikkel kGrainRate, 0, -1, async, 0, -1,
       giSigmoRise, giSigmoFall, 0, ka_d_ratio,
       kgdur, kamp, -1, kwavfreq, 0.5, -1, -1,
       awavfm, -1, -1, giCosine, 1, 1, 1, -1,
       krandommask, kwaveform1, kwaveform2,
```

```
        kwaveform3, kwaveform4, iwaveamptab,
        asamplepos1, asamplepos2, asamplepos3,
        asamplepos4,  kwavekey1, kwavekey2,
        kwavekey3, kwavekey4, 100

  out a1
  endin
```

15.4 Clock Synchronisation

Most granular synthesisers use an internal clock to trigger generation of new grains. To control the exact placement of grains in time, we might sometimes need to manipulate this clock. Grain displacement can be done simply with the kdistribution parameter of partikkel, offsetting individual grains in time within a period of 1/grainrate.

Listing 15.24 Using the kdistribution parameter of partikkel, individual grains are displaced in time within a time window of 1/grainrate. The stochastic distribution can be set with the idisttab table. The following is an excerpt of code that can be used for example with the example in listing 15.4. See also Figure 15.6 for an illustration

```
...
kdistribution line 0, p3-1, 1
idisttab ftgen 0, 0, 32768, 7, 0, 32768, 1
...
a1 partikkel kGrainRate, kdistribution, idisttab, ...
```

For more elaborate grain clock patterns and synchronisations we can manipulate the internal clock directly. The partikkel opcode uses a sync input and sync output to facilitate such clock manipulation, and these can also be used to synchronise several partikkel generators if need be. The internal clock is generated by a ramping value, as is common in many digital metronomes and oscillators. The internal ramping value is updated periodically, and the exact increment determines the steepness of the ramp. The ramping value normally starts at zero, and when it exceeds 1.0 a clock pulse is triggered, and the ramp value is reset to zero. The sync input to partikkel lets us directly manipulate the ramping value, and offsetting this value directly affects the time until the next clock pulse. Figure 15.7 provides an illustration of this mechanism. Sync output from partikkel is done via the helper opcode partikkelsync. This opcode is linked to a particular partikkel instance by means of an opcode ID. The partikkelsync instance then has internal access to the grain scheduler clock of a partikkel with the same ID. It outputs the clock pulse and the internal ramping value, and these can be used to directly drive other partikkel instances, or to synchronise external clocks using the same ramping technique. The ramping value is the phase of the clock, and this

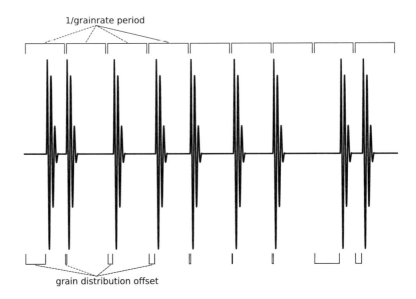

Fig. 15.6 Grain distribution, showing periodic grain periods on top, random displacements at bottom

can be used to determine whether we should nudge the clock up or down towards the nearest beat.

Listing 15.25 Soft synchronisation between two `partikkel` instances, using `partikkelsync` to get the clock phase and clock tick, then nudging the second clock up or down towards the nearest clock tick from `partikkel` instance 1

```
instr 1
iamp = ampdbfs(-12)
kamp linen iamp, 0.01, p3, 1
asamplepos1 = 0
kGrainRate1 = 7
async = 0.0 ; (disable external sync)
kGrainDur = 1.0
kgdur1 = (kGrainDur*1000)/kGrainRate1
ka_d_ratio = 0.1
kwavfreq1 = 880
kwavekey1 = 1
awavfm = 0 ; (FM disabled)
id1 = 1
a1 partikkel kGrainRate1, 0, -1, async, 0, -1,
    giSigmoRise, giSigmoFall, 0, ka_d_ratio, kgdur1,
```

```
      1, -1, kwavfreq1, 0.5, -1, -1, awavfm,  -1, -1,
      giCosine, 1, 1, 1, -1, 0, giSine, giSine, giSine,
      giSine, -1, asamplepos1, asamplepos1, asamplepos1,
      asamplepos1, kwavekey1, kwavekey1, kwavekey1,
      kwavekey1, 100, id1

async1, aphase1 partikkelsync, id1

kphaSyncGravity line 0, p3, 0.7
aphase2 init 0
asyncPolarity limit (int(aphase2*2)*2)-1, -1, 1
asyncStrength =
    abs(abs(aphase2-0.5)-0.5)*asyncPolarity
; Use the phase of partikkelsync instance 2 to find
; sync polarity for partikkel instance 2.
; If the phase of instance 2 is less than 0.5,
; we want to nudge it down when synchronizing,
; and if the phase is > 0.5 we
; want to nudge it upwards.
async2in = async1*kphaSyncGravity*asyncStrength

kGrainRate2 = 5
kgdur2 = (kGrainDur*1000)/kGrainRate2
kwavfreq2 = 440
id2 = 2
a2 partikkel kGrainRate2, 0, -1, async2in, 0, -1,
      giSigmoRise, giSigmoFall, 0, ka_d_ratio, kgdur2,
      1, -1, kwavfreq2, 0.5, -1, -1, awavfm, -1, -1,
      giCosine, 1, 1, 1, -1, 0, giSine, giSine, giSine,
      giSine, -1, asamplepos1, asamplepos1 asamplepos1,
      asamplepos1, kwavekey1, kwavekey1, kwavekey1,
      kwavekey1, 100, id2

async2, aphase2 partikkelsync, id2

; partikkel instance 1 outputs to the left
; instance 2 outputs to the right
```

Listing 15.26 Gradual synchronisation. As in listing 15.25, but here we also adjust the grain rate of `partikkel` instance 2 to gradually approach the rate of clock pulses from instance 1. This leads to a quite musical rhythmic gravitation, attracting instance 2 to the pulse of instance 1.

```
instr 1
iamp = ampdbfs(-12)
kamp linen iamp, 0.01, p3, 1
asamplepos1 = 0
```

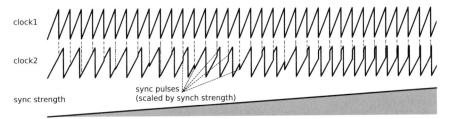

clock1

clock2

sync strength   sync pulses
                        (scaled by synch strength)

Fig. 15.7 Soft synchronisation between two clocks, showing the ramp value of each clock and sync pulses from clock 1 nudging the phase of clock 2. Sync pulses are scaled by sync strength, in this case increasing strength, forcing clock 2 to synchronise with clock 1

```
kGrainRate1 = 7
async = 0.0 ; (disable external sync)
kGrainDur = 1.0
kgdur1 = (kGrainDur*1000)/kGrainRate1
ka_d_ratio = 0.1
kwavfreq1 = 880
kwavekey1 = 1
awavfm = 0 ; (FM disabled)
id1 = 1
a1 partikkel kGrainRate1, 0, -1, async, 0, -1,
    giSigmoRise, giSigmoFall, 0, ka_d_ratio, kgdur1,
    1, -1, kwavfreq1, 0.5, -1, -1, awavfm,  -1, -1,
    giCosine, 1, 1, 1, -1, 0, giSine, giSine, giSine,
    giSine, -1, asamplepos1, asamplepos1, asamplepos1,
    asamplepos1, kwavekey1, kwavekey1, kwavekey1,
    kwavekey1, 100, id1

async1, aphase1 partikkelsync, id1

kphaSyncGravity linseg 0, 2, 0, p3-5, 1, 1, 1
aphase2 init 0
ksync2 init 0
asyncPolarity limit (int(aphase2*2)*2)-1, -1, 1
asyncStrength =
    abs(abs(aphase2-0.5)-0.5)*asyncPolarity
; Use the phase of partikkelsync instance 2 to find
; sync polarity for partikkel instance 2.
; If the phase of instance 2 is less than 0.5,
; we want to nudge it down when synchronizing,
; and if the phase is > 0.5
; we want to nudge it upwards.
async2in = async1*kphaSyncGravity*asyncStrength
```

```
; adjust grain rate of second partikkel instance
; to approach that of the first instance
krateSyncGravity = 0.0005
ksyncPulseCount init 0
ksync1 downsamp async1, ksmps
ksync1 = ksync1*ksmps
; count the number of master clock pulses
; within this (slave)clock period
ksyncPulseCount = ksyncPulseCount + ksync1
ksyncRateDev init 0
ksyncStrength downsamp asyncStrength
; sum of deviations within this (slave)clock period
ksyncRateDev = ksyncRateDev + (ksyncStrength*ksync1)

; adjust rate only on slave clock tick
if ksync2 > 0 then
; if no master clock ticks, my tempo is too high
if ksyncPulseCount == 0 then
krateAdjust = -krateSyncGravity
; if more than one master clock tick,
; my tempo is too low
elseif ksyncPulseCount > 1 then
krateAdjust = krateSyncGravity
; if exactly one master clock tick,
; it depends on the phase value at the time
; when the master clock tick was received
elseif ksyncPulseCount == 1 then
krateAdjust = ksyncRateDev*krateSyncGravity*0.02
endif

; Reset counters on (slave)clock tick
ksyncPulseCount = 0
ksyncRateDev = 0
endif

kGrainRate2 init 2
kGrainRate2 = kGrainRate2 +
        (krateAdjust*kGrainRate2*0.1)
kgdur2 = (kGrainDur*1000)/kGrainRate2
kwavfreq2 = 440
id2 = 2
a2 partikkel kGrainRate2, 0, -1, async2in, 0, -1,
    giSigmoRise, giSigmoFall, 0, ka_d_ratio, kgdur2,
    1, -1, kwavfreq2, 0.5, -1, -1, awavfm, -1, -1,
    giCosine, 1, 1, 1, -1, 0, giSine, giSine, giSine,
```

```
        giSine, -1, asamplepos1, asamplepos1, asamplepos1,
        asamplepos1, kwavekey1, kwavekey1, kwavekey1,
        kwavekey1, 100, id2

async2, aphase2 partikkelsync, id2
ksync2 downsamp async2, ksmps
ksync2 = ksync2*ksmps
```

15.5 Amplitude Modulation and Granular Synthesis

Amplitude modulation is inherent in all granular processing, because we fade individual snippets of sound in and out, i.e. modulating the amplitude of the grains. Under certain specific conditions, we can create the same spectrum with granular synthesis as we can with amplitude modulation (AM) (figs. 15.8 and 15.9). This goes to show an aspect of the flexibility of granular synthesis, and can also help us understand some of the artefacts that can occur in granular processing. Let's look at a very specific case: if the grain envelope is sinusoidal and the grain duration is exactly 1/grainrate (so that the next grain starts exactly at the moment where the previous grain stops), and the waveform inside grains has a frequency which has an integer ratio to the grain rate (so that one or more whole cycles of the waveform fit exactly inside a grain), the result is identical to amplitude modulation with a sine wave where the modulation frequency equals the grain rate and the carrier frequency equals the source waveform frequency. For illustrative purposes, we call denote grain rate by gr and waveform frequency by wf. We observe a partial at gr, with sidebands at $gr+wf$ and $gf-wf$.

Now, if we dynamically change the frequency of the source waveform, the clean sidebands of traditional AM will start to spread, creating a cascade of sidebands (Fig. 15.10)

At the point where the waveform frequency attains an integer multiple of the grain rate, the cascading sidebands will again diminish, approaching a new steady state where granular synthesis equals traditional AM (see listing 15.27 and Fig. 15.11). The non-integer ratio of $gr{:}wf$ amounts to a periodic phase reset of the AM carrier. This may seem like a subtle effect since the phase reset happens when the carrier has zero amplitude. The difference is significant, as can be heard in the output of listing If we want to avoid these artefacts, we can create an equivalent effect by cross-fading between two source waveforms, each being of a frequency in integer relationship to the grain rate (listing 15.28 and Fig. 15.12).

Listing 15.27 Granular synthesis equals traditional AM when the ratio of grain rate to source waveform frequency is of an integer ratio (grain rate 200 Hz, waveform frequency 400 Hz). Here we gradually go from an integer to a non-integer ratio (sweeping the waveform frequency from 400 Hz to 600 Hz, then from 600 Hz to 800 Hz), note the spreading of the sidebands. The example ends at an integer ratio again (source waveform frequency 800 Hz)

```
instr 1
```

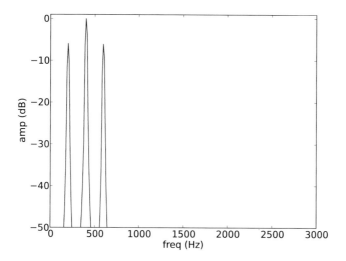

Fig. 15.8 FFT of AM with modulator frequency 200 Hz and carrier frequency 400 Hz

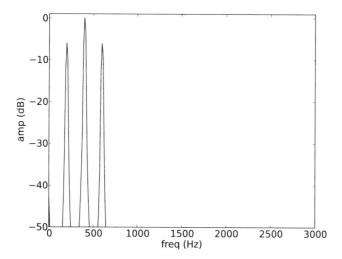

Fig. 15.9 Granular synthesis with grain rate 200 Hz and source waveform frequency 400 Hz. We use a sine wave as the source waveform and a sinusoid grain shape

```
kamp linen 1, 0.1, p3, 0.1
kamp = kamp*ampdbfs(-3)
asamplepos1 = 0
```

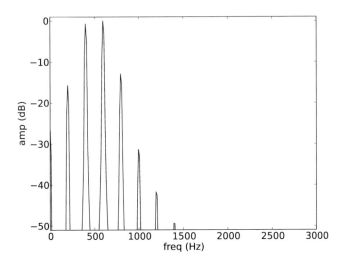

Fig. 15.10 Granular synthesis with grain rate 200 Hz and source waveform frequency 500 Hz. We use a sine wave as the source waveform and a sigmoid grain shape. Note the extra sidebands added due to the non-integer relationship between grain rate and source waveform frequency

```
kGrainRate = p4
async = 0.0 ; (disable external sync)
kGrainDur = 1.0
kgdur = (kGrainDur*1000)/kGrainRate
ka_d_ratio = 0.5
ipitch1 = p4*2
ipitch2 = p4*3
ipitch3 = p4*4
kwavfreq linseg ipitch1, 1, ipitch1, 2,
      ipitch2, 1, ipitch2, 2, ipitch3, 1, ipitch3
kwavekey1 = 1
iwaveamptab ftgentmp 0, 0, 32, -2, 0, 0, 1,0,0,0,0
awavfm = 0 ; (FM disabled)
a1 partikkel kGrainRate, 0, -1, async, 0, -1,
    giSigmoRise, giSigmoFall, 0, ka_d_ratio, kgdur,
    kamp, -1, kwavfreq, 0.5, -1, -1, awavfm,
    -1, -1, giCosine, 1, 1, 1, -1, 0,
    giSine, giSine, giSine, giSine, iwaveamptab,
    asamplepos1,asamplepos1,asamplepos1,asamplepos1,
    kwavekey1, kwavekey1, kwavekey1, kwavekey1, 100
out a1
endin
```

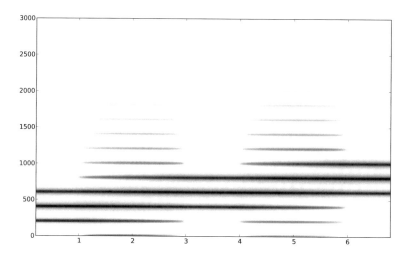

Fig. 15.11 Spectrogram of the waveform produced by listing 15.27, note the extra sidebands during source waveform frequency sweep.

Listing 15.28 Similar to the previous example, but avoiding the spreading sidebands by cross-fading between source waveforms of integer frequency ratios (crossfading a source waveform of frequency 400 Hz with a waveform of frequency 600 Hz, then cross-fading with another waveform with frequency 800 Hz)

```
instr 1
kamp linen 1, 0.1, p3, 0.1
kamp = kamp*ampdbfs(-3)
asamplepos1 = 0
; invert phase to retain constant
; power during crossfade
asamplepos2 = 0.5
asamplepos3 = 0
kGrainRate = p4
async = 0.0 ; (disable external sync)
kGrainDur = 1.0
kgdur = (kGrainDur*1000)/kGrainRate
ka_d_ratio = 0.5
awavfm = 0 ; (FM disabled)
kwavfreq = 1
kwavekey1 = p4*2
kwavekey2 = p4*3
kwavekey3 = p4*4
; crossface by using wave mix masks
iwaveamptab ftgentmp 0, 0, 32, -2, 0, 0, 0,0,0,0,0
kamp1 linseg 1, 1, 1, 2, 0, 4, 0
kamp2 linseg 0, 1, 0, 2, 1, 1, 1, 2, 0, 2, 0
```

```
kamp3 linseg 0, 4, 0, 2, 1, 1, 1
tablew kamp1, 2, iwaveamptab
tablew kamp2, 3, iwaveamptab
tablew kamp3, 4, iwaveamptab
a1 partikkel kGrainRate, 0, -1, async, 0, -1,
    giSigmoRise, giSigmoFall, 0, ka_d_ratio, kgdur,
    kamp, -1, kwavfreq, 0.5, -1, -1, awavfm,
    -1, -1, giCosine, 1, 1, 1, -1, 0,
    giSine, giSine, giSine, giSine, iwaveamptab,
    asamplepos1,asamplepos2,asamplepos3,asamplepos1,
    kwavekey1, kwavekey2, kwavekey3, kwavekey1, 100
out a1
```

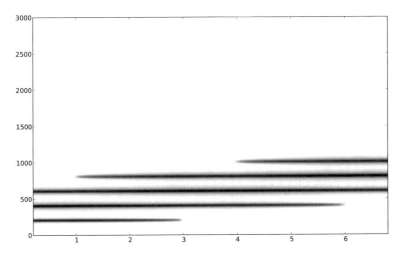

Fig. 15.12 Spectrogram of the waveform produced by listing 15.28, cleaner transition with no extra sidebands

15.6 Pitch Synchronous Granular Synthesis

In this section we will look at the technique *pitch synchronous granular synthesis* (PSGS) and use this for formant shifting of a sampled sound. As we have seen in Section 15.1.2, we can use a high grain rate with a periodic grain clock to constitute pitch. We have also seen that the pitch of the waveform inside grains can be used to create formant regions. Here we will use pitch tracking to estimate the fundamental frequency of the source waveform, and link the grain clock to this fundamental frequency. This means that the pitch constituted by the granular process will be the same as for the source sound, while we are free to manipulate the grain pitch for the purpose of shifting the formants of the synthesised sound. Shifting the formants of a sampled sound is perceptually similar to changing the size of the acoustic object generating the sound, and in this manner we can create the illusion of a "gender change" (or change of head and throat size) for recorded vocal sounds. The amplitude modulation effects considered in Section 15.5 can be minimised by using a slightly longer grain size. In our example, we use a grain size of 2/grainrate, although this can be tuned further.

Listing 15.29 Pitch synchronous granular synthesis used as a formant shift effect

```
giSoundfile ftgen 0, 0, 0, 1,"Vocphrase.wav",0,0,0

instr 1
kamp adsr 0.0001, 0.3, 0.5, 0.5
kamp = kamp*ampdbfs(-6)
isoundDur = filelen("Vocphrase.wav")
asamplepos1 phasor 1/isoundDur
aref loscil 1, 1, giSoundfile, 1
kcps, krms pitchamdf aref, 100, 800
kGrainRate limit kcps, 100, 800
async = 0.0 ; (disable external sync)
kGrainDur = 2.0
kgdur = (kGrainDur*1000)/kGrainRate
ka_d_ratio = 0.5
kwavfreq = semitone(p4)
kwavekey1 = 1/isoundDur
awavfm = 0 ; (FM disabled)
a1 partikkel kGrainRate, 0, -1, async, 0, -1,
    giSigmoRise, giSigmoFall, 0, ka_d_ratio, kgdur,
    kamp, -1, kwavfreq, 0.5, -1, -1, awavfm,
    -1, -1, giCosine, 1, 1, 1, -1, 0, giSoundfile,
    giSoundfile, giSoundfile, giSoundfile, -1,
    asamplepos1,asamplepos1,asamplepos1,asamplepos1,
    kwavekey1, kwavekey1, kwavekey1, kwavekey1, 100
out a1
endin

schedule(1,0,7,0)
schedule(1,7,7,4)
schedule(1,14,7,8)
schedule(1,21,7,-4)
schedule(1,28,7,-8)
```

15.7 Morphing Between Classic Granular Synthesis Types

Many of the classic types of granular synthesis described by Roads [109] originally required a specialised audio synthesis engine for each variant of the technique. Each granular type has some specific requirements not shared with other varieties of the technique. For example, a glisson generator needs the ability to create a specific pitch trajectory during the span of each grain; a pulsar generator needs grain masking and parameter linking; a trainlet generator needs an internal synthesiser to provide the desired parametrically controlled harmonic spectrum of the source waveform; and so on. The `partikkel` opcode in Csound was designed specifically to overcome the impracticalities posed by the need to use a different audio generator for each granular type. All of the specific needs associated with each granular type were implemented in a combined super-generator capable of all time-based granular synthesis types. Due to a highly optimised implementation, this is not overly processor intensive, and a high number of `partikkel` generators can run simultaneously in real-time. The following is a brief recounting of some classic granular synthesis types not already covered in the text, followed by a continuous parametric morph through all types.

15.7.1 Glissons

Glisson synthesis is a straightforward extension of basic granular synthesis in which the source waveform for each grain has an independent frequency trajectory. The grain or glisson creates a short glissando (see Fig. 15.13).

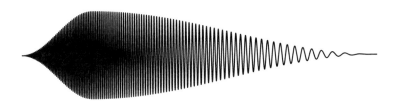

Fig. 15.13 A glisson with a downward glissando during the grain

With `partikkel`, we can do this by means of pitch masking, with independent masks for the start and end frequencies (the `iwavfreqstarttab` and `iwavfreqendtab` parameters of the opcode). Since grain masking gives control over individual grains, each grain can have a separate pitch trajectory. Due to this flexibility we can also use statistical control over the grain characteristics.

15.7.2 Grainlets

Grainlet synthesis is inspired by ideas from wavelet synthesis. We understand a wavelet to be a short segment of a signal, always encapsulating a constant number of cycles. Hence the duration of a wavelet is always inversely proportional to the frequency of the waveform inside it. Duration and frequency are thus linked (through an inverse relationship). Grainlet synthesis as described by [109] allows a generalisation of the linkage between different synthesis parameters. Some exotic combinations mentioned by Roads are duration/space, frequency/space and amplitude/space. The space parameter refers to the placement of a grain in the stereo field or the spatial position in a 3D multichannel set-up.

15.7.3 Trainlets

Trainlets differ from other granular synthesis techniques in that they require a very specific source waveform for the grains. The waveform consists of a bandlimited impulse train as shown in Figure 15.14.

Fig. 15.14 Bandlimited trainlet pulse

A trainlet is specified by:

- Base frequency.
- Number of harmonics.
- Harmonic balance (chroma): the energy distribution between high- and low-frequency harmonics.

The `partikkel` opcode has an internal impulse train synthesiser to enable creation of these specific source waveforms. This is controlled by the `ktraincps`, `knumpartials` and `kchroma` parameters. To enable seamless morphing between trainlets and other types of granular synthesis, the impulse train generator has been implemented as a separate source waveform, and the mixing of source waveforms is done by means of grain masks (the `iwaveamptab` parameter to `partikkel`). See Section 15.3.2 for more details about waveform mixing.

15.7.4 Pulsars

Pulsar audio synthesis relates to the phenomenon of fast-rotating neutron stars (in astronomy, a pulsar is short for a *pulsating radio star*), which emit a beam of electromagnetic radiation. The speed of rotation of these stars can be as high as several hundred revolutions per second. A stationary observer will then observe the radiation as a pulse, appearing only when the beam of emission points toward the observer. In the context of audio synthesis, Roads [109] uses the term pulsar to describe a sound particle consisting of an arbitrary waveform (the pulsaret) followed by a silent interval. The total duration of the pulsar is called the pulsar period, while the duration of the pulsaret is called the duty cycle (see Fig. 15.15. The pulsaret itself can be seen as a special kind of grainlet, where pitch and duration are linked. A pulsaret can be contained by an arbitrary envelope, and the envelope shape affects the spectrum of the pulsaret due to the amplitude modulation effects inherent in applying the envelope to the signal. Repetitions of the pulsar signal form a pulsar train. We can use parameter linkage to create the pulsarets and amplitude masking of grains to create a patterned stream of pulsarets, making a pulsar train).

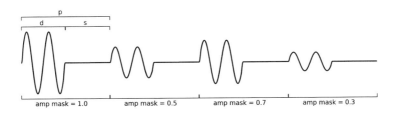

Fig. 15.15 A pulsar train consisting of pulsarets with duty cycle *d*, silent interval *s* and pulsar period *p*. Amplitude masks are used on the pulsarets

15.7.5 Formant Synthesis

As we have seen in Sections 15.1.2 and 15.5, we can use granular techniques to create a spectrum with controllable formants. This can be utilised to simulate vocals or speech, and also other formant-based sounds. Several variants of particle-based formant synthesis (FOF, Vosim, Window Function Synthesis) have been proposed [109]. As a generalisation of these techniques we can say that the base pitch is constituted by the grain rate (which is normally periodic), the formant position is determined by the pitch of the source waveform inside each grain (commonly a sine wave), and the grain envelope controls the formant's spectral shape. The Csound manual for `fof2` contains an example of this kind of formant synthesis. We can also find a full listing of formant values for different vowels and voice types at http://csound.github.io/docs/manual/MiscFormants.html. With `partikkel` we can approximate the same effect by using the four source waveforms, all set to sine waves, with a separate frequency trajectory for each source waveform (Fig. 15.16). The formant frequencies will be determined by the source waveform frequencies. We can use waveform-mixing techniques as described in Section 15.3.2 to adjust the relative amplitudes of the formants. We will not have separate control over the bandwidth of each formant, since the same grain shape will be applied to all source waveforms. On the positive side, we only use one grain generator instance so the synthesiser will be somewhat less computationally expensive, and we are able to gradually morph between formant synthesis and other types of granular synthesis.

Listing 15.30 Formant placement by transposition of source waveforms, here moving from a bass 'a' to a bass 'e'

```
kwavekey1 linseg 600, 1, 600, 2, 400, 1, 400
kwavekey2 linseg 1040, 1, 1040, 2, 1620, 1, 1620
kwavekey3 linseg 2250, 1, 2250, 2, 2400, 1, 2400
kwavekey4 linseg 2450, 1, 2450, 2, 2800, 1, 2800
```

Listing 15.31 Relative level of formants controlled by wave-mix masking table. We use `ftmorf` to gradually change between the different masking tables

```
iwaveamptab ftgentmp 0, 0, 32, -2, 0, 0,
     1, 0, 0, 0, 0
iwaveamptab1 ftgentmp 0, 0, 32, -2, 0, 0,
     1, ampdbfs(-7), ampdbfs(-9), ampdbfs(-9), 0
iwaveamptab2 ftgentmp 0, 0, 32, -2, 0, 0,
     1, ampdbfs(-12), ampdbfs(-9), ampdbfs(-12), 0
iwavetabs ftgentmp 0, 0, 2, -2,
     iwaveamptab1, iwaveamptab2
kwavemorf linseg 0, 1, 0, 2, 1, 1, 1
ftmorf kwavemorf, iwavetabs, iwaveamptab
```

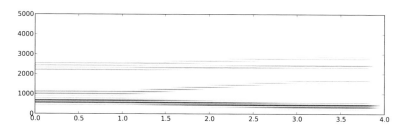

Fig. 15.16 Sonogram of the formants created by listings 15.30 and 15.31

15.7.6 Morphing Between Types of Granular Synthesis

To show how to move seamlessly between separate granular types, we will give an example that morphs continuously through: sampled waveform - single cycle waveform - glissons - trainlets - pulsars - formant synthesis - asynchronous granular - waveform mixing. The morph is done in one single continuous tone where only the parameters controlling the synthesis change over time. The best way to study this example is to start listening to the produced sound, as the code to implement it is necessarily somewhat complex. The main automated parameters are:

- source waveform
- source waveform pitch
- phase (time pointer into the source waveform)
- amplitude
- grain rate
- clock sync
- grain shape
- pitch sweep
- trainlet parameters
- grain masking (amplitude, channel, wave mix).

Listening to the sound produced by listing 15.32, we can hear the different granular types at these times (in seconds into the soundfile):

- 0 s: sampled waveform
- 5 s: single cycle waveform
- 7 s: glissons
- 14 s: trainlets
- 25 s: pulsars
- 35 s: formant synthesis
- 42 s: asynchronous granular
- 49 s: waveform mixing.

Listing 15.32 Morphing through different variations of granular synthesis

```
nchnls = 2
0dbfs = 1

; load audio files
giVocal ftgen 0, 0, 0, 1, "Vocphrase.wav", 0, 0, 0
giChoir ftgen 0, 0, 0, 1, "Choir.wav", 0, 0, 0
giCello ftgen 0, 0, 0, 1, "Cello.wav", 0, 0, 0
giVibLine ftgen 0, 0, 0, 1, "VibDist.wav", 0, 0, 0

; classic waveforms
giSine ftgen 0,0,65537,10,1
giCosine ftgen 0,0,8193,9,1,1,90
```

```
giTri ftgen 0,0,8193,7,0,2048,1,4096,-1,2048,0

; grain envelope tables
giSigmoRise ftgen 0,0,8193,19,0.5,1,270,1
giSigmoFall ftgen 0,0,8193,19,0.5,1,90,1
giExpFall ftgen 0,0,8193,5,1,8193,0.00001
giTriangleWin ftgen 0,0,8193,7,0,4096,1,4096,0

; asynchronous clock UDO
opcode probabilityClock, a, k
kdens xin
setksmps 1
krand rnd31 1, 1
krand = (krand*0.5)+0.5
ktrig = (krand < kdens/kr ? 1 : 0)
atrig upsamp ktrig
xout atrig
endop

instr 1

; * use instrument running time
; as the morphing "index"
kmorftime timeinsts

; * source waveform selection automation
; single-cycle waveforms must
; be transposed differently
; than sampled waveforms,
; hence the kwaveXSingle variable
kwaveform1 = (kmorftime < 30 ? giVocal : giSine)
kwave1Single = (kmorftime < 30 ? 0 : 1)
kwaveform2 = (kmorftime < 51 ? giSine : giChoir)
kwave2Single = (kmorftime < 51 ? 1 : 0)
kwaveform3 = (kmorftime < 51 ? giSine : giCello)
kwave3Single = (kmorftime < 51 ? 1 : 0)
kwaveform4 = (kmorftime < 51 ? giSine : giVibLine)
kwave4Single = (kmorftime < 51 ? 1 : 0)
kwaveform1 = (kmorftime > 49 ? giVocal : kwaveform1)
kwave1Single = (kmorftime > 49 ? 0 : kwave1Single)

; * get source waveform length
; (used when calculating transposition
;  and time pointer)
kfildur1 = tableng(kwaveform1) / sr
```

```
kfildur2 = tableng(kwaveform2) / sr
kfildur3 = tableng(kwaveform3) / sr
kfildur4 = tableng(kwaveform4) / sr

; * original pitch for each waveform,
; use if they should be transposed individually
kwavekey1 linseg 1, 30, 1, 4, 600, 3, 600, 2,
     400, 11, 400, 0, 1
kwavekey2 linseg 440, 30, 440, 4, 1040, 3, 1040, 2,
     1620, 12, 1620, 1, semitone(-5)
kwavekey3 linseg 440, 30, 440, 4, 2250, 3, 2250, 2,
     2400, 12, 2400, 1, semitone(10)
kwavekey4 linseg 440, 30, 440, 4, 2450, 3, 2450, 2,
     2800, 12, 2800, 1, semitone(-3)

; set original key dependant on waveform length
; (only for sampled waveforms,
;  not for single cycle waves)
kwavekey1 = (kwave1Single > 0 ?
     kwavekey1 : kwavekey1/kfildur1)
kwavekey2 = (kwave2Single > 0 ?
     kwavekey2 : kwavekey2/kfildur2)
kwavekey3 = (kwave3Single > 0 ?
     kwavekey3 : kwavekey3/kfildur3)
kwavekey4 = (kwave4Single > 0 ?
     kwavekey4 : kwavekey4/kfildur4)

; * time pointer (phase) for each source waveform.
isamplepos1 = 0
isamplepos2 = 0
isamplepos3 = 0
isamplepos4 = 0

; phasor from 0 to 1,
; scaled to the length of the source waveform
kTimeRate = 1
asamplepos1 phasor kTimeRate / kfildur1
asamplepos2 phasor kTimeRate / kfildur2
asamplepos3 phasor kTimeRate / kfildur3
asamplepos4 phasor kTimeRate / kfildur4

; mix initial phase and moving phase value
; (moving phase only for sampled waveforms,
; single cycle waveforms use static samplepos)
asamplepos1 = asamplepos1*(1-kwave1Single) +
```

```
    isamplepos1
asamplepos2 = asamplepos2*(1-kwave2Single) +
    isamplepos2
asamplepos3 = asamplepos3*(1-kwave3Single) +
    isamplepos3
asamplepos4 = asamplepos4*(1-kwave4Single) +
    isamplepos4

; * amplitude
kdb linseg -3, 8, -3, 4, -10, 2.5, 0, 0.5, -2, 10,
    -2, 3, -6, 2, -6, 1.3, -2, 1.5, -5, 4, -2
kenv expseg 1, p3-0.5, 1, 0.4, 0.001
kamp = ampdbfs(kdb) * kenv

; * grain rate
kGrainRate linseg 12, 7, 12, 3, 8, 2, 60, 5,
    110, 22, 110, 2, 14

; * sync
kdevAmount linseg 0, 42, 0, 4, 1, 2, 1, 2, 0
async probabilityClock kGrainRate
async = async*kdevAmount

; * distribution
kdistribution = 0.0
idisttab ftgentmp 0, 0, 16, 16, 1, 16, -10, 0

; * grain shape
kGrainDur linseg 2.5, 2, 2.5, 5, 1.0, 5, 5.0, 4,
    1.0, 1, 0.8, 5, 0.2, 10, 0.8, 7, 0.8, 3,
    0.1, 5, 0.1, 1, 0.2, 2, 0.3, 3, 2.5
kduration = (kGrainDur*1000)/kGrainRate
ksustain_amount linseg 0, 16, 0, 2, 0.9,
    12 ,0.9, 5, 0.2
ka_d_ratio linseg 0.5, 30, 0.5, 5, 0.25, 4, 0.25, 3,
    0.1, 7, 0.1, 1, 0.5
kenv2amt linseg 0, 30, 0, 5, 0.5

; * grain pitch
kwavfreq = 1
awavfm = 0 ;(FM disabled)

; * pitch sweep
ksweepshape = 0.75
iwavfreqstarttab ftgentmp 0, 0, 16, -2, 0, 0, 1
```

```
iwavfreqendtab ftgentmp 0, 0, 16, -2, 0, 0, 1
kStartFreq randh 1, kGrainRate
kSweepAmount linseg 0, 7, 0, 3, 1, 1, 1, 4, 0
kStartFreq = 1+(kStartFreq*kSweepAmount)
tablew kStartFreq, 2, iwavfreqstarttab

; * trainlet parameters
; amount of parameter linkage between
; grain dur and train cps
kTrainCpsLinkA linseg 0, 17, 0, 2, 1
kTrainCpsLink = (kGrainDur*kTrainCpsLinkA)+
      (1-kTrainCpsLinkA)
kTrainCps = kGrainRate/kTrainCpsLink
knumpartials = 16
kchroma linseg 1, 14, 1, 3, 1.5, 2, 1.1

; * masking
; gain masking table, amplitude for
; individual grains
igainmasks ftgentmp 0, 0, 16, -2, 0, 1, 1, 1
kgainmod linseg 1, 19, 1, 1, 1, 3, 0.5, 1,
      0.5, 6, 0.5, 7, 1
; write modified gain mask,
; every 2nd grain will get a modified amplitude
tablew kgainmod, 3, igainmasks

; channel masking table,
; output routing for individual grains
; (zero based, a value of 0.0
;   routes to output 1)
ichannelmasks ftgentmp 0, 0, 16, -2, 0, 3,
      0.5, 0.5, 0.5, 0.5

; create automation to modify channel masks
; the 1st grain moving left,
; the 3rd grain moving right,
; other grains stay at centre.
kchanmodL linseg 0.5, 25, 0.5, 3, 0.0, 5, 0.0, 4, 0.5
kchanmodR linseg 0.5, 25, 0.5, 3, 1.0, 5, 1.0, 4, 0.5
tablew kchanmodL, 2, ichannelmasks
tablew kchanmodR, 4, ichannelmasks

; amount of random masking (muting)
; of individual grains
krandommask linseg 0, 22, 0, 7, 0, 3, 0.5, 3, 0.0
```

```
; wave mix masking.
; Set gain per source waveform per grain,
; in groups of 5 amp values, reflecting
; source1, source2, source3, source4,
; and the 5th slot is for trainlet amplitude.
iwaveamptab ftgentmp 0, 0, 32, -2, 0, 0, 1,0,0,0,0

; vocal sample
iwaveamptab1 ftgentmp 0, 0, 32, -2, 0, 0, 1,0,0,0,0
; sine
iwaveamptab2 ftgentmp 0, 0, 32, -2, 0, 0, 0,1,0,0,0
; trainlet
iwaveamptab3 ftgentmp 0, 0, 32, -2, 0, 0, 0,0,0,0,1
; formant 'a'
iwaveamptab4 ftgentmp 0, 0, 32, -2, 0, 0,
     1, ampdbfs(-7), ampdbfs(-9), ampdbfs(-9), 0
; formant 'e'
iwaveamptab5 ftgentmp 0, 0, 32, -2, 0, 0,
     1, ampdbfs(-12), ampdbfs(-9), ampdbfs(-12), 0
iwavetabs ftgentmp 0, 0, 8, -2,
     iwaveamptab1, iwaveamptab2, iwaveamptab3,
     iwaveamptab4, iwaveamptab5, iwaveamptab1,
     iwaveamptab2, iwaveamptab1
kwavemorf linseg 0, 4, 0, 3, 1, 4, 1, 5, 2, 14, 2, 5,
     3, 2, 3, 2, 4, 8, 4, 1, 5, 1, 6, 1, 6, 1, 7
ftmorf kwavemorf, iwavetabs, iwaveamptab

; generate waveform crossfade automation
; (only enabled after 52 seconds, when we
; want to use the 2D X/Y axis
; method to mix sources)
kWaveX linseg 0, 52,0, 1,0, 1,1, 1,1
kWaveY linseg 0, 52,0, 1,1, 1,1, 1,0

if kmorftime < 52 kgoto skipXYwavemix
; calculate gain for 4 sources from XY position
kwgain1 limit ((1-kWaveX)*(1-kWaveY)), 0, 1
kwgain2 limit (kWaveX*(1-kWaveY)), 0, 1
kwgain3 limit ((1-kWaveX)*kWaveY), 0, 1
kwgain4 limit (kWaveX*kWaveY), 0, 1
tablew kwgain1, 2, iwaveamptab
tablew kwgain2, 3, iwaveamptab
tablew kwgain3, 5, iwaveamptab
tablew kwgain4, 4, iwaveamptab
```

```
skipXYwavemix:

a1,a2,a3,a4,a5,a6,a7,a8 partikkel kGrainRate,
   kdistribution, idisttab, async, kenv2amt,
   giExpFall, giSigmoRise, giSigmoFall,
   ksustain_amount, ka_d_ratio,
   kduration, kamp, igainmasks, kwavfreq, ksweepshape,
   iwavfreqstarttab, iwavfreqendtab, awavfm,  -1, -1,
   giCosine, kTrainCps, knumpartials, kchroma,
   ichannelmasks, krandommask, kwaveform1, kwaveform2,
   kwaveform3, kwaveform4, iwaveamptab,
   asamplepos1, asamplepos2, asamplepos3, asamplepos4,
   kwavekey1, kwavekey2, kwavekey3, kwavekey4, 100

outs a1, a2
endin

schedule(1,0,56.5)
```

15.8 Conclusions

This chapter has explored granular synthesis, and the combinations of synthesis parameters that most strongly contribute to the different types of granular synthesis. Most significant among these are the grain rate and pitch along with grain envelope and source waveform. We can also note that the perceptual effect of changing the value of one parameter sometimes strongly depends on the current value of another parameter. One example is the case with grain pitch, which gives a pitch change when the grain rate is low, and a formant shift when the grain rate is high. We have also looked at the application of granular techniques for processing a live audio stream, creating granular delays and reverb effects. Furthermore we looked at manipulation of single grains to create intermittencies, filtering, subharmonics, spatial effects and waveform mixing. Techniques for clock synchronisation have also been shown, enabling soft (gradual) or hard synchronisation between different `partikkel` instances and/or other clock sources. The relationship between AM and granular synthesis was considered, and utilised in the pitch synchronous granular synthesis technique. Finally, a method of parametric morphing between different types of granular synthesis was shown. Hopefully, the potential of granular synthesis as an abundant source of sonic transformations has been sketched out, encouraging the reader to further experimentation.

Chapter 16
Physical Models

Abstract This chapter will introduce the area of physical modelling synthesis, which has attracted much interest recently. It will discuss the basic ways which can be used to model and simulate the sound-producing parts of different instruments. It begins with an introduction to waveguides, showing how these can be constructed with delay lines. It then discusses modal synthesis, which is another approach that is based on simulating resonances that exist in sound-producing objects. Finally, it presents an overview of finite difference methods, which can produce very realistic results, but with considerable computational costs.

16.1 Introduction

Humans have been making and playing musical instruments for thousands of years, and have developed ways of controlling the sounds they make. One of the problems in computer-realised instruments is the lack of knowledge of how they behave, and in particular how the control parameters interact. This leads to the idea of mimicking the physical instrument, so we may understand the controls, but of course takes the model beyond possible reality, for example by having unreasonable ranges or impossible sizes or materials.

Thus we have collections of sounds derived from or inspired by physical arte-facts. Within this category there are a wide variety of levels of accuracy and speed. Here we explore separately models based on waveguides, detailed mathematical modelling and models inspired by physics but approximated for speed.

16.2 Waveguides

It is well known that sound travelling in a perfect medium is governed by the wave equation; in one spatial dimension that is $\frac{\partial^2 u}{\partial t^2} = c^2 \frac{\partial^2 u}{\partial x^2}$. This equation actually has a

© Springer International Publishing Switzerland 2016
V. Lazzarini et al., *Csound*, DOI 10.1007/978-3-319-45370-5_16

simple solution. It is easy to see that $f(x-ct)$ is a solution, for an arbitrary function f. Similarly $g(x+ct)$ is a solution. Putting these together we see that $f(x-ct) + g(x+ct)$ is a very general solution.

In order to interpret this let us concentrate on the f function. As time increases the shape will stay the same, but moved forward in space, with the constant c being the speed of movement. Considering the case of a perfect string, this corresponds to a wave travelling in the positive direction. It is easy to see that the g component is also a wave travelling in the negative direction. It is this solution and interpretation that give rise to the waveguide concept.

16.2.1 Simple Plucked String

If we consider a finite string held at both ends, such as is found in physical string instruments, we can apply the waveguide idea directly. It is simplest to consider a plucked string, so the initial shape of the string is a triangle divided equally between the functions f and g. On releasing the pluck the triangle will move in both directions. To continue the model we need to know what happens at the held ends. Experimentation shows that the displacement is inverted and reflected, so the wave travels back. But realistically some of the energy is lost, and converted into the sound we wish to hear. A simple model for this loss in a low-pass filter, such as an averaging one $y_n = (x_n + x_{n-1})/2$, which can be placed at one end of the string (see also Section 12.2.2 for an analysis of this particular filter). To listen we take a point on the string and copy the displacement value to the output. If we use two simple delay lines to model the left- and right-moving waves we obtain a diagram like Fig. 16.1.

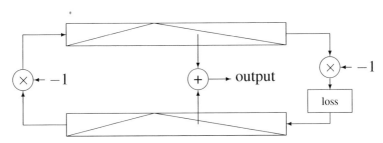

Fig. 16.1 Diagram of a simple pluck model

But what pitch does this produce? That can be calculated by considering the history of an initial impulse in one of the delay lines. If the delay in each line is N samples it will take $2N$ samples, plus any delay incurred in the loss filter. In the case of a simple averaging filter the group delay is half a sample, so we can play pitches at $sr/(2N + 0.5)$ Hz. For low frequencies the errors are small, a 50 Hz target and CD quality sampling rate we can choose 49.97 Hz or 50.09 Hz with the delay lines 441 or 440 long. But for higher frequencies the error is very great; at 2,000 Hz we need to chose between 1,960 and 2,151 Hz. This can be solved by introducing another all-pass filter into the signal path which can have an adjustable delay to tune it accurately. Generally speaking the parameters of the model are the delay line lengths, the initial pluck shape, the loss and tuning filters, and the location of the reading to the delay lines for the output.

It is possible to encode all this within Csound, using opcodes `delayr` and `delayw`, and filter opcodes.

Listing 16.1 Simple waveguide string model

```
<CsoundSynthesizer>
<CsInstruments>
/***********************
asig String kamp,ifun,ils,ipos,ipk
kamp - amplitude
ifun - fundamental freq
ils - loss factor
ipos - pluck position
ipk - pickup position
******************/
opcode String,a,kiii
 setksmps 1
 kamp,ifun,ipos,ipk xin
 ain1 init 0
 ain2 init 0
 idel = 1/(2*ifun)
 kcnt line 0, p3, p3
 if kcnt < idel then
  ainit1 linseg  0,idel*ipos, 1, idel*(1-ipos),0
  ainit2 linseg  0,idel*(1-ipos),-1, idel*ipos,0
 else
  ainit1=0
  ainit2=0
 endif
 awg1 delayr idel
 apick1 deltap idel*(1-ipk)
     delayw  ain1+ainit1
 awg2 delayr idel
 apick2 deltap idel*ipk
     delayw  ain2+ainit2
 ain1 = (-awg2 + delay1(-awg2))*0.5
 ain2 = -awg1
```

```
      xout (apick1+apick2)*kamp
endop

instr 1
 asig String p4,p5,0.3,0.05
 out asig
endin

</CsInstruments>
<CsScore>
i1 0 1 10000 220
i1 + 1 10000 440
i1 + 1 10000 330
i1 + 1 10000 275
i1 + 1 10000 220
</CsScore>
</CsoundSynthesizer>
```

However there are a number of packaged opcodes that provide variants on the theme. The `pluck` opcode implements a much simplified model, the Karplus and Strong model, with one delay and random (white noise) initial state [57]. Fuller implementations can be found in the opcodes `wgpluck` and `wgpluck2`, which use slightly different filters and tuning schemes.

It should be noted what these models do not provide: in particular instrument body, sympathetic string vibration and vibration angle of the string. Body effects can be simulated by adding one of the reverberation methods to the output sound. A limited amount of sympathetic sound is provided by `repluck`, which includes an audio excitement signal, and `streson`, which models a string as a resonator with no initial pluck.

A mandolin has pairs of strings which will be at slightly different pitches; the waveguide model can be extended to this case as well, as in opcode `mandol`.

Closely related to plucked strings are bowed strings, which use the same model except the bow inserts a sequence of small plucks to insert energy into the string. Csound offers the opcode `wgbow` for this. There are additional controls such as where on the string the bowing happens, or equivalently where in the delay line to insert the plucks, and how hard to press and move the bow, which is the same as controlling the mini-plucks.

Tuning

As discussed above, the waveguide model can only be tuned to specific frequencies at $sr/(2N+0.50)$ Hz, where N is the length in samples of each delay line. In order to allow a fractional delay to be set for fine-tuning of the model, we can add an all-pass filter to the feedback loop. As discussed in Chapter 13, these processors only affect the timing (phase) of signals, and leave their amplitude unchanged. With a simple first-order all-pass filter, we can add an extra delay of less than one sample, so that the waveguide fundamental has the correct pitch.

The form of this all-pass filter is

$$y(t) = c(x(t) - y(t-1)) + x(t-1) \tag{16.1}$$

where $x(t)$ is the input, $y(t-1)$ and $x(t-1)$ are the input and output delayed by one sample, respectively. The coefficient c is determined from the amount of fractional delay d ($0 < d < 1$) required:

$$c = \frac{1-d}{1+d} \tag{16.2}$$

The total delay in samples will then be $N_{wdelay} + 0.5 + d$, where N_{wdelay} is the total waveguide delay length. To fine-tune the model, we need to

1. Obtain the waveguide delay length N_{wdelay}:

$$N_{wdelay} = \left\lfloor \frac{sr}{f_0} - 0.5 \right\rfloor \tag{16.3}$$

where $\lfloor . \rfloor$ is the floor function.
2. Calculate d:

$$d = \frac{sr}{f_0} - (N_{wdelay} + 0.5) \tag{16.4}$$

3. Insert the all-pass filter into the model, using the coefficient c as above.

Finally, when we come to implement this, it is easier to aggregate the two delay lines used in the waveguide into one single block, removing the explicit reflection at one end. Because the reflections cancel each other, we should also remove the other one in the feedback loop. We can still initialise and read from the delay line as before, but now taking account of this end-to-end joining of the two travelling directions. By doing this, we can manipulate its size more easily. Also, with a single delay line, we do not require sample-by-sample processing (ksmps=1), although we will need to make sure that the minimum delay required is not less than the orchestra ksmps.

A fine-tuned version of the String UDO is shown below. It uses the Ap UDO, which implements the all-pass tuning filter.

Listing 16.2 String model with an all-pass tuning filter

```
/*********************
asig Ap ain,ic
ain - input signal
ic - all-pass coefficient
*******************/
opcode Ap,a,ai
 setksmps 1
 asig,ic xin
```

```
 aap init 0
 aap = ic*(asig - aap) + delay1(asig)
 xout aap
endop


/**********************
asig String kamp,ifun,ils,ipos,ipk
kamp - amplitude
ifun - fundamental freq
ils - loss factor
ipos - pluck position
ipk - pickup position
******************/
opcode String,a,kiii
 kamp,ifun,ipos,ipk xin
 aap init 0
 idel = 1/ifun
 ides = sr*idel
 idtt = int(ides-0.5)
 ifrc = ides - (idtt + 0.5)
 ic = (1-ifrc)/(1+ifrc)
 kcnt line 0, p3, p3
 if kcnt < idel then
  ainit linseg 0,ipos*idel/2,-1,
               (1-ipos)*idel,1,
               ipos*idel/2,0
 else
  ainit=0
 endif
 awg delayr idtt/sr
 apick1 deltap idel*(1-ipk)
 apick2 deltap idel*ipk
 afdb = Ap((awg + delay1(awg))*0.5, ic)
        delayw  afdb+ainit
     xout (apick1+apick2)*kamp
endop
```

16.2.2 Wind Instruments

Blowing into a cylindrical tube is in many ways similar to the string model above.
In this case, there is a pressure wave and an associated displacement of air, but the
governing equation and general solution is the same. The main difference is at the
ends of the tube. At a closed end the displacement wave just reverses its direction

with the same negation of the string. The open end is actually very similar to the string case as the atmosphere outside the tube is massive and the wave is reflected back up the tube without the negation, although with energy loss.

The other component is the insertion of pressure into the tube from some reed, fipple or mouthpiece. Modelling this differentiates the main class of instrument. A basic design is shown in Fig. 16.2, where a reed input for a clarinet is depicted, with a waveguide consisting of one open and one closed end. In this case, because of the mixed boundaries, the fundamental frequency will be one octave lower compared to an equivalent string waveguide.

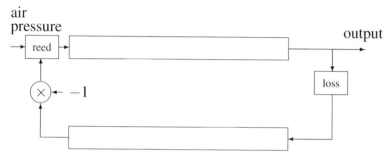

Fig. 16.2 Diagram of a simple wind instrument

Examples of cylindrical tubular instruments include flutes, clarinets and trumpets. To complete these instruments we need to consider the action of blowing.

The simplest model is a single reed as used in a clarinet. A reed opens by bending away from the rest position under pressure from the breath, mitigated by the pressure in the tube. The exact way in which this happens is non-linear and depends on the reed stiffness. A sufficiently accurate table lookup for this was developed by Perry Cook [27], and this is used in the csound `wgclar` opcode. There are a range of parameters, controlling things such as reed stiffness, amount of noise and the time is takes to start and stop blowing.

We can modify the string example to create a clarinet waveguide instrument by adding a reed mode as shown in Fig. 16.2, as well as a better low-pass filter to simulate the loss at the bell end. The output of this instrument has an inherent DC offset, which we can block using a `dcblock2` filter.

Listing 16.3 Clarinet waveguide model

```
<CsoundSynthesizer>
<CsOptions>
</CsOptions>
<CsInstruments>
isiz = 16384
ifn ftgen 1,0,isiz,-7,0.8,0.55*isiz,-1,0.45*isiz,-1
```

```
/*******************************
idel LPdel ifo,ifc
idel - lowpass delay (samples)
ifo - fund freq
ifc - cutoff freq
*******************************/
opcode LPdel,i,ii
 ifo,ifc xin
 ipi = $M_PI
 itheta = 2*ipi*ifc/sr
 iomega = 2*ipi*ifo/sr
 ib = sqrt((2 - cos(itheta))^2 -1) - 2 + cos(itheta)
 iden = (1 + 2*ib*cos(iomega) + ib*ib)
 ire = (1. + ib + cos(iomega)*(ib+ib*ib))/iden
 img =  sin(iomega)*(ib + ib*ib)/iden
 xout -taninv2(img,ire)/iomega
endop


/********************************
asig Reed ain,kpr,kem,ifn
ain - input (feedback) signal
kpr - pressure amount
kem - embouch pos (0-1)
ifn - reed transfer fn
********************************/
opcode Reed,a,akki
 ain,kpr,kem,ifn xin
 apr linsegr 0,.005,1,p3,1,.01,0
 asig = ain-apr*kpr-kem
 awsh tablei .25*asig,ifn,1,.5
 asig *= awsh
 xout asig
endop


/********************
asig Ap ain,ic
ain - input signal
ic - all-pass coefficient
******************/
opcode Ap,a,ai
 setksmps 1
 asig,ic xin
 aap init 0
 aap = ic*(asig - aap) + delay1(asig)
```

```
 xout aap
endop

/******************************
asig Pipe kamp,ifun,ipr,iem,ifc
kamp - amplitude
ifun - fundamental
ipr - air pressure
iem - embouch pos
ifc - lowpass filter factor
*****************************/
opcode Pipe,a,kiiii
 kamp,ifun,ipr,ioff,ifc xin
 awg2 init 0
 aap init 0
 ifun *= 2
 ifc = ifun*ifc
 ilpd = LPdel(ifun,ifc)
 ides = sr/ifun
 idtt = int(ides - ilpd)
 ifrc = ides - (idtt + ilpd)
 ic = (1-ifrc)/(1+ifrc)
 awg1 delayr idtt/sr
 afdb = Ap(tone(awg1,ifc), ic)
      delayw Reed(-afdb,ipr,ioff,1)
 xout dcblock2(awg1*kamp)
endop

instr 1
 asig Pipe p4,p5,p6,p7,p8
      out asig
endin

</CsInstruments>
<CsScore>
i1 0 1 10000 440 0.7 0.7 6
i1 + 1 10000 880 0.9 0.7 5
i1 + 1 10000 660 0.7 0.7 4.5
i1 + 1 10000 550 0.8 0.6 3
i1 + 1 10000 440 0.7 0.6 3.5
</CsScore>
</CsoundSynthesizer>
```

Note that the use of an IIR filter such as `tone` for the loss makes the fine-tuning of the waveguide somewhat more complex. However, the all-pass filter approach can still be used to fix this. All we need to do is to calculate the exact amount of delay that is added by the filter, which is dependent on the cut-off frequency f_c. With this, the phase response of the filter will tell us what the delay is at a given fundamental. The i-time UDO `LPdel` calculates this for the specific case of the `tone` filter, which is defined as

$$y(t) = (1+b)x(t) - by(t-1) \tag{16.5}$$

$$b = \sqrt{(2 - \cos(\theta))^2 - 1} - 2 + \cos(\theta) \tag{16.6}$$

where $\theta = 2\pi f_c/sr$. With this in hand, we can proceed to determine the waveguide length, and the necessary fractional delay. In `LPDel`, we calculate the phase response for this filter at the fundamental frequency and return it as a time delay in samples. This is then subtracted from the desired length for this fundamental. The integral part of this is the delay line size; the fractional part determines the all-pass coefficient.

The flute model is similar to the clarinet except excitation is based on a non-linear jet that fluctuates over the lip hole and the pipe is open at both ends. Again the Csound opcode `wgflute` implements this with a detailed architecture due to Cook. An opcode modelled on the brass family `wgbrass` uses an excitation controlled by the mass, spring constant and damping of the lip, but is otherwise similar.

What has not been considered so far is how different pitches are played. There are two approaches; the simpler is to take inspiration from the slide trombone and simply change the length of the tube (delay line). As long as the lowest pitch of an instrument is known, this is easy to implement. It is indeed the method used in the Csound opcodes. The alternative is tone-holes, as usually found on woodwind instruments. These can be modelled by splitting the delay line and inserting a filter system to act for the energy loss. This is explored in detail in research papers [113, 78].

16.2.3 More Waveguide Ideas

The concept of the waveguide is an attractive one and has been used outside strings and tubes. Attempts have been made to treat drum skins as a mesh of delays with scattering at the nodes where they meet. There are a number of problems with this, lsuch the connectivity of the mesh (Fig. 16.3 and what happens at the edge of a circular membrane. There are a number of research papers about this but so far the performance has not been anywhere near real-time, and other approaches, typically noise and filters, have proved more reliable. However this remains a possible future method [61, 4, 3].

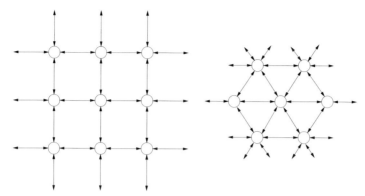

Fig. 16.3 Two possible drum meshes

16.3 Modal Models

In Section 12.4 the idea of constructing a sound from its sinusoidal partials was discussed. A related idea for a physical object is to determine its basic modes of vibration and how they decay and to create a physical model based on adding these modes together, matching the initial excitation. The resulting system can be very impressive (see [107] for example), but it does require a significant amount of analysis and preprocessing for an arbitrarily shaped object.

This method is particularly useful for generating the sound from a regularly shaped object such as a wooden or metallic block, where the modes can be precalculated or estimated. Using a rich excitement we can get a usable sound. This example shows the general idea. The sound decay needs to be long to get the effect.

Listing 16.4 Marimba model using modal resonances

```
<CsoundSynthesizer>
<CsInstruments>
gicos    ftgen 0, 0,8192,11,1

/********************
asig MyMarimba idur,iamp,ifrq,ibias
idur - time to decay
iamp - amplitude
ifrq - fundamental freq
******************/
opcode MyMarimba,a,iii
  idur,ifrq,iamp xin
  ; envelope
  k1       expseg  .0001,.03,iamp,idur-.03,.001
  ; anticlick
```

```
   k25       linseg  1,.03,1,idur-.03,3
   ; power to partials
   k10       linseg  2.25,.03,3,idur-.03,2
   a1        gbuzz   k1*k25,ifrq,k10,0,35,gicos
   a2        reson   a1,500,50,1      ;filt
   a3        reson   a2,1500,100,1    ;filt
   a4        reson   a3,2500,150,1    ;filt
   a5        reson   a4,3500,150,1    ;filt
   a6        balance a5,a1
             xout    a6
endop

instr 1
  ifq  =           cpspch(p4)
  asig MyMarimba   20,ifq,p5
       out         asig
endin
</CsInstruments>
<CsScore>
i1      0         10      8.00      30000
i1      4         .       8.02      30000
i1      8         .       8.04      30000
i1      12        .       8.05      30000
i1      16        .       8.07      30000
i1      20        .       8.09      30000
i1      24        .       8.11      30000
i1      28        .       9.00      30000
i1      32        .       8.00      10000
i1      32        .       8.04      10000
i1      32        .       8.07      10000
</CsScore>
</CsoundSynthesizer>
```

In Csound this technique is used in three opcodes, marimba, vibes and gogobel, all build from four modal resonances but with more controls that the simple marimba shown above.

It is possible to use the modal model directly, supported by the opcode mode. This is a resonant filter that can be applied to an impulse or similarly harmonically rich input to model a variety of percussive sounds. For example we may model the interaction between a physical object and a beater with the following UDO.

Listing 16.5 Mode filter example

```
opcode hit,a,iiiii
  ihard, ifrq1, iq1, ifrq2, iq2 xin
  ashock  mpulse  3,0 ;; initial impulse
  ; modes of beater-object interaction
```

```
  aexc1   mode ashock,ifrq1,iq1
  aexc2   mode ashock,ifrq2,iq2
  aexc   =      ihard*(aexc1+aexc2)/2
  ;"Contact" condition : when aexc reaches 0,
  ; the excitator looses contact with the
  ; resonator, and stops influencing it
  aexc    limit aexc,0,3*ihard
  xout   aexc
endop
```

This can be applied to the following object.

Listing 16.6 Modal resonances example

```
opcode ring,a,aiiii
  aecx, ifrq1, iq1, ifrq2, iq2  xin
  ares1   mode aexc,ifrq1,iq1
  ares2   mode aexc,ifrq2,iq2
  ares   =      (ares1+ares2)/2
  xout aexc+ares
endop
```

All that remains is to use suitable frequencies and Q for the interactions. There are tables of possible values in an appendix to the Csound manual, but a simple example might be

```
instr 1
  astrike  hit   ampdb(68),80, 8, 180, 3
  aout   ring astrike, 440, 60, 630, 53
  out   aout
endin
```

More realistic outputs can be obtained with more uses of the `mode` resonator in parallel, and a more applicable instrument might make use of parameters to select the parameters for the resonators.

16.4 Differential Equations and Finite Differences

In the earlier discussion of a physical model of a string it was stated that it was for a perfect string; that is for a string with no stiffness or other properties such as shear. If we think about the strings of a piano, the stiffness is an important part of the instrument and so the two wave solution of a waveguide is not valid. The differential equation governing a more realistic string is

$$\frac{\partial^2 u}{\partial t^2} = c^2 \frac{\partial^2 u}{\partial x^2} - \kappa^2 \frac{\partial^4 u}{\partial x^4} \tag{16.7}$$

where κ is a measure of stiffness. To obtain a solution to this one creates a discrete grid in time and space, time step size being one sample, and approximates the derivative by difference equations, where $u(x_n, t)$ is the displacement at point x_n at time t:

$$\frac{\partial u}{\partial x} \rightarrow \frac{u_{n+1} - u_n}{\delta x}$$
$$\frac{\partial^2 u}{\partial x^2} \rightarrow \frac{u_{n+1} - 2u_n + u_{n-1}}{\delta x^2} \tag{16.8}$$

and similarly for the other terms. For example the equation

$$\frac{\partial u}{\partial t} = \frac{\partial u}{\partial x} \tag{16.9}$$

could become

$$u(x_n, t+1) = u(x_n, t) + \frac{1}{\delta x} u(x_{n+1}, t) + \left(1 - \frac{1}{\delta x}\right) u(x_n, t) \tag{16.10}$$

Returning to our stiff strings this mechanism generates an array of linear equations in $u(n, t)$ which, subject to some stability conditions on the ratio of the time and space steps, represents the evolution in time of the motion of the string. This process, a finite difference scheme, is significantly slower to use than the waveguide model, but is capable of very accurate sound. Additional terms can be added to the equation to represent energy losses and similar properties.

Forms such as this are hard and slow to code in the Csound language. Producing usable models is an active research area [11], and GPU parallelism is being used to alleviate the performance issues.

There are three opcodes in Csound that follow this approach while delivering acceptable performance, modelling a metal bar, a prepared piano string and a reverberating plate.

A thin metal bar can be considered as a special case of a stiff string, and a finite difference scheme set up to calculate the displacement of the bar at a number of steps along it. The ends of the bar can be clamped, pivoting or free, and there are parameters for stiffness, quality of the strike and so on. The resulting opcode is called barmodel.

A mathematically similar but sonically different model is found in the opcode prepiano which incorporates three stiff strings held at both ends that interact, and with optional preparations of rattles and rubbers to give a Cageian prepared piano. The details can be found in [12].

The third opcode is for a two-dimensional stiff rectangular plate excited by an audio signal and allowed to resonate. This requires a two-dimensional space grid to advance time which adds complexity to the difference scheme but is otherwise similar to the strings and bars. The opcode is called platerev and allows a defined stiffness and aspect ratio of the plate as well as multiple excitations and listening points. The added complexity makes this much slower than real-time.

16.5 Physically Inspired Models

Not quite real physical modelling, there is a class of instrument models introduced by Cook [28] that are inspired by the physics but do not solve the equations, but rather use stochastic methods to produce similar sounds. Most of these models are percussion sounds.

Consider a maraca. Inside the head there are a number of seeds. As the maraca is moved the seeds fly in a parabola until they hit the inner surface, exciting the body with an impulse. A true physical model would need to follow every seed in its motion. But one could just think that the probability of a seed hitting the gourd at a certain time depends on the number of seeds in the instrument and how vigorously it is shaken. A random number generator suitably driven supplies this, and the model just needs to treat the gourd as a resonant filter. That is the thinking behind this class.

In Csound there are a number of shaken and percussive opcodes that use this method; `cabasa`, `crunch`, `sekere`, `sandpaper`, `stix`, `guiro`, `tambourine`, `bamboo`, `dripwater`, and `sleighbells`. The differences are in the body resonances and the excitations.   The inspiration from physics is not limited to percussion. It is possible to model a siren, or a referee's whistle, in a similar way. In the whistle the pea moves round the body in response to the blowing pressure, and the exit jet is sometimes partially blocked by the pea as it passes. The model uses the pressure to control the frequency of blocking, with some stochastic dither to give an acceptable sound.

16.6 Other Approaches

There is an alternative way of making physical models where all interactions are seen as combinations of springs, masses and dampers. In many ways this is similar to the differential equation approach and may end in computations on a mesh. Examples of these ideas can be found in the literature such as [33] or the CORDIS-ANIMA project [60].

16.6.1 Spring-Mass System

While there are no Csound opcodes using the CORDIS-ANIMA abstraction, spring-mass instruments can be designed using similar principles. A simple example is given by a spring-mass system. If we can set up a number of equations that describe the movement of the system, in terms of position (in one dimension) and velocity, then we can sample it to obtain our audio signal (Fig. 16.4).

A force acting on a mass in such a system, whose displacement is x, can be defined by

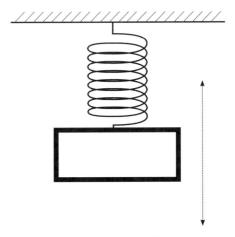

Fig. 16.4 Spring-mass system, whose position is sampled to generate an output signal

$$F = -k \times x \tag{16.11}$$

where k is the spring constant for a mass on a spring. From this, and Newton's second law ($F = ma$), we can calculate the acceleration a in terms of the displacement position x and mass m:

$$a = \frac{-k \times x}{m} \tag{16.12}$$

Finally, to make the system *move* we need to propagate the changes in velocity and position. These are continuous, so we need a discretisation method to do this. The simplest of these is Euler's method:

$$y_{n+1} = y_n + y'(t) \times h \tag{16.13}$$

with $t_{n+1} = t_n + h$.

In other words, the next value of a function after a certain step is the combination of the current one plus a correction that is based on its derivative and the step. For the velocity v, this translates to

$$v_{n+1} = v_n + a \times h \tag{16.14}$$

as the acceleration is the derivative of the velocity. For the position x, we have

$$x_{n+1} = x_n + v \times h \tag{16.15}$$

That works for an ideal system, with no dampening. If we want to dampen it, then we introduce a constant d, $0 < d < 1$, to the velocity integration:

$$v_{n+1} = v_n \times d + a \times h \tag{16.16}$$

and we have everything we need to make a UDO to model this system and sample the position of the mass as it oscillates:

Listing 16.7 A spring-mass UDO

```
/ * * * * * * * * * * * * * * * * * * * * * * * * * *
asig Masspring ad,ik,idp,ih
ad - external displacement
ik - spring constant
idp - dampening constant
ims - mass
* * * * * * * * * * * * * * * * * * * * * * * * * */
opcode Masspring,a,aiiii
 setksmps 1
 ad,ik,idp,ims xin
 av init 0
 ax init 0
 ih = 1/sr
 ac = -(ik*ax+ad)/ims
 av = idp*av+ac*ih
 ax = ax+av*ih
 xout ax
endop
```

Note that h is set to the sampling period ($1/sr$), which is the step size of the audio signal. To run the opcode, we need to set up the spring constant, the mass of the object and the dampening. The first two determine the pitch of the sound, whereas the final parameter sets how long the sound will decay.

Since this is a simple harmonic system, we can use the well-known relationship between period T_0, mass m and spring constant k:

$$T_0 = 2\pi\sqrt{\frac{m}{k}} \tag{16.17}$$

which can be reworked to give a certain model mass for an arbitrary f_0 and spring constant k:

$$m = k \times \left(\frac{1}{2\pi f_0}\right)^2 \tag{16.18}$$

The value of k will influence the amplitude of the audio signal, as it can be interpreted as the amount of stiffness in the spring. Increasing it has the effect of reducing the output level for a given f_0. The actual amplitude will also vary with the mass, increasing for smaller masses for the same k. The following example shows an instrument that is designed to use the model. It makes an initial displacement (lasting for one k-cycle) and then lets the system carry on vibrating.

Listing 16.8 An instrument using the spring-mass UDO

```
instr 1
```

```
asig = p4
ifr = p5
idp = p6
ik = 1.9
im = ik*(1/(2*$M_PI*ifr))^2
ams Masspring asig,ik,idp,im
out dcblock(ams)
asig = 0
endin
schedule(1,0,5,0dbfs/2,440,0.9999)
```

The input into the system is very simple, like a single strike. Other possibilities can be experimented with, including using audio files, to which the system will respond like a resonator.

16.6.2 Scanned Synthesis

Scanned synthesis [130] is another method that can be cast as a physical modelling technique. From this perspective, it can be shown to be related to the idea of a network of masses, springs and dampers. The principle behind it is a separation between a dynamic physical (spatial) system and the reading (scanning) of it to produce the audio signal. These two distinct components make up the method.

The configuration of the physical system can be done in different ways. In the Csound implementation, this is done by defining a set of masses, which can be connected with each other via a number of springs. This creates a spring-mass network, whose components have the basic characteristics of mass, stiffness, dampening and initial velocity. This is set into motion, making it a dynamic system.

The fundamental layout of the network is set up by the connections between the masses. This is defined in the scanning matrix M, which is an $N \times N$ matrix, where N is the number of masses in the system, containing zeros and ones. If $M_{i,j}$ is set to one, we will have a connection between the mass elements indexed by i and j. A zero means that there is no connection.

For instance, with $N = 8$, the following matrix has all the elements connected in a line, making it an open unidirectional string:

$$M = \begin{pmatrix} 0\ 1\ 0\ 0\ 0\ 0\ 0\ 0 \\ 0\ 0\ 1\ 0\ 0\ 0\ 0\ 0 \\ 0\ 0\ 0\ 1\ 0\ 0\ 0\ 0 \\ 0\ 0\ 0\ 0\ 1\ 0\ 0\ 0 \\ 0\ 0\ 0\ 0\ 0\ 1\ 0\ 0 \\ 0\ 0\ 0\ 0\ 0\ 0\ 1\ 0 \\ 0\ 0\ 0\ 0\ 0\ 0\ 0\ 1 \\ 0\ 0\ 0\ 0\ 0\ 0\ 0\ 0 \end{pmatrix} \tag{16.19}$$

As can be seen, mass 0 is connected to mass 1, which is connected to mass 2 etc. $M_{i,i+1} = 1$ for $1 \leq i < N - 1$. Linked at right angles to each mass there are centring springs to which some dampening can be applied. By adding the backwards connections from the end of the string, $M_{i+1,i} = 1$, we can make it bidirectional (Fig. 16.5). By adding two further connections ($N - 1$ to 0, and 0 to $N - 1$), we can make a circular network. Many other connections are possible.

Fig. 16.5 Bidirectional string network with eight masses (represented by balls) connected by springs. There are also vertical springs to which dampening is applied.

Csound will take matrices such as these as function tables, where a two-dimensional matrix is converted to a vector in row-major order (as sequences of rows, $V_{iN+j} = M_{i,j}$):

$$V = (M_{0,0}, M_{0,1}, ..., M_{0,N-1}, M_{1,0}, ..., M_{N-1,N-1})$$

The following code excerpt shows how connection matrix could be created for a bidirectional string, using the table-writing opcode `tabw_i`:

```
ii init 0
while ii < iN-1 do
   tabw_i 1,ii*iN+ii+1,ift
   tabw_i 1,(ii+1)*iN+ii,ift
   ii += 1
od
```

where `ift` is the function table holding the matrix in vector form, and `iN` is the number of masses in the network. The opcode that sets up the model and puts it into motion is `scanu`:

```
scanu ifinit, iupd, ifvel, ifmass,
         ifcon, ifcentr, ifdamp,
         kmass, kstif, kcentr, kdamp,
         ileft, iright, kx, ky,
         ain, idisp, id
```

`ifinit`: function table number with the initial position of the masses. If negative, it indicates the table is used as a hammer shape.

`iupd`: update period, determining how often the model calculates state changes.

`ifvel`: initial velocity function table.

`ifmass`: function table containing the mass of each object.

`ifcon`: connection matrix ($N \times N$ size table).

`ifcentr`: function table with the centring force for each mass. This acts orthogonally to the connections in the network.

`ifdamp`: function table with the dampening factor for each mass

`kmass`: mass scaling.

`kstif`: spring stiffness scaling.

`kdamp`: dampening factor scaling.

`ileft, iright`: position of left/right hammers (`init < 0`).

`kx`: position of a hammer along the string ($0 \leq kx \leq 1$), (`init < 0`).

`ky`: hammer power (`init < 0`).

`ain`: audio input that can be used to excite the model.

`idisp`: if 1, display the evolution of the masses.

`id`: identifier to be used by a scanning opcode. If negative, it is taken to be a function table to which the waveshape created by the model will be written to, so any table-reading opcode can access it.

All tables, except for the matrix one, should have the same size as the number of masses in the network. To make sound out of the model, we define a scanning path through the network, and read it at a given rate. For each such model, we can have more than one scanning operation occurring simultaneously:

```
ar scans kamp, kfreq, iftr, id[,iorder]
```

where `kfr` is the scanning frequency, `iftr` is the function table determining the trajectory taken, and `id` is the model identifier. The output amplitude is scaled by `kamp`. The last, optional, parameter controls the interpolation order, which ranges from 0, the default, to fourth-order. Note also that other opcodes can access the waveform produced by the model if that is written to a function table.

The trajectory function table is a sequence of mass positions, 0 to $N-1$, in an N-sized table, which will be used as the scanning order by the opcode. The following example demonstrates a `scanu - scans` pair, using a bi-directional string network with 1024 masses. The scan path is linear up and down the string. The scanning frequency and the model update rate are also controlled as parameters to the instrument

Listing 16.9 Scanned synthesis example

```
giN = 1024
ginitf = ftgen(0,0,giN,7,0,giN/2,1,giN/2,0)
gimass = ftgen(0,0,giN,-7,1,giN,1)
gimatr = ftgen(0,0,giN^2,7,0,giN^2,0)
gicntr = ftgen(0,0,giN,-7,0,giN,2)
gidmpn = ftgen(0,0,giN,-7,0.7,giN,0.9)
givelc = ftgen(0,0,giN,-7,1,giN,0)
gitrjc = ftgen(0,0,giN,-7,0,giN/2,giN-1,giN/2,0)

instr 1
 ii init 0
 while ii < giN-1 do
```

```
    tabw_i 1,ii*giN+ii+1,gimatr
    tabw_i 1,(ii+1)*giN+ii,gimatr
    ii += 1
 od
 asig init 0
 scanu ginitf,1/p6,
        givelc,gimass,
        gimatr,gicntr,gidmpn,
        1,.1,.1,-.01,0,0,
        0,0,asig,1,1
 a1   scans p4,p5,gitrjc,1,3
      out a1
endin
schedule(1,0,10,0dbfs/20,220,100)
```

Scanned synthesis is a complex, yet very rich method for generating novel sounds. Although not as completely intuitive as the simpler types of physical modelling, it is very open to exploration and interactive manipulation.

16.6.3 ... and More

Csound has a rich collection of opcodes including ones not described above, but that owe their existence to the thinking of this chapter. For example there are opcodes for a simplified scanned synthesis and for a faster implementation lacking just one little-used option. One might also explore wave terrain synthesis [14], which predates scanned synthesis with a fixed environment. There is much to explore here; with physical models there is the possibility to drive them in non-physical ways.

16.7 Conclusions

This chapter has introduced the basic ideas behind creating sounds based on the physics of real instruments. The simplest style is based on the general solution of the simple wave equation, and that generates controllable instruments in the string and wind groups. Modal modelling can produce realistic sounds for struck objects. For a greater investment in computer time it is possible to use finite difference schemes to model more complex components of an instrument, such as stiffness and non-simple interaction with other objects. We also briefly considered a class of stochastic instruments inspired by physics, and the chapter was completed by a look at other approaches such as spring-mass models and scanned synthesis.

Part V
Composition Case Studies

Chapter 17
Iain McCurdy: Csound Haiku

Abstract This chapter examines the set of pieces *Csound Haiku* and describes the most interesting features of each one. The Csound code is intentionally written with a clarity, simplicity and concision that hopefully allows even those with no knowledge of the Csound language to garner some insight into the workings of the pieces. This is intended to confront the conscious estericism of some live-coding performances. The pieces are intended to be run in real time thereby necessitating the use of efficient synthesis techniques. Extensive use of real-time event generation is used, in fact the traditional Csound score is not used in any of the pieces. It was my preference in these pieces to devise efficient synthesis cells that would allow great polyphony rather than to create an incredibly complex and CPU-demanding, albeit possibly sonically powerful, synthesis engine that would only permit one or two simultaneous real-time instances. The pieces' brevity in terms of code should make their workings easily understandable. These illustrations will highlight the beauty to be found in simplicity and should also provide technical mechanisms that can be easily transplanted into other projects.

17.1 Introduction

Csound Haiku is a set of nine real-time generative pieces that were composed in 2011 and which are intended to be exhibited as a sound installation in the form of a book of nine pages. Each page represents one of the pieces and turning to that piece will start that piece and stop any currently playing piece. The pages themselves contain simple graphical elements reflecting compositional processes in the piece overlaid with the code for the piece typed using a mechanical typewriter onto semi-transparent paper. The use of a typewriter is intended to lend the code greater permanence than if it were stored digitally and to inhibit the speed of reproduction that is possible with digitally held code through copy and paste. Each of the pieces can be played for as long or as short a time as desired. The experience should be like tuning in and out of a radio station playing that piece indefinitely. The first moment

© Springer International Publishing Switzerland 2016
V. Lazzarini et al., *Csound*, DOI 10.1007/978-3-319-45370-5_17

we hear is not the beginning of the piece and the last moment we hear is not the ending – they are more like edit points. None of the nine pieces make use of the Csound score but instead they generate note events from within the orchestra. The Csound score has traditionally been a mainstay of Csound work so this approach is more reflective of modern approaches to how Csound can be used and allows Csound to compete in the area of event generation and the use of patterns with software more traditionally associated with this style of working such as SuperCollider [86] and Pure Data [103]. To be able to print the complete code for each piece on a single side of paper necessitated that that code be concise and this was in fact one of the main compositional aims of the project. Running each piece comfortably in real time also demanded that restrictions were employed in the ambitions of the Csound techniques used. This spilled over into a focussing of the musical forces employed: great orchestrations of disparate sound groups were ruled out.

17.2 Groundwork

The first task was to identify simple and economic, yet expressive synthesising 'cells' that would characterise each piece. Techniques that proved to be too expensive in their CPU demands or that would limit polyphony were quickly rejected. In addition a reasonable degree of CPU headroom was sought in order to accommodate the uncertainty inherent in a generative piece. When the set of pieces was combined as an installation a brief overlapping of pieces was allowed as pages were turned. This period of time when two pieces were briefly playing would inevitably produce a spike in CPU demand. Across the nine pieces synthesis techniques that suggest natural sounds have been employed, machine-like and overly synthetic-sounding synthesis was avoided. Similarly, when sound events are triggered, their timings, durations and densities have been controlled in such a way as to suggest natural gestures and perhaps activation by a human hand. In particular the opcodes `rspline` and `jspline` were found to be particularly useful in generating natural gestures with just a few input arguments. The various random number generators used in the pieces are seeded by the system clock, which means that each time a piece is run the results will be slightly different. This is an important justification for the pieces to be played live by Csound and not just to exist as fixed renderings. Csound Haiku was written using Csound 5. A number of syntactical innovations in Csound 6 would allow further compression of the code for each piece but the code has not since been updated. In fact the clarity of traditional longhand Csound code is preferred as this helps fulfil the pieces' dictatic aim. A large number of synthesiser cells proved to be of interest but ultimately this number was whittled down to nine. The next stage involved creating 'workbenches' for each of the synthesiser cells. A simple GUI was created using Csound's FLTK opcodes (see Fig. 17.1) so the range of timbres possible for each cell could be explored. The FLTK opcodes are not recommended for real-time use on account of threading issues that can result in interruptions in real-time audio, besides there are now more reliable options built in to a number of Csound's frontends. Nonetheless they are still useful for sketching ideas and remaining independent of a frontend.

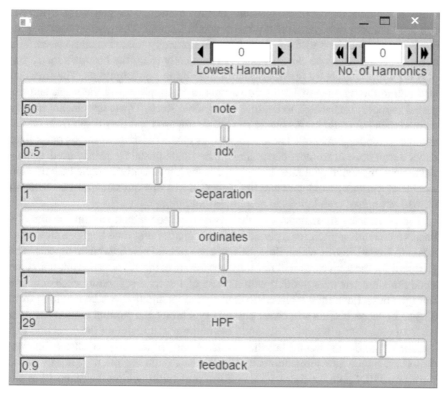

Fig. 17.1 An example of one of the FLTK 'workbenches' that were used to explore ideas for synthesiser cells

For each synthesis idea that proved capable of a range of interesting sounds, a set of critical values was created establishing the useful ranges for parameters and combinations of parameters that would produce specific sounds. These sets of values provided the starting points for the construction of the final pieces, sometimes defining initialisation values for opcodes and at other times defining range values for random functions.

17.3 The Pieces

In this section, we will consider each piece separately, looking at its key elements, and how they were constructed.

17.3.1 Haiku I

The two principle elements in this piece are the rich tone quality of the basic synthesis cell and the design of the glissandi that recur through the piece. The synthesis cell used comes more or less ready-made within Csound. The gbuzz opcode produces a rich tone derived from a stack of harmonically related cosine waves. The overall spectral envelope can be warped dynamically using the opcode's kmul parameter which musically provides control over timbre or brightness. The user can also define the number of harmonic partials to be employed in the stack and the partial number from which to begin the harmonic stack. These options are merely defined then left static in this piece; 75 partials are used and they begin from partial number 1: the fundamental. These two settings characterise the result as brass-like: either a trombone or trumpet depending on the fundamental frequency used. Essential to how this sound generator is deployed is the humanisation: applied the random wobble to various parameters and the slow changes in expression created through the use of gradual undulations in amplitude and timbre.

The piece uses six instances of the gbuzz opcode which sustain indefinitely. Each voice holds a pitch for a period of time and after that time begins a slow glissando to a new pitch at which time the sequence of held note/glissando repeats. The glissandi for the six voices will not fully synchronise; the six notes begin, not at the same time, but one after another with a pause of 1 second separating each entrance from the next. Each time a glissando is triggered its duration can be anything from 22.5 to 27 seconds. This will offset possible synchrony and parallel motion between the six coincident glissandi. Glissando beginnings will maintain the 1 second gap established at the beginning but their conclusions are unlikely to follow this pattern. Each time a new glissando is triggered a new pitch is chosen randomly in the form of a MIDI note number with a uniform probability within the range 12 to 54. The glissando will start from the current note and move to the new note following the shape of an inflected spline: a stretched 'z' shape if rising and a stretched 's' shape if falling. Glissandi can therefore cover a large or small or even a zero interval. This further contributes to undermining any synchrony between the six voices and provides varying beating effects as the six glissandi weave across one another. When the piece begins firstly the reverb instrument is triggered to be permanently active using the alwayson opcode. An instrument called start_long_notes plays for zero performance-time seconds, so therefore only acts at its initialisation time to trigger six iterations of the instrument "trombone" that synthesises sound. This arrangement is shown in Fig. 17.2.

The glissandi are created using the transeg opcode. This opcode generates a breakpoint envelope with the curvature of each segment being user-definable. Each glissando is constructed as a two-segment envelope. The curvature of the first segment will always be 2 and the curvature of the second segment -2. This ensures that, regardless of start and end pitches, each glissando will begin slowly, picking up speed to a maximum halfway through its duration before slowing again and smoothly drawing to a halt at its destination pitch. This can be observed graphically in Fig. 17.3.

Fig. 17.2 The instrument schematic of Haiku I. The six connections between "START LONG NOTES" and "TROMBONE" represent six real-time score events being sent

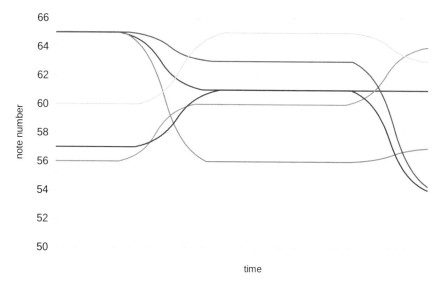

Fig. 17.3 A snapshot of the pitches of the six voices across one minute of the piece. Note that sometimes voices coalesce onto a unison pitch. Independent modulations of the fine tuning of each voice will prevent these being in absolute unison

The timbre of each note experiences random modulation by means of `rspline` random spline generators using the following line:

```
kmul rspline 0.3,0.82,0.04,0.2
```

The first two input arguments define the amplitude limits of the random function and the third and fourth terms define the limits of the rate of modulation. The function produced will actually exceed the amplitude limits slightly but in this instance this will not be a catastrophic problem. When `kmul` is around its minimum value the timbre produced will resemble that of a brass instrument played softly, at its maximum the timbre will more closely resemble the brilliance and raspiness of a brass instrument played fortissimo. Once pitch glissando is added the listener will probably relate the tone produced to that of a trombone. The `rspline` function generator for each voice will follow its own independent path. This means that individual voices will rise out of the texture whenever their timbre brightens and submerge back into it when their timbre darkens. This is an important feature in providing each voice with its own personality, the sense that each voice is an individual player rather than a note of a chord played on a single instrument. The independence of the glissandi from voice to voice also reinforces this individuality. Amplitude, fine pitch control (detuning) and panning position are also modulated using random spline functions:

```
kamp rspline 0.02,3,0.05,0.1
kdtn jspline 0.05,0.4,0.8
kpan rspline 0,1,0.1,1
```

The combined result of these random modulations enriched by also being passed through the `screverb` reverb opcode provides timbral interest on a number of levels. Faster random modulations, those of 0.2 Hz or higher, provide a wobble that could be described as "humanising". Slower modulations of 0.1 Hz or less are perceived as an intentional musical expression such as crescendo.

Listing 17.1 Csound Haiku I

```
ksmps = 32
nchnls = 2
0dbfs = 1

seed  0
gicos ftgen 0,0,131072,11,1
gasendL,gasendR init   0
event_i  "i", "start_long_notes", 0, 0
alwayson "reverb"

instr start_long_notes
 event_i "i","trombone",0,60*60*24*7
 event_i "i","trombone",1,60*60*24*7
 event_i "i","trombone",2,60*60*24*7
 event_i "i","trombone",3,60*60*24*7
 event_i "i","trombone",4,60*60*24*7
 event_i "i","trombone",5,60*60*24*7
endin

instr trombone
 inote     random          54,66
 knote     init            int(inote)
 ktrig     metro           0.015
 if ktrig==1 then
  reinit   retrig
 endif
 retrig:
 inote1    init      i(knote)
 inote2    random    54, 66
 inote2    =         int(inote2)
 inotemid =          inote1+((inote2-inote1)/2)
```

```
idur       random  22.5,27.5
icurve  =          2
timout   0,idur,skip
knote     transeg inote1,idur/2,icurve,inotemid,
                                idur/2,-icurve,inote2
skip:
rireturn
kenv     linseg  0,25,0.05,p3-50,0.05,25,0
kdtn     jspline 0.05,0.4,0.8
kmul     rspline 0.3,0.82,0.04,0.2
kamp     rspline 0.02,3,0.05,0.1
a1       gbuzz   kenv*kamp, \
  cpsmidinn(knote)*semitone(kdtn),75,1,kmul^1.75,gicos
kpan      rspline 0,1,0.1,1
a1, a2  pan2      a1,kpan
outs      a1,a2
gasendL =         gasendL+a1
gasendR =         gasendR+a2
endin

instr reverb
 aL, aR reverbsc          gasendL,gasendR,0.85,10000
 outs     aL,aR
 clear   gasendL, gasendR
endin
```

17.3.2 *Haiku II*

This piece explores polyrhythms and fragmentation and diminution of a rhythmic theme. The source rhythmic theme is defined in a function table as

```
giseq ftgen 0,0,-12,-2,1,1/3,1/3,1/3,1,1/3,1/3,1/3,
                              1/2,1/2,1/2,1/2
```

This rhythmic theme translates into conventional musical notation as shown in Fig. 17.4.

Fig. 17.4 The rhythmic motif stored in the GEN 2 function table `giseq`. Note that the GEN routine number needs to be prefixed by a minus sign in order to inhibit normalisation of the values it stores

Fragments of the motif are looped and overlaid at different speeds in simple ratio with one another. The index from which the loop will begin will be either 0, 1, 2, 3, 4 or 5 as defined by `istart`. An index of zero will point to the first crotchet of the motif as a start point. A value of 6 from the `random` statement will be very unlikely. The fractional part of all `istart` values will be dumped in the `seqtime` line to leave an integer. The index value at which to loop back to the beginning of the loop is defined in a similar fashion by `iloop`. `iloop` will be either 6, 7, 8, 9, 10, 11 or 12. An `iloop` value of 12 will indicate that looping back will occur on the final quaver of the motif. Finally the tempo of the motivic fragment can be scaled via the `itime_unit` variable using an integer value of either 2, 3 or 4. This modification is largely responsible for the rhythmic complexity that is generated; without this feature the rhythmic relationships between layers are rather simple.

Listing 17.2 Mechanism for reading loop fragments from the rhythmic theme.

```
itime_unit  random   2,5
istart      random   0,6
iloop       random   6,13
ktrig_in    init     0
ktrig_out   seqtime  int(itime_unit)/3,int(istart),
                                int(iloop),0,giseq
```

The synthesis cell for this piece uses additive synthesis complexes of harmonically related sinusoids created through GEN 9 function tables. What is different is that the lower partials are omitted and the harmonics that are used are non-sequential. These waveforms are then played back as frequencies that assume the lowermost partial is the fundamental. This will require scaling down of frequencies given to an oscillator, for example the first partial of one of the function tables, `giwave6`, is partial number 11; if we desire a note of 200 Hz, then the oscillator should be given a frequency of 200/11.

Listing 17.3 GEN 9 table used to create a pseudo-inharmonic spectrum

```
giwave6 ftgen 0,0,131072,9,  11,1,0,  19,1/10,0,
                  28,1/14,0,  37,1/18,0,  46,1/22,0,
                  55,1/26,0,  64,1/30,0,  73,1/34,0
```

If wavetables are played back at low frequencies this raises the possibility of quantisation artefacts, particularly in the upper partials. This can be obviated by using large table sizes and interpolating oscillator opcodes such as `poscil` or `poscil3`. The resulting timbre from this technique is perceived as being more inharmonic than harmonic, perhaps resembling a resonating wooden or metal bar. These sorts of timbres can easily be creating using techniques of additive synthesis using individual oscillators, but the method employed here will prove to be much more efficient as each iteration of the timbre will employ just one oscillator. Seven wavetables are created in this way which describe seven different timbral variations. Each time a rhythmic loop begins, one of these seven wavetables will be chosen at random for that loop. The durations of the notes played are varied from note to note. This creates a sense of some note being damped with the hand and others being allowed to resonate. This is a technique common when playing something like a cowbell. The duration values are derived from a probability histogram created using GEN 17. The following table produces the histogram shown in Fig. 17.5.

```
gidurs ftgen 0,0,-100,-17,0,0.4,50,0.8,90,1.5
```

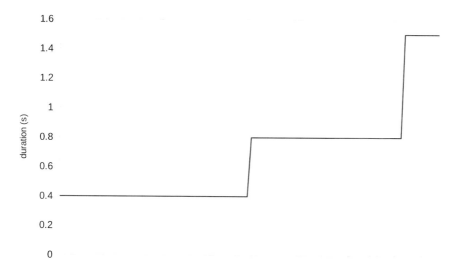

Fig. 17.5 The length of a horizontal line corresponding to a particular value defines the probability of that value. A duration of 0.4 s will be most common, a duration of 1.5 s least common

The instrument named "start_sequences" triggers a new rhythmic loop every 4 seconds. Each rhythmic loop then sustains for 48 seconds, itself generating note events used to trigger the instrument "play_note". The complete schematic is shown in Fig. 17.6.

Fig. 17.6 The connections leading out of "START SEQUENCES" and "PLAY SEQUENCE" are real-time event triggers. Only "PLAY NOTE" generates audio.

Listing 17.4 Csound Haiku II

```
ksmps = 64
nchnls = 2
0dbfs = 1
```

```
giampscl ftgen 0,0,-20000,-16,1,20,0,1,19980,-5,1
giwave1 ftgen 0,0,131073,9, 6,1,0, 9,1/10,0, 13,1/14,0,
 17,1/18,0, 21,1/22,0, 25,1/26,0, 29,1/30,0, 33,1/34,0
giwave2 ftgen 0,0,131073,9, 7,1,0, 10,1/10,0, 14,1/14,0,
 18,1/18,0, 22,1/22,0, 26,1/26,0, 30,1/30,0, 34,1/34,0
giwave3 ftgen 0,0,131073,9, 8,1,0, 11,1/10,0, 15,1/14,0,
 19,1/18,0, 23,1/22,0, 27,1/26,0, 31,1/30,0, 35,1/34,0
giwave4 ftgen 0,0,131073,9, 9,1,0, 12,1/10,0, 16,1/14,0,
 20,1/18,0, 24,1/22,0, 28,1/26,0, 32,1/30,0, 36,1/34,0
giwave5 ftgen 0,0,131073,9, 10,1,0, 13,1/10,0, 17,1/14,0,
 21,1/18,0, 25,1/22,0, 29,1/26,0, 33,1/30,0, 37,1/34,0
giwave6 ftgen 0,0,131073,9, 11,1,0, 19,1/10,0, 28,1/14,0,
 37,1/18,0, 46,1/22,0, 55,1/26,0, 64,1/30,0, 73,1/34,0
giwave7 ftgen 0,0,131073,9, 12,1/4,0, 25,1,0, 39,1/14,0,
 63,1/18,0, 87,1/22,0, 111,1/26,0, 135,1/30,0, 159,1/34,0
giseq ftgen 0,0,-12,-2, 1,1/3,1/3,1/3,1,1/3,1/3,1/3,1/2,
                                           1/2,1/2,1/2
gidurs   ftgen  0,0,-100,-17, 0,0.4, 50,0.8, 90,1.5
gasendL  init   0
gasendR  init   0
gamixL   init   0
gamixR   init   0

girescales ftgen 0,0,-7,-2,6,7,8,9,10,11,12
alwayson          "start_sequences"
alwayson          "sound_output"
alwayson          "reverb"
seed              0

opcode tonea,a,aii
 setksmps  1
 ain,icps,iDecay xin
 kcfenv       transeg     icps*4,iDecay,-8,1,1,0,1
 aout         tone        ain, kcfenv
 xout         aout
endop

instr start_sequences
 ktrig  metro            1/4
 schedkwhennamed ktrig,0,0,"play_sequence",0,48
endin

instr            play_sequence
 itime_unit random 2, 5
 istart    random  0, 6
```

```
 iloop       random   6, 13
 ktrig_in    init     0
 ktrig_out   seqtime  int(itime_unit)/3, int(istart),
                               int(iloop), 0, giseq
 inote     random   48, 100
 ienvscl   =        ((1-(inote-48)/(100-48))*0.8)+0.2
 ienvscl   limit    ienvscl,0.3,1
 icps      =        cpsmidinn(int(inote))
 ipan      random   0, 1
 isend     random   0.3, 0.5
 kamp      rspline  0.007, 0.6, 0.05, 0.2
 kflam     random   0, 0.02
 ifn       random   0, 7
 schedkwhennamed ktrig_out,0,0,"play_note",kflam,0.01,
                  icps,ipan,isend,kamp,int(ifn),ienvscl
endin

instr        play_note
 idurndx     random   0, 100
 p3          table    idurndx, gidurs
 ijit        random   0.1, 1
 acps        expseg   8000, 0.003, p4, 1, p4
 aenv        expsega  0.001,0.003,ijit^2,(p3-0.2-0.002)*p9,
                               0.002,0.2,0.001,1,0.001
 adip        transeg  1, p3, 4, 0.99
 iampscl     table    p4, giampscl
 irescale    table    p8, girescales
 idtn        random   0.995,1.005
 a1   oscili  p7*aenv*iampscl,
                 (acps*adip*idtn)/(6+irescale),
                 giwave1+p8
 adlt        rspline  1, 10, 0.1, 0.2
 aramp       linseg   0, 0.02, 1
 acho        vdelay   a1*aramp, adlt, 40
 icf         random   0, 2
 kcfenv      transeg  p4+(p4*icf^3), p9, -8, 1, 1, 0, 1
 a1          tonex    a1, kcfenv
 a1, a2      pan2     a1,p5
 outs    a1,a2
 gamixL      =        gamixL + a1
 gamixR      =        gamixR + a2
 gasendL     =        gasendL + (a1*(p6^2))
 gasendR     =        gasendR + (a2*(p6^2))
endin
```

```
instr   sound_output
 a1,a2 reverbsc gamixL, gamixR, 0.01, 500
 a1     =       a1*100
 a2     =       a2*100
 a1     atone   a1, 250
 a2     atone   a2, 250
 outs     a1, a2
 clear    gamixL, gamixR
endin

instr   reverb
 aL, aR reverbsc gasendL, gasendR, 0.75, 4000
 outs     aL, aR
 clear    gasendL, gasendR
endin
```

17.3.3 Haiku III

This piece uses for its synthesis cell the `wguide2` opcode which implements a double waveguide algorithm with first-order low-pass filters applied to each delay line (Fig. 17.7). The simple algorithm employed by the `wguide1` opcode tends to produce plucked-string-like sound if excited by a short impulse. The Csound manual describes the sound produced by `wguide2` as being akin to that of a struck metal plate, but perhaps it is closer to the sound of a plucked piano or guitar string with the string partially damped somewhere along its length by, for example, a finger. This arrangement of waveguides produces resonances on a timbrally rich input signal that express inharmonic spectra when the frequencies of the two individual waveguides from which it is formed are not in simple ratio with one another. Care must be taken that the sum of the two feedback factors does not exceed 0.5; the reason this figure is not 1 is because the output of each waveguide is fed back into the input of both waveguides.

At the beginning of each note, a short impulse sound is passed into the wguide2 network. This sound is simply a short percussive harmonic impulse. Its duration is so short that we barely perceive its pitch, instead we hear a soft click or a thump – this is our model of a 'pluck' or a 'strike'. The pitch of the impulse changes from note to note and this imitates the hardness of the strike, with higher pitches imitating a harder strike. Note events are triggered using a metronome but the rate of this metronome is constantly changing as governed by a function generated by the `rspline` opcode. A trick is employed to bias the function in favour of lower values: a random function between zero and 1 is first created, then it is squared (resulting in the bias towards lower values) and finally this new function is re-scaled according to the desired range of values. This sequence is shown in the code snippet below:

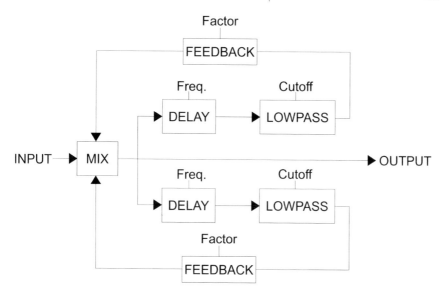

Fig. 17.7 The `wguide2` opcode implements a double waveguide with the output of each delay line being passed through a low-pass filters before begin fed back into the input. Crucially the feedback from each waveguide is fed back into both waveguides, not just itself.

```
krate    rspline 0,1,0.1,2
krate    scale    krate^2,10,0.3
```

Natural undulations of pulsed playing will result provided that the four input arguments for `rspline` are carefully chosen (Fig. 17.8).

Simultaneously with this, `rspline` random functions generate frequencies for the two waveguides contained within wguide2, These frequencies are only applied as initialisation-time values at the beginning of note events. This method is chosen as the sound of continuous glissando throughout each note was not desired. The shapes of the `rsplines` are still heard in how the spectra of note events evolve note to note. Panning and amplitude are also modulated by `rspline` functions and by combining all these independent `rsplines` we create movements in sound that are strongly gestural.

Listing 17.5 Csound Haiku III

```
ksmps = 32
nchnls = 2
0dbfs = 1

giImpulseWave    ftgen              0,0,4097,10,1,1/2,1/4,1/8
gamixL,gamixR,gasendL,gasendR init 0
seed             0
gitims                ftgen                0,0,128,-7,1,100,0.1
```

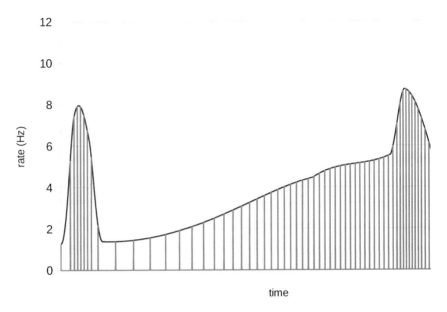

Fig. 17.8 The continuous curve shows the frequency value that is passed to the metro opcode, the vertical lines indicate where note triggers occur.

```
alwayson          "start_sequences"
alwayson          "spatialise"
alwayson          "reverb"

instr     start_sequences
 krate    rspline          0, 1, 0.1, 2
 krate    scale            krate^2,10,0.3
 ktrig    metro            krate
 koct     rspline          4.3, 9.5, 0.1, 1
 kcps     =                cpsoct(koct)
 kpan     rspline          0.1, 4, 0.1, 1
 kamp     rspline          0.1, 1, 0.25, 2
 kwgoct1  rspline          6, 9, 0.05, 1
 kwgoct2  rspline          6, 9, 0.05, 1
 schedkwhennamed ktrig,0,0,"wguide2_note",0,4,kcps,
                               kwgoct1,kwgoct2,kamp,kpan
endin

instr     wguide2_note
 aenv        expon      1,10/p4,0.001
 aimpulse poscil      aenv-0.001,p4,giImpulseWave
 ioct1       random     5, 11
```

```
ioct2      random    5, 11
aplk1      transeg   1+rnd(0.2), 0.1, -15, 1
aplk2      transeg   1+rnd(0.2), 0.1, -15, 1
idmptim    random    0.1, 3
kcutoff    expseg    20000,p3-idmptim,20000,idmptim,200,1,200
awg2       wguide2   aimpulse,cpsoct(p5)*aplk1,
              cpsoct(p6)*aplk2,kcutoff,kcutoff,0.27,0.23
awg2       dcblock2  awg2
arel       linseg    1, p3-idmptim, 1, idmptim, 0
awg2       =         awg2*arel
awg2       =         awg2/(rnd(4)+3)
aL,aR      pan2      awg2,p8
gasendL    =         gasendL+(aL*0.05)
gasendR    =         gasendR+(aR*0.05)
gamixL     =         gamixL+aL
gamixR     =         gamixR+aR
endin


instr      spatialise
adlytim1 rspline 0.1, 5, 0.1, 0.4
adlytim2 rspline 0.1, 5, 0.1, 0.4
aL         vdelay  gamixL, adlytim1, 50
aR         vdelay  gamixR, adlytim2, 50
outs       aL, aR
gasendL    =         gasendL+(aL*0.05)
gasendR    =         gasendR+(aR*0.05)
clear      gamixL, gamixR
endin


instr      reverb
aL, aR     reverbsc gasendL,gasendR,0.95,10000
outs       aL, aR
clear      gasendL,gasendR
endin
```

17.3.4 Haiku IV

This piece focusses on forming periodic gestural clusters within which the sound spectra morphs in smooth undulations. Its synthesis cell is based on Csound's hsboscil opcode. This opcode generates a tone comprising of a stack of partials, each spaced an octave apart from its nearest neighbour. The scope of the spectrum above and below the fundamental is defined in octaves (up to a limit of eight) and this complex of partials is then shaped by a spectral window function typically providing emphasis on the central partial (fundamental). The spectral window can be shifted up or down, thereby introducing additional higher partials when shifted up and additional lower partials when shifted down. In this piece the spectral window is shifted up and down quite dramatically and quickly once again using an `rspline` function and it is this technique that lends the piece a shifting yet smooth and glassy character. A sonogram of an individual note is shown in Fig. 17.9. Using a logarithmic scale, octaves appear equally spaced and it is clearly observable how the spectral envelope descends from its initial position, ascends slightly then descends again before ascending for a second time.

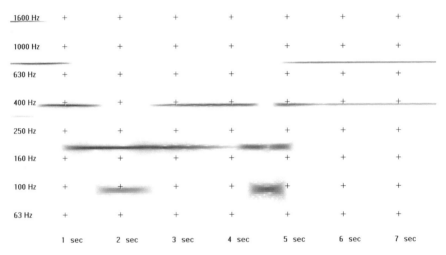

Fig. 17.9 Spectrogram of an `hsboscil` gesture

On its own this sound is rather plain, so it is ring-modulated. Ring modulation adds sidebands – additional partials – which undermine the pure organ-like quality of `hsboscil`'s output. The ring-modulated version and the unmodulated source are mixed using a dynamic cross-fade. Microtonal pitch and panning are modulated using LFOs following sine shapes to evoke a sense of spinning or vibrato, but the amplitudes and rates of these LFOs are continuously varied using `rspline` functions so that the nature of this rotational movement constantly changes, suggesting acceleration or deceleration and expansion or contraction of the rotational displacement. Notes are triggered in groups of four on average every 12 seconds to create the desired density of texture. The actual time gap between clusters varies about this mean and note onsets within each cluster can temporally smear by as much as 2 seconds. These steps ensure a controlled amount of variation in how the piece progresses and sustains interest while not varying excessively from the desired textural density and pace.

Listing 17.6 Csound Haiku IV

```
ksmps   = 32
nchnls  = 2
0dbfs   = 1

gisine           ftgen    0, 0, 4096, 10, 1
gioctfn          ftgen    0, 0, 4096, -19, 1,0.5,270,0.5
gasendL,gasendR init     0
ginotes          ftgen    0,0,-100,-17,0,8.00,10,8.03,
        15,8.04,25,8.05,50,8.07,60,8.08,73,8.09,82,8.11
seed     0
alwayson "trigger_notes"
alwayson "reverb"

instr    trigger_notes
 krate   rspline      0.05, 0.12, 0.05, 0.1
 ktrig   metro        krate
 gktrans trandom      ktrig,-1, 1
 gktrans =            semitone(gktrans)
 idur    =            15
 schedkwhen ktrig, 0, 0, "hboscil_note", rnd(2), idur
 schedkwhen ktrig, 0, 0, "hboscil_note", rnd(2), idur
 schedkwhen ktrig, 0, 0, "hboscil_note", rnd(2), idur
 schedkwhen ktrig, 0, 0, "hboscil_note", rnd(2), idur
endin

instr           hboscil_note
 ipch           table    int(rnd(100)),ginotes
 icps = cpspch(ipch)*i(gktrans)*semitone(rnd(0.5)-0.25)
 kamp           expseg   0.001,0.02,0.2,p3-0.01,0.001
 ktonemoddep jspline    0.01,0.05,0.2
 ktonemodrte jspline    6,0.1,0.2
 ktone          oscil    ktonemoddep,ktonemodrte,gisine
 kbrite         rspline  -2,3,0.0002,3
 ibasfreq       init     icps
 ioctcnt        init     2
 iphs           init     0
 a1             hsboscil kamp,ktone,kbrite,ibasfreq,gisine,
                                  gioctfn,ioctcnt,iphs
 amod  oscil  1, ibasfreq*3.47, gisine
 arm            =        a1*amod
 kmix expseg   0.001, 0.01, rnd(1),
               rnd(3)+0.3, 0.0018
 a1             ntrpol   a1, arm, kmix
 a1             pareq    a1/10, 400, 15, .707
 a1             tone     a1, 500
```

```
kpanrte      jspline   5, 0.05, 0.1
kpandep      jspline   0.9, 0.2, 0.4
kpan         oscil     kpandep, kpanrte, gisine
a1,a2        pan2      a1, kpan
a1           delay     a1, rnd(0.1)
a2           delay     a2, rnd(0.1)
kenv         linsegr   1, 1, 0
a1           =         a1*kenv
a2           =         a2*kenv
outs     a1, a2
gasendL      =         gasendL+a1/6
gasendR      =         gasendR+a2/6
endin

instr   reverb
aL, aR reverbsc gasendL,gasendR,0.95,10000
outs        aL, aR
clear       gasendL,gasendR
endin
```

17.3.5 Haiku V

This piece is formed from two simple compositional elements: a continuous motor rhythm formed from elements resembling struck wooden bars, and periodic punctuating gestures comprising of more sustained metallic sounds. The two sound types are created from the same synthesising instrument. The differences in their character are created by their dramatically differing durations and also by sending the instrument a different set of p-field values which govern key synthesis parameters. The synthesis cell is based around the phaser2 opcode. This opcode implements a series of second-order all-pass filters and is useful in creating a stack of inharmonically positioned resonances. Higher frequencies will decay more quickly than lower ones and this allows phaser2 to imitate struck resonating objects which would display the same tendency. The filters are excited by a single-cycle wavelet of a sine wave. The period of this excitation signal (the reciprocal of the frequency) can be used to control the brightness of the filtered sound – a wavelet with a longer period will tend to excite the lower resonances more than the higher ones. The duration of time for which the synthesis cell will resonate is controlled mainly by the phaser2's feedback parameter. The more sustained metallic sounds are defined by having a much higher value for feedback (0.996) than the shorter sounds (0.9). The continuous motor rhythm is perceived as a repeating motif of four quavers with the pitches of the four quavers following their own slow glissandi. This introduces the possiblity of contrary motion between the different lines traced by the four quavers of the motif.

Listing 17.7 Csound Haiku V

```
ksmps = 32
nchnls = 2
0dbfs = 1

gisine  ftgen    0, 0, 4096, 10, 1
gasendL init     0
gasendR init     0
seed     0
alwayson "start_sequences"
alwayson "reverb"

instr       start_sequences
 iBaseRate random              1, 2.5
 event_i "i","sound_instr",0,3600*24*7,iBaseRate,0.9,
                                0.03,0.06,7,0.5,1
 event_i "i","sound_instr",1/(2*iBaseRate),3600*24*7,
                     iBaseRate,0.9,0.03,0.06,7,0.5,1
 event_i "i","sound_instr",1/(4*iBaseRate),3600*24*7,
                     iBaseRate,0.9,0.03,0.06,7,0.5,1
 event_i "i","sound_instr",3/(4*iBaseRate),3600*24*7,
                     iBaseRate,0.9,0.03,0.06,7,0.5,1
ktrig1    metro             iBaseRate/64
 schedkwhennamed ktrig1,0,0,"sound_instr",1/iBaseRate,
    64/iBaseRate,iBaseRate/16,0.996,0.003,0.01,3,0.7,1
 schedkwhennamed ktrig1,0,0,"sound_instr",2/iBaseRate,
    64/iBaseRate,iBaseRate/16,0.996,0.003,0.01,4,0.7,1
ktrig2    metro             iBaseRate/72
 schedkwhennamed ktrig2,0,0,"sound_instr",3/iBaseRate,
    72/iBaseRate,iBaseRate/20,0.996,0.003,0.01,5,0.7,1
 schedkwhennamed ktrig2,0,0,"sound_instr",4/iBaseRate,
    72/iBaseRate,iBaseRate/20,0.996,0.003,0.01,6,0.7,1
endin

instr       sound_instr
 ktrig     metro             p4
 if ktrig=1 then
  reinit PULSE
 endif
 PULSE:
 ioct      random    7.3,10.5
 icps      init      cpsoct(ioct)
 aptr      linseg    0,1/icps,1
 rireturn
```

```
a1          tablei     aptr, gisine, 1
kamp        rspline    0.2, 0.7, 0.1, 0.8
a1          =          a1*(kamp^3)
kphsoct     rspline    6, 10, p6, p7
isep        random     0.5, 0.75
ksep        transeg    isep+1, 0.02, -50, isep
kfeedback   rspline    0.85, 0.99, 0.01, 0.1
aphs2 phaser2 a1,cpsoct(kphsoct),0.3,p8,p10,isep,p5
iChoRate    random     0.5,2
aDlyMod     oscili     0.0005,iChoRate,gisine
acho vdelay3 aphs2+a1, (aDlyMod+0.0005+0.0001)*1000,100
aphs2       sum        aphs2,acho
aphs2       butlp      aphs2,1000
kenv        linseg     1, p3-4, 1, 4, 0
kpan        rspline    0, 1, 0.1, 0.8
kattrel     linsegr    1, 1, 0
a1, a2      pan2       aphs2*kenv*p9*kattrel, kpan
a1          delay      a1, rnd(0.01)+0.0001
a2          delay      a2, rnd(0.01)+0.0001
ksend       rspline    0.2, 0.7, 0.05, 0.1
ksend       =          ksend^2
outs        a1*(1-ksend), a2*(1-ksend)
gasendL     =          gasendL+(a1*ksend)
gasendR     =          gasendR+(a2*ksend)
endin

instr       reverb
 aL, aR     reverbsc   gasendL,gasendR,0.85,5000
 outs       aL, aR
 clear      gasendL, gasendR
endin
```

17.3.6 Haiku VI

This piece imagines an instrument with six resonating strings as the sound-producing model. The strings are modelled using the wguide1 opcode (Fig. 17.10), which provides a simple waveguide model consisting of a delay line fed into a first-order low-pass filter with a feedback loop encapsulating the entire unit.

The waveguides are excited by sending short-enveloped impulses of pink noise into them. This provides a softer and more realistic 'pluck' impulse than would be provided by a one-sample click. When a trigger ktrig is generated a new pluck impulse is instigated by forcing a reinitialisation of the envelope that shapes the pink noise.

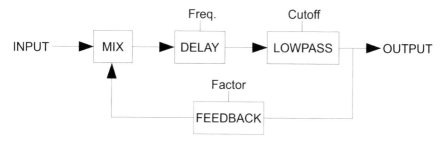

Fig. 17.10 The schematic of the `wguide1` opcode

Listing 17.8 Mechanism to create a retriggerable soft 'pluck' impulse

```
ktrig   metro   krate
if ktrig==1 then
 reinit   update
endif
update:
aenv    expseg   0.0001,0.02,1,0.2,0.0001,1,0.0001
apluck pinkish   aenv
rireturn
```

The six waveguides are essentially sent the same impulse, but for each waveguide the audio of the pluck is delayed by a randomly varying amount between 50 and 250 milliseconds. This offsetting will result in a strumming effect as opposed to the six waveguides being excited in perfect sync. The strumming occurs periodically at a rate that varies between 0.005 and 0.15 Hz. The initial pitches of each waveguide follow the conventional tuning of a six-string guitar (E-A-G-D-B-E) but they each experience an additional and continuous detuning determined by `rspline` opcodes as if each string is being continuously and independently detuned. The detuning of the six waveguides in relation to when pluck impulses occur is shown in Fig. 17.11.

Each waveguide output then experiences a slowly changing and random auto-panning before being fed into a reverb. The full schematic of steps is shown in Fig. 17.12. The use of a large amount of reverb with a long reverberant tail effectively creates pitch clusters as the waveguides slowly glissando, and therefore results in interesting 'beating' effects.

Listing 17.9 Csound Haiku VI

```
ksmps   = 32
nchnls  = 2
0dbfs   = 1

gasendL init     0
gasendR init     0
```

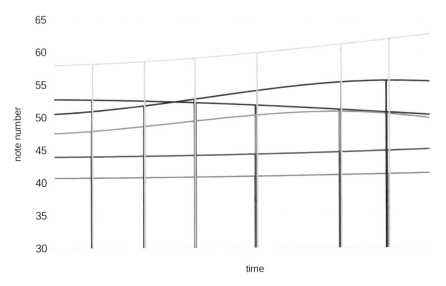

Fig. 17.11 The slow modulations of the pitches of the six waveguides are shown by the six continuous curves. The occurrence of the pluck impulses is shown by the vertical lines. Their misalignment indicates the strumming effect that has been implemented

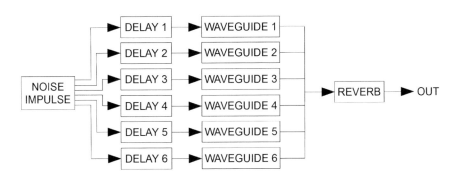

Fig. 17.12 Schematic showing the flow of audio through the main elements used in *Haiku VI*

```
seed      0
alwayson "trigger_6_notes_and_plucks"
alwayson "reverb"

instr    trigger_6_notes_and_plucks
 event_i "i","string",0,60*60*24*7,40
 event_i "i","string",0,60*60*24*7,45
 event_i "i","string",0,60*60*24*7,50
 event_i "i","string",0,60*60*24*7,55
 event_i "i","string",0,60*60*24*7,59
```

```
event_i "i","string",0,60*60*24*7,64
krate   rspline    0.005, 0.15, 0.1, 0.2
ktrig   metro      krate
if ktrig==1 then
 reinit     update
endif
update:
aenv    expseg     0.0001, 0.02, 1, 0.2, 0.0001, 1, 0.0001
apluck  pinkish    aenv
rireturn
koct    randomi    5, 10, 2
gapluck butlp      apluck, cpsoct(koct)
endin

instr           string
adlt            rspline 50, 250, 0.03, 0.06
apluck          vdelay3 gapluck, adlt, 500
adtn            jspline 15, 0.002, 0.02
astring wguide1 apluck,cpsmidinn(p4)*semitone(adtn),\
                                   5000,0.9995
astring         dcblock astring
kpan            rspline 0, 1, 0.1, 0.2
astrL, astrR pan2    astring, kpan
outs      astrL, astrR
gasendL       =          gasendL+(astrL*0.6)
gasendR       =          gasendR+(astrR*0.6)
endin

instr           reverb
aL, aR          reverbsc gasendL,gasendR,0.85,10000
outs      aL, aR
clear     gasendL, gasendR
endin
```

17.3.7 *Haiku VII*

This piece shares a method of generating rhythmic material with Haiku II. Again a sequence of durations is stored in a GEN 2 function table and a loop fragment is read from within this sequence. These durations are converted into a series of triggers using the seqtime opcode. The key difference with the employment of this technique in this piece, besides the rhythmic sequence being different, is that the tempo is much slower to the point where the gaps between notes become greatly dramatised. This time the rhythmic sequence stored in a function table is

```
giseq ftgen  0,0,-12,-2,
                     3/2,2,3,1,1,3/2,1/2,3/4,5/2,2/3,2,1
```

The duration values start from the fifth parameter field, 3/2. The duration value of 2/3 is inserted to undermine the regularity and coherence implied by other values which are more suggestive of 'simple' time. Two synthesis cells are used; the first uses a similar technique to that used in *Csound Haiku II* – the use of a GEN 9 table to imitate an inharmonic spectrum – but this time the sound has a longer duration and uses a stretched percussive amplitude and low-pass filter envelope to imitate a resonating bell sound. The second synthesis cell mostly follows the pitch of the first and is generally much quieter, functioning as a sonic shadow of the first. It is generated using a single instance of gbuzz whose pitch is modulated by a high-frequency jspline function, making the pitch less distinct and adding an 'airiness'. This type of sound is often described as a 'pad' although in this example it could more accurately be described as a distant screech. To add a further sense of space the gbuzz sound passes entirely through the reverb before reaching Csound's audio output. In performance the piece will slowly randomly cross-fade between the GEN 9 bell-like sound and the gbuzz pad sound. The flow of instruments is shown in Fig. 17.13.

Fig. 17.13 Instrument schematic of *Csound Haiku VII*: "TRIGGER SEQUENCE" triggers notes in "TRIGGER NOTES" which in turn triggers notes in "LONG BELL" and "GBUZZ LONG NOTE" which then synthesise the audio

The long bell sounds are triggered repeatedly using schedkwhen whereas the gbuzz long note is triggered by event_i and only plays a single note during the same event. The rhythm of events that the long bell follows will be determined by looping a fragment of the values in the giseq table but the pitch can change owing to the influence of the k-rate function kSemiDrop upon kcps:

```
kSemiDrop line  rnd(2),  p3,  -rnd(2)
kcps        =       cpsmidinn(inote+int(kSemiDrop))
```

The use of the int() function upon kSemiDrop means that changes in pitch will only occur in semitone steps and the choice of values in defining the line function means that kSemiDrop can only be a falling line. rnd(2) creates a fractional value in the range 0 to 2 and rnd(-2) creates a value in the range 0 to -2. This sense of a descending chromatic scale is key to the mood of this piece and can also be observed in the representation of pitch shown in Fig. 17.14. This balance between the use of descending chromaticism and its undermining through the use of other intervals leaves the internal workings of the piece tantalising but slightly enigmatic to the listener.

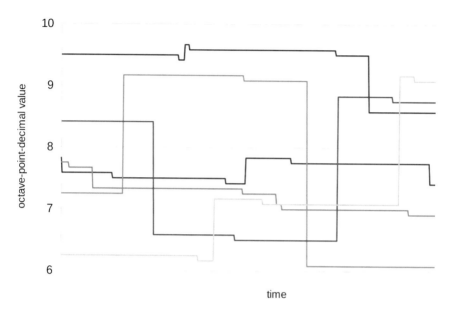

Fig. 17.14 The pitches of six voices are indicated by the six lines. The predominance of descending semitone steps can be seen. The larger intervals are produced when the start of a new sequence occurs

Listing 17.10 Csound Haiku VII

```
ksmps = 32
nchnls = 2
0dbfs = 1

giampscl ftgen 0,0,-20000,-16,1,20,0,1,19980,-30,0.1
giwave ftgen 0,0,4097,9, 3,1,0, 10,1/10,0, 18,1/14,0,\
 26,1/18,0, 34,1/22,0, 42,1/26,0, 50,1/30,0, 58,1/34,0
gicos    ftgen    0,0,131072, 11, 1
giseq    ftgen    0,0,-12, \
                  -2,3/2,2,3,1,1,3/2,1/2,3/4,5/2,2/3,2,1
gasendL  init     0
gasendR  init     0
seed     0
alwayson "trigger_sequence"
alwayson "reverb"
```

```
instr          trigger_sequence
 ktrig         metro              0.2
 schedkwhennamed ktrig,0,0,"trigger_notes",0,30
 kcrossfade  rspline        0, 1, 0.01, 0.1
 gkcrossfade =              kcrossfade^3
endin

instr          trigger_notes
 itime_unit random     2, 10
 istart        random     0, 6
 iloop         random     6, 13
 ktrig_out   seqtime    int(itime_unit),int(istart),
         int(iloop),0,giseq
 idur          random     8, 15
 inote         random     0, 48
 inote         =          (int(inote))+36
 kSemiDrop   line       rnd(2), p3, -rnd(2)
 kcps          =          cpsmidinn(inote+int(kSemiDrop))
 ipan          random     0, 1
 isend         random     0.05, 0.2
 kflam         random     0, 0.02
 kamp          rspline  0.008, 0.4, 0.05, 0.2
 ioffset       random     -0.2, 0.2
 kattlim       rspline  0, 1, 0.01, 0.1
 schedkwhennamed ktrig_out,0,0,"long_bell",kflam,idur,
  kcps*semitone(ioffset), ipan, isend, kamp
 event_i  "i","gbuzz_long_note",0,30,cpsmidinn(inote+19)
endin

instr     long_bell
 acps      transeg  1, p3, 3, 0.95
 iattrnd random    0, 1
 iatt    =         (iattrnd>(p8^1.5)?0.002:p3/2)
 aenv expsega 0.001,iatt,1,p3-0.2-iatt,0.002,0.2,0.001
 aperc     expseg   10000, 0.003, p4, 1, p4
 iampscl table     p4, giampscl
 ijit      random    0.5, 1
 a1        oscili    p7*aenv*iampscl*ijit*(1-gkcrossfade),
                     (acps*aperc)/2,giwave
 a2        oscili    p7*aenv*iampscl*ijit*(1-gkcrossfade),
                (acps*aperc*semitone(rnd(.02)))/2,giwave
 adlt      rspline  1, 5, 0.4, 0.8
 acho      vdelay    a1, adlt, 40
 a1        =         a1-acho
 acho      vdelay    a2, adlt, 40
```

```
a2         =           a2-acho
icf        random      0, 1.75
icf        =           p4+(p4*(icf^3))
kcfenv     expseg      icf, 0.3, icf, p3-0.3, 20
a1         butlp       a1, kcfenv
a2         butlp       a2, kcfenv
a1         butlp       a1, kcfenv
a2         butlp       a2, kcfenv
outs       a1, a2
gasendL =              gasendL+(a1*p6)
gasendR =              gasendR+(a2*p6)
endin

instr      gbuzz_long_note
kenv       expseg  0.001, 3, 1, p3-3, 0.001
kmul       rspline 0.01, 0.1, 0.1, 1
kNseDep    rspline 0,1,0.2,0.4
kNse       jspline kNseDep,50,100
agbuzz     gbuzz   gkcrossfade/80,p4/2*semitone(kNse),
                                   5,1,kmul*kenv,gicos
a1         delay   agbuzz, rnd(0.08)+0.001
a2         delay   agbuzz, rnd(0.08)+0.001
gasendL =              gasendL+(a1*kenv)
gasendR =              gasendR+(a2*kenv)
endin

instr      reverb
aL,aR      reverbsc gasendL,gasendR,0.95,10000
outs       aL,aR
clear      gasendL, gasendR
endin
```

17.3.8 Haiku VIII

This piece creates waves of pointillistic gestures within which individual layers and voices are distinguishable by the listener. The principal device for achieving this is the use of aleatoric but discrete duration values for the note events. This means that there are very long note events and very short note events but not a continuum between them; this allows the ear to group events into streams or phrases emanating from the same instrument. Radical modulation of parameters that could break this perceptive streaming, such as pitch, is avoided. A sequence of note triggering from instrument to instrument is employed. The instrument 'start_layers' is triggered from within instrument 0 (the orchestra header area) using an event_i statement for

zero seconds meaning that it will carry out initialisation-time statements only and
then cease. The i-time statements it carries out simply start three long instances of
the instrument "layer". Adding or removing streams of notes in the piece can easily
be achieved by simply adding or removing iterations of the `event_i` statement
in the instrument "start_layers". Each layer then generates streams of notes using
`schedkwhen` statements which trigger the instrument "note". The synthesis cell in
this piece again makes use of the `gbuzz` opcode but the sound is more percussive.
The fundamental pitch of a stream of notes generated by a layer changes only very
occasionally as dictated by a random sample and hold function generator:

```
knote randomh 0,12,0.1
```

The frequency with which new values for `knote` will be generated is 0.1 Hz;
each note will persist for 10 seconds before changing. The number of partials employed by `gbuzz` in each note of the stream and the lowest partial present are chosen randomly upon each new note. The number of partials can be any integer from
1 to 500 and the lowest partial can be any integer from 1 to 12. This generates quite
a sense of angularity from note to note but the perceptual grouping from note to
note is maintained by virtue of the fact that the fundamental frequency changes so
infrequently. The waveform used by `gbuzz` is not the usual cosine wave but instead
is a sine wave with a weak fifth partial:

```
giwave ftgen 0,0,131072,10,1,0,0,0,0.05
```

A distinctive colouration is added by employing occasional pitch bends on notes.
The likelihood that a note will bend is determined using interrogation of a random
value, the result of which will dictate whether a conditional branch implementing
the pitch bend should be followed:

```
iprob     random    0,1
if iprob<=0.1 then
  irange  random    -8,4
  icurve  random    -4,4
  abend   linseg    1,p3,semitone(irange)
  aperc   =         aperc*abend
endif
```

The range of random values is from 0 to 1 and the conditional threshold is 0.1.
A similar approach is employed to apply a frequency-shifting effect to just some of
the notes. The frequency shifting is carried out using the `hilbert` transformation
opcode. This effect is reserved for the longer notes for reasons of efficiency – the
effect would be less perceivable in shorter notes. The code that conditionally adds
frequency shifting is shown below:

```
iprob2              random              0,1
if iprob2<=0.2&&p3>1 then
  kfshift  transeg 0,p3,-15,rnd(200)-100
  ar,ai    hilbert a1
  asin     oscili  1, kfshift, gisine, 0
```

```
acos      oscili  1, kfshift, gisine, 0.25
amod1     =       ar*acos
amod2     =       ai*asin
a1        =       ((amod1-amod2)/3)+a1
endif
```

Using GEN 17 we create a step function that defines the possible time gaps between consecutive notes in a layer and their probabilities. The table given below will create the distribution shown in Fig. 17.15:

```
gigaps ftgen 0,0,-100,-17, 0,32,5,2,45,1/2,70,1/8,90,2/9
```

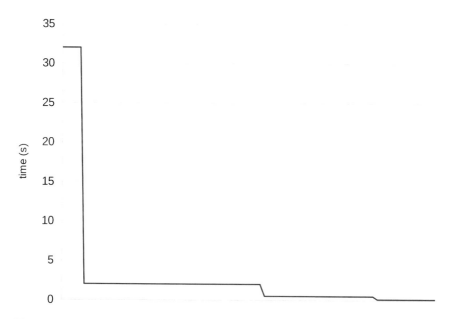

Fig. 17.15 Time gap distribution: the length of a horizontal line corresponding to a particular value defines the probability of that value

A new time gap value between consecutive events will be chosen once every second using an indexing variable created using a random sample and hold function. Normalised indexing is used with the table opcode so that the indexing range for the entire table ranges from 0 to 1. A trigger that will trigger notes is created using metro, the frequency of which will be the reciprocal of the time gap. Random values are selected and triggers are generated using the code snippet shown below:

```
kndx   randomh  0,1,1
kgap   table    kndx,gigaps,1
ktrig metro     1/kgap
```

From the GEN 17 distribution table `gigaps` we can see that the majority of time gaps will be 2 seconds. 5% of the time the gap duration value will be 32 seconds but it is very unlikely this gap value will persist for an entire trigger period as new gap values are generated every second. The instrument "layer" triggers the instrument "note" and it determines its own duration, again by randomly selecting a value from the distribution table `gidurs` (Fig. 17.16):

```
gidurs ftgen 0,0,-100,-17,  0,0.4,  85,4
```

Fig. 17.16 Note duration distribution

85% of note durations will be 0.4 seconds and only 15% will be 4 seconds but the but the longer notes' predominance will seem more significant simply on account of their longer persistence.

Listing 17.11 Csound Haiku VIII

```
ksmps = 32
nchnls = 2
0dbfs = 1

gigaps ftgen 0,0,-100,-17,0,32,5,2,45,1/2,70,1/8,90,2/9
gidurs  ftgen      0,0,-100,-17,  0,0.4,  85,4
giwave  ftgen      0,0,131072,10,1,0,0,0,0.05
gisine  ftgen      0,0,4096,10,1
gasendL init       0
gasendR init       0
seed       0
```

```
event_i  "i","start_layers",0,0
alwayson "reverb"

instr     start_layers
 event_i "i","layer",0,3600*24*7
 event_i "i","layer",0,3600*24*7
 event_i "i","layer",0,3600*24*7
endin

instr   layer
 kndx  randomh      0,1,1
 kgap  table        kndx,gigaps,1
 ktrig metro        1/kgap
 knote randomh      0,12,0.1
 kamp  rspline      0,0.1,1,2
 kpan  rspline      0.1,0.9,0.1,1
 kmul  rspline      0.1,0.9,0.1,0.3
 schedkwhen ktrig,0,0,"note",rnd(0.1),0.01,int(knote)*3,\
                                     kamp,kpan,kmul
endin

instr      note
 iratio   =         int(rnd(20))+1
 p3       table  rnd(1), gidurs, 1
 aenv     expseg 1, p3, 0.001
 aperc    expseg 5, 0.001, 1, 1, 1
 iprob    random 0, 1
 if iprob<=0.1 then
  irange  random -8, 4
  icurve  random -4, 4
  abend   linseg 1, p3, semitone(irange)
  aperc   =         aperc*abend
 endif
 kmul     expon  abs(p7), p3, 0.0001
 a1       gbuzz  p5*aenv, cpsmidinn(p4)*iratio*aperc,\
              int(rnd(500))+1,rnd(12)+1,kmul,giwave
 iprob2   random 0,1
 if iprob2<=0.2\&&p3>1 then
  kfshift transeg 0, p3, -15, rnd(200)-100
  ar,ai  hilbert a1
  asin   oscili 1, kfshift, gisine, 0
  acos   oscili 1, kfshift, gisine, 0.25
  amod1  =         ar*acos
  amod2  =         ai*asin
  a1     =         ((amod1-amod2)/3)+a1
```

```
endif
a1        butlp     a1, cpsoct(rnd(8)+4)
a1,a2     pan2      a1, p6
a1        delay     a1, rnd(0.03)+0.001
a2        delay     a2, rnd(0.03)+0.001
outs      a1, a2
gasendL =           gasendL+a1*0.3
gasendR =           gasendR+a2*0.3
endin

instr     reverb
aL,aR     reverbsc gasendL, gasendR, 0.75, 10000
outs      aL, aR
clear     gasendL, gasendR
endin
```

17.3.9 Haiku IX

This piece is based around sweeping arpeggios that follow the intervals of the harmonic series. The overlapping of steps of these arpeggios and the long attack and decay times of their amplitude envelopes means that the result is more that of a shifting spectral texture. The sequence of instruments used to generate arpeggios and notes is shown in Fig. 17.17.

Fig. 17.17 Instrument schematic of Csound Haiku IX

The waveform used by the partials of these harmonic series sweeps is not a pure sine wave but is itself a stack of sinusoidal elements from the harmonic series. The notes of the arpeggios will often overlap so this use of a rich waveform will provide dense clustering of partials as the arpeggio sweeps. Each arpeggio plays for 25 seconds and the fundamental frequency ibas of this arpeggio is chosen randomly at the start of this 25 seconds and does not change thereafter. The fundamental is defined using the following steps:

```
ibas    random  0,24
ibas    =       cpsmidinn((int(ibas)*3)+24)
```

Before conversion to a value in hertz, ibas can be an integer value within the set 24, 27, 30, 96. As these are MIDI note numbers the interval between adjacent

possible fundamentals is always a minor third. This arrangement partly lends the
piece the mood attributed to a diminished arpeggio. The values for the frequency of
generation of arpeggios (`krate` in instrument "trigger_arpeggio"), the rate of note
generation within an arpeggio (`krate` in instrument "arpeggio") and the duration
of individual notes within an arpeggio have been carefully chosen to allow for the
generation of rich textures with many overlapping notes but also for the occasional
possibility of significant pauses and silence between arpeggios. To permit these extremes, but rarely in close succession, allows the piece to maintain interest over a
number of minutes.

Listing 17.12 Csound Haiku IX

```
ksmps   = 32
nchnls = 2
0dbfs   = 1

gasendL    init       0
gasendR    init       0
giwave     ftgen      0,0,128, 10, 1, 1/4, 1/16, 1/64
giampscl1 ftgen       0,0,-20000,-16,1,20,0,1,19980,-20,0.01
seed       0
alwayson "trigger_arpeggio"
alwayson "reverb"

instr    trigger_arpeggio
krate    randomh             0.0005, 0.2, 0.04
ktrig    metro               krate
schedkwhennamed ktrig,0,0,"arpeggio",0,25
endin

instr    arpeggio
 ibas    random      0, 24
 ibas    =           cpsmidinn((int(ibas)*3)+24)
 krate   rspline     0.1, 3, 0.3, 0.7
 ktrig   metro       krate
 kharm1 rspline      1, 14, 0.4, 0.8
 kharm2 random       -3, 3
 kharm   mirror      kharm1+kharm2, 1, 23
 kamp    rspline     0, 0.05, 0.1, 0.2
 schedkwhen ktrig,0,0,"note",0,4,ibas*int(kharm),kamp
endin

instr       note
 aenv       linsegr   0, p3/2, 1, p3/2, 0, p3/2, 0
 iampscl    table     p4, giampscl1
```

```
asig       oscili    p5*aenv*iampscl, p4, giwave
adlt       rspline   0.01, 0.1, 0.2, 0.3
adelsig    vdelay    asig, adlt*1000, 0.1*1000
aL,aR      pan2      asig+adelsig, rnd(1)
outs       aL, aR
gasendL    =         gasendL+aL
gasendR    =         gasendR+aR
endin

instr    reverb
aL, aR   reverbsc gasendL,gasendR,0.88,10000
outs       aL, aR
clear      gasendL,gasendR
endin
```

17.4 Conclusions

The *Csound Haiku* pieces provide a demonstration of Csound's ability to jettison the traditional score in favour of notes being generated in the orchestra in real time. Instruments no longer need be regarded merely as individual synthesisers; their roles can be as generators of notes for other instruments or as generators of global variables for use by multiple iterations of later instruments. It has also been shown that these note-generating instruments can be chained in series to multiply the note structures formed. Some of the techniques used for note generation can be described as algorithmic composition and to this end a number of Csound's opcodes for random value and function generation have been employed. Extensive use was made of rspline and jspline for the generation of random spline functions. These opcodes were found to be particularly strong at generating natural flowing gestures when applied to a wide range of synthesis and note generation parameters. These pieces also exemplify the richness of some of Csound's opcodes for sound synthesis and also the brevity of code with which they can be deployed.

Chapter 18
Øyvind Brandtsegg: Feedback Piece

Abstract This chapter presents a case study of a live electronics piece where the only sound source is audio feedback captured with two directional microphones. The performer controls the timbre by means of microphone position. Algorithms for automatic feedback reduction are used, and timbral colouring added by using delays, granular effects, and spectral panning.

18.1 Introduction

The piece is based on audio feedback as the only sound generator. It was originally inspired by the works of Alvin Lucier and Agostino Di Scipio, and how they use audio feedback to explore the characteristics of a physical space. Audio from the speakers is picked up by microphones (two shotgun/supercardioid microphones) and treated with a slow feedback suppression technique. In my feedback piece, a performer holds these two microphones and moves around in the concert space, exploring the different resonant characteristics. Different parts of the room will have resonances at specific frequencies, and the microphone position in relation to the speakers will also greatly affect the feedback potential.

Some subtle colouring effects are added, for example delays to extend the tail of the changing timbres and also to help to get feedback at lower sound levels. Spectral-panning techniques are used to spread timbral components to different locations in the room. During the later part of the piece, granular effects are used to further shape and extend the available sonic palette, still maintaining the concept of using microphone feedback as the only sound source. The overall form of the piece is set by a timed automation of the effects treatment parameters, and the actual sonic content is determined by the performer moving the microphones around the room where the piece is performed. The relationship between material and form is thus explored, in the context of the particular performance venue. Example recordings of the piece can be found at [21, 20].

© Springer International Publishing Switzerland 2016
V. Lazzarini et al., *Csound*, DOI 10.1007/978-3-319-45370-5_18

18.2 Feedback-Processing Techniques

As the signal from audio feedback picked up by microphones in the room is the only sound source for the piece, the digital treatment of the signal is significant for allowing the instrument a certain amount of timbral plasticity. The performer ultimately shapes the sound by way of microphone positioning, but the potential for timbral manipulation of the feedback lies in the adaptive filters. Some of these filters selectively reduce feedback at specific frequencies, while others maintain or increase the potential for feedback. Before A/D conversion, the microphone signal is gently compressed, as the signal level picked up by the microphone can vary greatly between the furthest corner of the room and a position directly in front of the speaker. A regular equalising stage is also used, manually tuned according to venue and speaker system, with the purpose of evening out the general frequency response of the system. The adaptive filters for digitally controlling the feedback consist of three different effects in series. The effects are

1. Autolevel
2. Adaptive spectral equalizer
3. Adaptive parametric equalizer

The autolevel effect measures the input level and compares this to a desired (target) level, then calculates a gain factor as the ratio between the two. Gain adjustment is bypassed if the input level is below a certain threshold (i.e. the background noise level). The gain factor is limited to a maximum value, and the rate of change for the gain factor is filtered so as to avoid abrupt glitches.

Listing 18.1 Autolevel effect, trying to maintain a constant output signal level, excluding sounds below the noise threshold

```
kRefLevel = ampdbfs(-10)
kLevelRate = 10
kMaxLevelFact = 10
kLowThreshold = ampdbfs(-25)
krmsOut init 1
kLevel init 1
aLeve init 1
krmsIn rms ain
kLevel divz kRefLevel, krmsIn, 1
kLevel = (krmsIn < kLowThreshold ? 1 : kLevel)
kLevel limit kLevel, 0, kMaxLevelFact
kLevel tonek kLevel, kLevelRate
aLevel interp kLevel*0.5
aout = ain * aLevel
```

In an audio signal generated by feedback from speaker to microphone, we will have a set of clearly defined partials and these will generally be the spectral components with the highest energy. The adaptive spectral equaliser uses an FFT (or more specifically a streaming phase vocoder, pvs opcodes) to analyze the spectrum, and then selectively lower the amplitudes of the loudest frequencies. This is done by creating a masking table, where each table index corresponds to a specific frequency (a bin of the pvs signal). The masking table is initially filled with all 1's. When we want to lower the energy of the *N* strongest spectral components, we iterate over the set of bins *N* times, each time looking for the bin with the highest energy and setting the masking table to zero for the index of the strongest component. We can then use the table as an adjustable spectral mask by means of the `pvsmaska` opcode, adjusting the amount of gain reduction for the selected bands with the kdepth parameter. To be dynamically updated the spectral profile, we run this process periodically at a selectable update rate. Each time we have an updated spectral profile, we use the `ftmorf` opcode to gradually change from the old profile to the new one.

Listing 18.2 Global ftables for the adaptive spectral equalizer

```
gifftsize = 256
giFftTabSize = (gifftsize / 2)+1
; amplitudes and frequencies for the pvs bins
gifna ftgen 0,0,giFftTabSize,7,0,giFftTabSize,0
gifnf ftgen 0,0,giFftTabSize,7,0,giFftTabSize,0
; all 1's
gil ftgen 0,0,giFftTabSize,7,1,giFftTabSize,1
; tables for storing and morphing spectral masks
gifnaMod ftgen 0,0,giFftTabSize,7,0,giFftTabSize,0
gifnaMod1 ftgen 0,0,giFftTabSize,7,0,giFftTabSize,0
gifnaMod2 ftgen 0,0,giFftTabSize,7,0,giFftTabSize,0
gifnaMorf ftgen 0,0,4,-2,gifnaMod2,gifnaMod1,
                          gifnaMod2,gifnaMod1
```

Listing 18.3 Adaptive spectral equalizer instrument code

```
fsin pvsanal ain,gifftsize,gifftsize/4,gifftsize,1
kflag pvsftw fsin,gifna,gifnf

kpvsNumBands = 4
kpvsAmpMod = 1
kpvsResponseTime = 1
kpvsSmoothTime = 0.5

kpvsResponseCps divz 1,kpvsResponseTime,1
kmetro metro kpvsResponseCps
kdoflag init 0
kdoflag = kdoflag + kmetro
kswitch init 1 ; count 1,2,1,2
kswitch = (kswitch == 1 ? \
      kswitch + kmetro :  kswitch - kmetro)
```

```
; copy pvs data from table to array
; modify amplitude of single bin
; repeat the above N number of times

if (kdoflag > 0) && (kflag > 0) then
kArr2[] init giFftTabSize-2
kArrM[] init giFftTabSize-2
copyf2array kArr2, gifna
copyf2array kArrM, gi1
kcount = 0

process:
kMa, kMaxIndx maxarray kArr2
kArr2[kMaxIndx] = 0
kArrM[kMaxIndx] = 0
kcount = kcount + 1
if kcount < kpvsNumBands then
kgoto process
endif

if kswitch == 1 then
copya2ftab kArrM, gifnaMod2
reinit morftable
else ; (if switch is 2)
copya2ftab kArrM, gifnaMod1
reinit morftable
endif
kdoflag = 0
endif

morftable:
iswitch = i(kswitch)
kinterp = kpvsSmoothTime*kpvsResponseTime
ikinterp = i(kinterp)
kmorfindx linseg iswitch, ikinterp, iswitch+1,
        1, iswitch+1
ftmorf kmorfindx, gifnaMorf, gifnaMod
rireturn

; modify and resynth
fsout pvsmaska fsin, gifnaMod, kpvsAmpMod
aout pvsynth fsout
```

The adaptive spectral effect can be thought of as an equaliser with a large number of static and narrow bands. It works quite effectively to reduce feedback and the adjustable response time allows a certain room for feedback to build up before being damped. In addition to this, a parametric equaliser coupled with a pitch tracker is used as an alternative and complementary means of reducing feedback. We could use the term "homing filter" to describe its behaviour. No claims are made about the efficiency and transparency of this method for any other purposes than use in this composition. The frequency of a band-stop filter is controlled by the pitch tracker. If the pitch tracker output jumps to a new frequency, the band-stop filter slowly approaches this frequency, possibly removing components that did not create feedback, as it travels across the frequency range to close in on the target frequency. In this respect it is in no way considered a "correct" feedback reducer, but acts as part of the complex but deterministic audio system for the composition.

Listing 18.4 Adaptive parametric equalizer. One band is shown, it is commonly used with at least three of these in series

```
kFiltFreq chnget "HomingRate"
kStrength1 chnget "FilterAmount1"
kFiltQ1 chnget "FilterQ1"

kamp rms a1
acps,alock plltrack a1, 0.3, 20, 0.33, 20, 2000,
            ampdbfs(-70)
kcps downsamp acps
kcps limit kcps, 20, 2000
kcps tonek kcps, kFiltFreq
kamp tonek kamp*5, kFiltFreq
kdamp = 1-(kamp*kStrength1)
kgain limit kdamp, 0.01, 1
a2 pareq a1, kcps, kgain, kFiltQ1
```

18.3 Coloring Effects

To enhance and prolong the feedback effects, a simple stereo delay with crossfeed between channels is used. The piece is played with two microphones, and each microphone is mainly routed to its own output channel. The crossfeed in the delay processing allows a certain amount of feedback from one microphone to bleed into the signal chain of the other microphone, creating a more enveloping spatial image. Granular processing with the Hadron Particle Synthesizer[1] is used to extend the sonic palette with noisy, intermittent, and pitch-shifted textures. Spectral-panning techniques are also used to spread the frequency components of each signal chain to several audio outputs, inspired by Peiman Koshravi's circumspectral panning techniques [58] which he in turn picked up from Denis Smalley [117].

---

[1] www.partikkelaudio.com

Listing 18.5 Spectral panner

```
sr = 44100
ksmps = 256
nchnls = 2
0dbfs=1

giSine ftgen  0, 0, 65536, 10, 1
gifftsize = 2048

iNumBins = gifftsize/2
; tables to spectral panning shapes
gipantab1 ftgen 0,0,iNumBins,7,0,iNumBins, 0
gipantab2 ftgen 0,0,iNumBins,7,1,iNumBins, 1

; spectral points (anchors) for panning curve
ifq1 = 120
iHzPrBin = sr/iNumBins ; Hz per bin
iBin1 = round(ifq1/iHzPrBin) ; bin number for this fq
iBin2 = round(ifq1*2/iHzPrBin)
iBin3 = round(ifq1*4/iHzPrBin)
iBin4 = round(ifq1*8/iHzPrBin)
iBin5 = round(ifq1*16/iHzPrBin)
iBin6 = round(ifq1*32/iHzPrBin)
iBin7 = round(ifq1*64/iHzPrBin)
iBin8 = round(ifq1*128/iHzPrBin)

; spectral panning shape A
gipantab1A ftgen 0,0,iNumBins,27, 0, 0, iBin1, 0, iBin8,
               1, iNumBins-1, 0
; complementary spectral panning shape
gipantab2A ftgen 0,0,iNumBins,27, 0, 0, iBin1, 1, iBin8,
               0, iNumBins-1, 1

; shape B
gipantab1B ftgen 0,0,iNumBins,27, 0, 0, iBin1, 0, iBin5,
               1, iBin8, 0, iNumBins-1, 1
gipantab2B ftgen 0,0,iNumBins,27, 0, 0, iBin1, 1, iBin5,
               0, iBin8, 1, iNumBins-1, 0

; shape C
gipantab1C ftgen 0,0,iNumBins,27, 0, 0, iBin1, 0, iBin4,
               1, iBin7, 0, iNumBins-1, 0
gipantab2C ftgen 0,0,iNumBins,27, 0, 0, iBin1, 1, iBin4,
               0, iBin7, 1, iNumBins-1, 1
```

```
  ; shape D
  gipantab1D ftgen 0,0,iNumBins,27, 0, 0, iBin1, 0, iBin3,
                 1, iBin5, 0, iBin7, 1, iNumBins-1, 1
  gipantab2D ftgen 0,0,iNumBins,27, 0, 0, iBin1, 1, iBin3,
                 0, iBin5, 1, iBin7, 0, iNumBins-1, 0

  ; morphing between shapes
  gimorftable1 ftgen 0,0,4,-2, gipantab1A, gipantab1B,
                 gipantab1C, gipantab1D
  gimorftable2 ftgen 0,0,4,-2, gipantab2A, gipantab2B,
                 gipantab2C, gipantab2D

instr 1
  ; spectral panning
  kDepth chnget "Depth"
  kMorph chnget "Morph"
  kMorphLfoAmp chnget "MorphLfoAmp"
  kMorphLfoFreq chnget "MorphLfoFreq"
  kLFO poscil 0.5, kMorphLfoFreq, giSine
  kLFO = (kLFO + 0.5)*kMorphLfoAmp
  kMorph = kMorph+kLFO
  kMorph limit kMorph, 0, 3

  ain inch 1
  fsin pvsanal ain,gifftsize,gifftsize/3,gifftsize,0
  fin1 pvsmix fsin,fsin ; just a simple copy
  fin2 pvsmix fsin,fsin

  ; morph between spectral panning shapes
  ftmorf kMorph, gimorftable1, gipantab1
  ftmorf kMorph, gimorftable2, gipantab2
  fin1 pvsmaska fin1, gipantab2, kDepth
  fin2 pvsmaska fin2, gipantab1, kDepth

  a1 pvsynth fin1
  a2 pvsynth fin2
  ; try to make up gain
  a1 = a1*ampdbfs(kDepth*3)
  a2 = a2*ampdbfs(kDepth*3)

  outs a1, a2
endin
```

18.4 Hosting and Interfacing

The processors are compiled as VST plug-ins using Cabbage, so as to be used with any VST host. I find it quite useful to rely on one of the standard DAWs or VST hosts to provide "bread and butter" functionality like signal i/o, metering, routing, and mixing. I have been using Reaper as the main host for this purpose. For practical purposes related to the full live rig (enabling several other pieces and improvisations), I have split the processing load between two computers, sending audio via ADAT between the two machines (Fig. 18.1). This allows the use of large audio buffer sizes for relaxed processing on the second computer while retaining the possibility for small buffer sizes and low latency on the primary computer. The delays and granular effects for the feedback piece are calculated on the second computer. The effects automation for the feedback piece was originally entered as an automation track in AudioMulch, and due to practical issues related to exporting said automation tracks, I have continued using that host for the automated effects in this piece. The adaptive filters and other feedback-conditioning processes run under Reaper on the first computer. If I were making a technical rig for this piece only, it would be quite possible to run it all on one single computer.

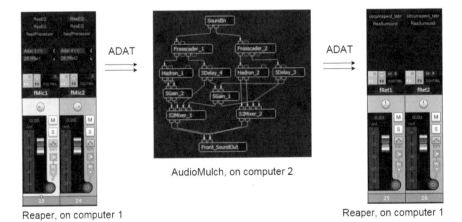

Reaper, on computer 1   Reaper, on computer 1

Fig. 18.1 Signal flow between the two computers, also showing the patching of effects within AudioMulch

18.5 Automation and Composed Form

The main form of the piece can be split into two sections, where the first one is free and the second is automated. Automation in this context relates to a timed script for changes to the effect parameters, gradually changing the instrument's timbre over time. The actual content and sonic gestures for the first section are not notated, and are indeed free to be performed differently in each performance of the piece. However, when I play it, I usually try to make two or three very similar sounds in the beginning of the piece, making the sound almost completely stop between each pair of gestures. This also serves as a way of sharing with the audience how this instrument works, showing the relationship between gestures and resulting sound. I then go on to vary the gestures and so develop nuances of the timbre. The free section has an unspecified duration, but usually continues until the sonic material has been sufficiently exposed. The automated part is then started by means of a trigger signal on a MIDI pedal, sending a play message to the automation host (see Fig. 18.2). During the automated part, the balance between a relatively clean feedback signal and a heavily processed signal is gradually changed. Automation envelopes for the parameters are shown in Fig. 18.3. The granular effects processing significantly changes the recognisable harmonic feedback tone into a scattering noise based texture. Parts of the granular processing are controlled by pitch tracking, so there is an additional layer of feedback between the pitch of the sound in the room and the specific parameters of the granular processing in Hadron, which again is sent out into the room. The automated section concludes by returning the balance of effects processing to a situation where the feedback signal is relatively clean. Performance-wise, I usually try to conclude the piece with a few repetitive sounds/gestures as a means of making a connection to the beginning, and reminding the listener of the basic premise for the piece.

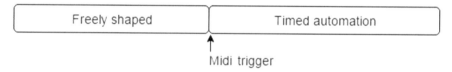

Fig. 18.2 Overall blockwise form

18.6 Spatial and Performative Considerations

The performer of the piece holds the two directional microphones, usually with wireless transmitters so as to increase mobility and reduce the clutter of cable when moving around the space. Microphone gain is controlled by means of EMG (muscle activity) sensors. When the performer tenses the arm muscles, the signal level is increased. This allows for expressive control, and is also a reasonable safety measure to ensure the feedback does not get out of hand (pun intended). An additional "free gift" of this performance setup is that the piece actually requires *physical effort* to perform. This is in contrast to many other forms of electronic music performance, where there is no direct relation between the physical effort and the produced sound.

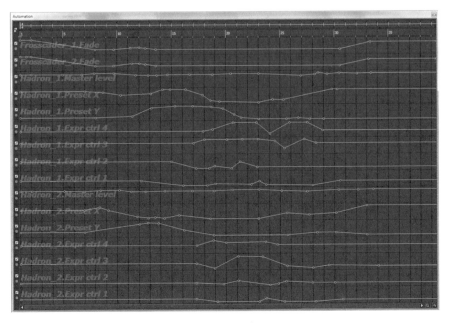

Fig. 18.3 Effects parameter automation in AudioMulch

From experience I also know that this can lend a certain dramatic aspect to the performance situation. The EMG system I personally use is a BodySynth built by Ed Severinghaus in the 1990s, but it would also be possible to make a similar system using an Arduino board with EMG sensors and a wireless expansion board.

An interesting aspect of the sound-producing mechanism created by this system is the complex relationship governing the availability and resonance of different fundamental frequencies. The distance between microphone and speaker effectively controls the potential fundamental frequencies, as a function of distance and the speed of sound. In addition, the resonances of the room allow some frequencies more aptitude for feedback, and these resonances change with the performer's position in the room. The position of the microphone in relation to the separate components of the speaker also greatly affects the feedback potential. The treble drivers are sometimes quite directional, allowing for precise control of feedback in the higher spectral register. One can also physically intervene in the feedback path by putting a hand between the speaker and the microphone, and in this manner balance the amount of high frequency content in the feedback timbre (see Fig. 18.4). Similarly, some speaker types radiate bass frequencies via backwards facing bass ports, so going up close to a speaker and "hugging it" may both muffle some direct sound and also put the microphones in as position where deeper frequencies are more pregnant. The software components controlling the feedback loop will also affect the spectral potential. Some of the components are specifically designed to reduce feedback, and others are designed to increase or maintain the feedback potential. The adaptive filters will continuously change in relation to the spectral content of the incoming sound, so it may not be possible to recreate the exact same sound with the same performative gesture at different points in time. The dynamic nature of the instrument thus may seem to create a very difficult situation for the performer, but keep in mind that there is nothing random in the signal chain. Even if it is governed by complex adaptive processes, it is completely deterministic, and thus ultimately controllable.

Fig. 18.4 Using a hand between speaker and microphone to obstruct the feedback path, affecting the spectral balance of the feedback signal

Chapter 19
Joachim Heintz: Knuth and Alma, Live Electronics with Spoken Word

Abstract This chapter exemplifies the usage of Csound in a live electronics setup. After introducing the general idea, two branches of working with spoken word as live input are shown. *Knuth* analyses the internal rhythm and triggers events at recognised accents, whereas *Alma* analyses sounding units of different sizes and brings them back in different modes. The concept is shown as a work in progress, with both realisations for improvisation and sketches for possible compositions.

19.1 Introduction

To compose a piece of music and to write a computer program seem to be rather different kinds of human activity. For music and composition, inspiration, intuition and a direct contact with the body[1] seem to be substantial, whereas clear ("cold") thinking, high abstraction and limitless control seem to be essential for writing a computer program. Arts must be dirty, programs must be clean. Arts must be immediate, programs are abstractions and formalisms.

Yet actually, there are various inner connections between composing and programming[2]. Programming should not be restricted to an act of technical application. Writing a program is a creative act, based on decisions which are at least in part intuitive ones. And on the other hand, the way contemporary music thinks, moves, forms is deeply connected to terms like parameters, series, permutations or substitutions. Whether it may be welcomed or not, abstractions and concepts are a substantial part of modern thinking in the arts, where they appear in manifold ways.

To program means to create a world in which the potential for change is inscribed. One of the most difficult choices in programming can be to decide on *this* shape instead of so many other possible shapes. Programming is always "on the

---

[1] Composing music, singing or playing an instrument is "embodied" in a similar way as reacting to rhythm, harmonies etc. on the listener's part.

[2] I have described some of these links in my article about composing before and with the computer [50].

V. Lazzarini et al., *Csound*, DOI 10.1007/978-3-319-45370-5_19

way"; it is not the one perfect form, but it articulates one of many possible expressions. Changes and transformations are the heart of programming; programs do not only allow change, they demand it.

So I would like to discuss here a compositional case study which is very much related to a concept rather than to a piece, and which is at the time of writing (December 2015) still at the beginning. I don't know myself whether it will evolve further, and if so, where to. I will show the basic idea first, then two different branches, and I will close with questions for future developments.

19.2 Idea and Set-up

There is a German children's song, called "Ich geh mit meiner Laterne". Children sing it in the streets in autumn, when dusk falls early, parading a paper-built lantern, with a candle in it. The song tells of the child going with the lantern in the same way as the lantern goes with the child, and establishes a relation between the stars in the sky and the lanterns here on earth.

I remembered this song after I wrote a couple of texts[3] and wondered how texts could be read in a different way, using live electronics. The set-up for this should be simple: in the same way as the children go anywhere with their lantern, the person who reads a text should be able to go anywhere with her live-electronics set-up. So I only use a small speaker, a microphone and a cheap and portable computer. The human speaker sits at a table, the loudspeaker is put on the table, too, and the human speaker goes with his electronics companion as his companion goes with him.

Within this set-up, I have worked on two concepts which focus on two different aspects of the spoken word. The one which I called *Knuth* deals with the internal rhythm of the language. The other, *Alma*, plays with the sounding units of the speaker's exclamations and thereby brings different parts of the past back into the present.

19.3 Knuth

At the core of *Knuth*, rhythmic analysis of speech is employed to provide its musical material. This triggers any pre-recorded or synthesised sound, immediately or with any delay, and feeds the results of the analysis into the way the triggered sounds are played back.

---

[3] www.joachimheintz.de/laotse-und-schwitters.html

19.3.1 Rhythm Analysis

The most significant rhythmical element of speech is the syllable. Metrics counts syllables and distinguishes marked and unmarked, long and short syllables. Drummers in many cultures learn their rhythms and when to beat the drums by speaking syllables together with drumming. Vinko Globokar composed his piece *Toucher* [47] as a modification of this practice, thereby musicalising parts of the *Galileo* piece of Bertolt Brecht [23].

But how can we analyse the rhythm of spoken words, for instance the famous Csound community speech example (Fig. 19.1), the "Quick Brown Fox"[4]? What we would like to get out is similar to our perception: to place a marker as soon as a vowel is recognised.

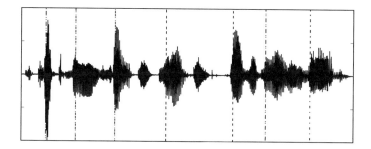

Fig. 19.1 "The quick brown fox jumps over the lazy dog" with desired recognition of internal rhythm

I chose a certain FFT application to get there. The Csound opcode `pvspitch` attempts to analyse the fundamental of a signal in the frequency domain [96]. Considering that a vowel is the harmonic part of speech, this should coincide with the task at hand. If no fundamental can be analysed, `pvspitch` returns zero as frequency. A threshold can be fed directly into the opcode to exclude everything which is certainly no peak because it is too soft. We only have to add some code to avoid repetitive successful analyses.

Listing 19.1 Rhythm detection by *Knuth*

```
/* get input (usually live, here sample) */
aLiveIn soundin "fox.wav"

/* set threshold (dB) */
kThreshDb = -20
```

---

[4] http://csound.github.io/docs/manual/examples/fox.wav

```
/* set minimum time between two analyses */
kMinTim = 0.2

/* initialise time which has passed since
   last detection */
kLastAnalysis init i(kMinTim)

/* count time since last detection */
kLastAnalysis += ksmps/sr

/* initialise the previous state of frequency
   analysis to zero hz */
kFreqPrev init 0

/* set fft size */
iFftSize = 512

/* perform fft */
fIn pvsanal aLiveIn,iFftSize,iFftSize/4,iFftSize,1

/* analyse input */
kFreq, kAmp pvspitch fIn, ampdb(kThreshDb)

/* ask for the new value being the first
   one jumping over kthresh */
if kFreqPrev == 0 &&
   kFreq > 0 &&
   kLastAnalysis > kMinTim then
 /* trigger subinstrument and pass
    analysed values */
  event "i", "whatever", 0, 1, kFreq, kAmp
 /* reset time */
 kLastAnalysis = 0
endif

/* update next previous freq to this freq */
kFreqPrev = kFreq
```

The result is shown in Fig. 19.2).

19.3.2 Possibilities

This analysis of rhythm internal to speech may not be sufficient for scientific purposes; yet it is close enough to the human recognition of emphases to be used in an

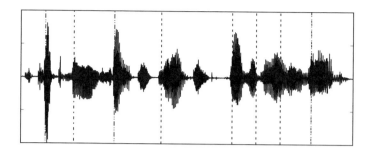

Fig. 19.2 "The quick brown fox jumps over the lazy dog" analysed by Knuth

artistic context. A basic implementation which I realised in CsoundQt (see Figure 19.3) offers an instrument for improvisation. The performer can select sounds which are triggered by *Knuth* in real time via MIDI. It is possible to mix samples and to vary their relative volumes. The basic analysis parameters (threshold, minimal time between two subsequent vowels, minimum/maximum frequency to be analysed) can be changed, too. And by request of some performers I introduced a potential delay, so that a detected vowel does not trigger a sound before a certain time. The detected base frequency is used to play back the samples at different speeds. In a similar way, a reverberation is applied, so that a vowel with a high frequency results in an upwards-transposed and more dry sound, whereas a low frequency results in a downwards-transposed and more reverberative sound.

Fig. 19.3 Knuth as an instrument for improvisation in CsoundQt

For a composition, or a conceptual piece, instead of an improvisation, an approach similar to Globokar's *Toucher* could already be implemented using an instrument based on the described one. Besides, instead of distinguishing vowels, *Knuth* could distinguish base frequencies, and/or the intensities of the accents. This could be linked with sounds, mixtures and structures, and all linkings could be changed in time.

19.4 Alma

Alma looks at the body of spoken words from a different perspective, playing with the speaker's vocalisations to recover elements of the past into the present.

19.4.1 Game of Times

Imagine someone who is reading a text. While they are reading, parts of what has already been read come back, in specific forms or modes. Parts of the past come back, thus confusing the perception of time as a flow, or succession. A *Game of Times* begins, and the text changes its face. Instead of a stream, always proceeding from past to future, it becomes a space in which all that has gone can come back and be here, in this moment, in the present. A line becomes a collection of fragments, and the fragments build up a new space, in which no direction is preferred. You can go back, you can go ahead, you can cease to move, you can jump, you can break down, you can rise again in a new mode of movement. But definitely, the common suggestion of a text as succession will experience strong irritations.

19.4.2 Speech as Different-Sized Pieces of Sounding Matter

So *Alma* is about the past, and she only works with the material the speaker has already uttered. But this material is not equivalent to all that has been recorded in a buffer. It would be unsatisfactory to play back some part of this past randomly: sometimes a syllable, sometimes the second half of a word, sometimes silence.

Moreover, the sounding matter must be analysed and selected. The choice for *Alma* is this: start by recognising sounding units of different sizes. A very small size of such a sounding unit approximately matches the phonemes; a middle size matches the syllables; a larger size matches whole words or even parts of sentences.

The number of sizes or levels is not restricted to three; there can be as many as desired, as many as are needed for a certain Game of Times. The method used to distinguish a unit is very easy: a sounding unit is considered as something which has a pause before and afterwards. The measurement is simply done by *rms*: if sounding material is below a certain *rms* threshold, it is considered as a pause. So the two parameters which determine the result are the threshold and the time span over which the *rms* value is measured. Figures 19.4–19.6 show three different results of sounding units in the "Quick Brown Fox" example, depending on threshold and time span.

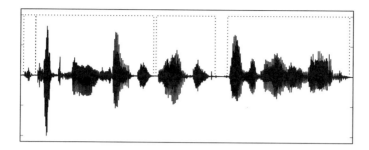

Fig. 19.4 Sounding units analysed with -40 dB threshold and 0.04 seconds rms time span: four large units

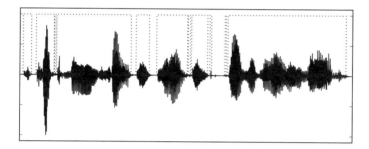

Fig. 19.5 Sounding units analysed with -40 dB threshold and 0.01 seconds rms time span: eleven units of very different sizes

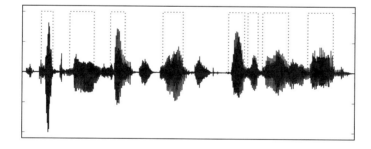

Fig. 19.6 Sounding units analysed with -20 dB threshold and 0.04 seconds rms time span: eight units of medium size

The basic code to achieve this is quite simple, reporting whether the state defined as silence has changed. Usually, these changes are then written in a table or array, indicating the start and end of sounding units.[5]

Listing 19.2 Analysis of sounding units by *Alma*

```
opcode IsSilence, k, akkj
 aIn, kMinTim, kDbLim, iHp xin
 /* rms */
 iHp = iHp == -1 ? 5 : iHp
 kRms rms aIn, iHp
 /* time */
 kTimeK init 0
 kTimeK += 1
 kTime = kTimeK/kr
 /* analyse silence as rms */
 kIsSilence = dbamp(kRms) < kDbLim ? 1 : 0
 /* reset clock if state changes */
 if changed(kIsSilence) == 1 then
  kNewTime = kTime
 endif
 /* output */
 if kIsSilence == 1 &&
    kTime > kNewTime+kMinTim then
  kOut = 1
 else
  kOut = 0
 endif
 xout kOut
endop

 /* minimal silence time (sec) */
 giMinSilTim = .04
 /* threshold (dB) */
 giSilDbThresh = -40
 /* maximum number of markers which can be written */
 giMaxNumMarkers = 1000
 /* array for markers */
 gkMarkers[] init giMaxNumMarkers

instr WriteMarker
 /* input (usually live) */
 aIn soundin "fox.wav"
 /* initialise marker number */
```

---

[5] To be precise, the units start and end a bit earlier, so half of the rms time span is subtracted from the current time.

```
kMarkerNum init 0
/* analyse silence */
kSil IsSilence aIn, giMinSilTim,
          giSilDbThresh, 1/giMinSilTim
/* store pointer positions */
if changed(kSil) == 1 then
 kPointer = times:k() - giMinSilTim/2
 gkMarkers[kMarkerNum] = kPointer
 kMarkerNum += 1
 endif
endin
```

In Figs. 19.4–19.6, we observe that the results only roughly correspond to the above-mentioned distinction between phonemes, syllables and words. As humans, we recognise these through complex analyses, whereas *Alma* deals with the spoken language only in relation to one particular aspect: sounding units, separated by "silence". It very much depends on the speaker, her way of connecting and separating the sounds, whether or not expected units are detected by *Alma*, or unexpected ones are derived. I like these surprises and consider them to be an integral part of the game.

19.4.3 Bringing Back the Past: Four Modes

Currently *Alma* can bring back the past in four different modes. None of these modes is a real "playback". All change or re-arrange the sounding past in a certain way. This is a short overview:

1. A large number of small speech particles create a sound which resembles a wave breaking on the shore. This is done via a special way of scratching, combined with a variable delay feedback unit. Depending mainly on the speed of scratching, the language is barely or not at all recognised.
2. Units in the overall size of syllables are put together in a new order, so that new words are created. Pauses between the syllables can be added, so that instead of words a more scattered image is created, one might say a landscape of syllables.
3. A short rhythm is given in proportions, for instance 1/2, 2/3, 1/3, 1/4, 3/4, 1. This rhythm controls the playback of isolated sound units, so that the natural, free rhythm of the language is left in favour of a metric rhythm. To avoid perfect repetitions, the rhythm is varied by applying permutations and different scalings of expansion/compression.
4. A sound snippet is transformed[6] into a bell-like sound which gently seems to speak. This sound can be of very different durations, starting from the original (= short) duration to a stretch factor of a thousand. Although it reproduces the

---

[6] via some FFT code which selects a number of prominent bins.

most prominent partials of the original sound, it sounds high and adds a pitched, slowly decaying sound to the overall image.

19.4.4 Improvisation or Composition

As for *Knuth*, one major application for *Alma* is to be an instrument for improvisation. Figure 19.7 shows the CsoundQt interface for a MIDI-driven improvisation.

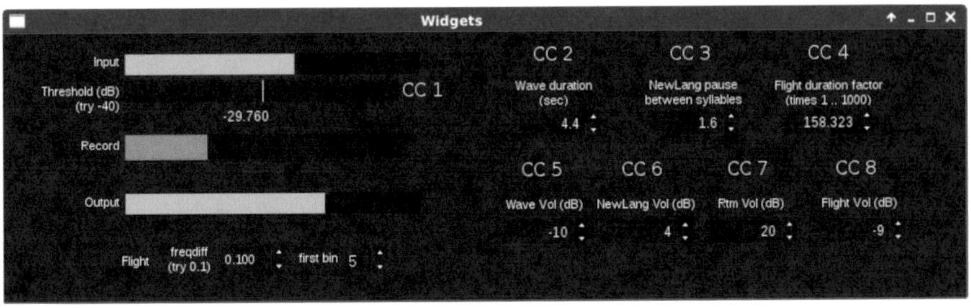

Fig. 19.7 Alma as an instrument for improvisation in CsoundQt

The microphone input is recorded in one huge buffer of arbitrary length, for instance ten minutes. The threshold can be adjusted in real time, as well as some parameters for the four modes. The markers for the modes are written in four tables, created automatically by an array which holds pairs of silence time span and maximum number of markers to be written.

Listing 19.3 Generation of marker tables in *Alma*

```
/* pairs of time span and number of markers */
giArrCreator[] fillarray .005, 100000, .02, 10000,
            .001, 500000, .01, 50000
instr CreateFtables
 iIndx = 0
 ;for each pair
 while iIndx < lenarray(giArrCreator) do
  iTableNum = giTableNumbers[iIndx/2]
   ;create a table
  event_i "f", iTableNum, 0, -iMarkers, 2, 0
   ;start an instance of instr WriteMarker
  iSilenceTime = giArrCreator[iIndx]
  schedule "WriteMarker", 0, p3, iTableNum,
                iMarkers, iSilenceTime
  iIndx += 2
 od
 turnoff
```

```
endin
```

As for compositional sketches, I am currently working on some studies which
focus on only one of the four modes. For the second mode for example, I took a text
by Ludwig Wittgenstein [132][7], separated it into parts of three, five, seven, nine or
eleven words, followed by pauses of one, two, three, four, five or six time units.[8]
Alma herself acts in a similar way, creating new "sentences" of five, seven, nine or
thirteen "words" and breaks of two, four, six or eight time units afterwards.[9]

Listing 19.4 *Alma* reading a text in her way

```
/* time unit in seconds */
giTimUnit = 1.2
/* number of syllables in a word */
giWordMinNumSyll = 1
giWordMaxNumSyll = 5
/* pause between words (sec) */
giMinDurWordPause = 0
giMaxDurWordPause = .5
/* possible number of words in a "sentence" */
giNumWordSent[] fillarray 5, 7, 9, 13
/* possible pauses (in TimUnits) after a sentence */
giDurSentPause[] fillarray 2, 4, 6, 8
/* possibe maximum decrement of volume (db) */
giMaxDevDb = -40

instr NewLangSentence

/* how many words */
iNumWordsIndx = int(random(0,
            lenarray:i(giNumWordSent)-.001))
iNumWords = giNumWordSent[iNumWordsIndx]

/* which one pause at the end */
iLenPauseIndx =
      int(random(0,lenarray:i(giDurSentPause)-.001))
iLenPause = giDurSentPause[iLenPauseIndx]

/* get current pointer position and
   make sure it is even (= end of a section) */
S_MarkerChnl sprintf "MaxMarker_%d", giMarkerTab
iCurrReadPos chnget S_MarkerChnl
```

---

[7] Numbers 683 and 691.

[8] A time unit is what the speaker feels as an inner pulse, so something around one second.

[9] I presented this study as part of my talk at the 3rd International Csound Conference in St. Petersburg, 2015.

```
iCurrReadPos = iCurrReadPos % 2 == 1 ?
              iCurrReadPos-1 : iCurrReadPos-2

/* make sure not to read negative */
if iCurrReadPos < 2 then
 prints {{  Not enough Markers available.
              Instr %d turned off.\n}}, p1
 turnoff
endif

/* set possible read position */
iMinReadPos = 1
iMaxReadPos = iCurrReadPos-1

/* actual time */
kTime timeinsts

/* time for next word */
kTimeNextWord init 0

/* word count */
kWordCount init 0

/* if next word */
if kTime >= kTimeNextWord then

 /* how many units */
 kNumUnits = int(random:k(giWordMinNumSyll,
                   giWordMaxNumSyll))

 /* call instrument to play units */
 kNum = 0
 kStart = 0
 while kNum < kNumUnits do

  /* select one of the sections */
  kPos = int(random:k(iMinReadPos, iMaxReadPos+.999))
  kPos = kPos % 2 == 1 ?
         kPos : kPos-1 ;get odd = start marker

  /* calculate duration */
  kDur = (table:k(kPos+1, giMarkerTab) -
         table:k(kPos, giMarkerTab)) * giBufLenSec

  /* reduce if larger than giSylMaxTim */
```

```
  kDur = kDur > giSylMaxTim ? giSylMaxTim : kDur

  /* get max peak in this section */
  kReadStart =
      table:k(kPos, giMarkerTab) * giBufLenSec ;sec
  kReadEnd = kReadStart + kDur
  kPeak GetPeak giBuf, kReadStart*sr, kReadEnd*sr

  /* normalisation in db */
  kNormDb = -dbamp(kPeak)

  /* calculate db deviation */
  kDevDb random giMaxDevDb, 0

  /* fade in/out */
  iFade = giMinSilTim/2 < 0.003 ?
          0.003 : giMinSilTim/2

  /* add giMinSilTim duration at the end */
  event "i", "NewLangPlaySnip", kStart,
      kDur+giMinSilTim, table:k(kPos, giMarkerTab),
      iFade, kDevDb+kNormDb

  /* but not for the start
     (so crossfade is possible) */
  kStart += kDur

  /* increase pointer */
  kNum += 1

 od

 /* which pause after this word */
 kPause random giMinDurWordPause, giMaxDurWordPause

 /* set time for next word to it */
 kTimeNextWord += kStart + kPause

 /* increase word count */
 kWordCount += 1

endif

/* terminate this instance and
   create a new one if number
```

```
   of words are generated */
 if kWordCount == iNumWords then
  event "i", "NewLangSentence", iLenPause, p3
  turnoff
 endif

endin

instr NewLangPlaySnip
 iBufPos = p4
 iFade = p5
 iDb = p6
 aSnd poscil3 1, 1/giBufLenSec, giBuf, iBufPos
 aSnd linen aSnd, iFade, p3, iFade
 out aSnd * ampdb(iDb)
endin
```

I think the main question for further steps with *Alma* is how the human speaker reacts to the accompaniment which comes out of the lantern. I hope I can go on exploring this field.[10]

19.5 Conclusions

This chapter showed an example of the usage of Csound for a live electronics set-up in both an improvisational and compositional environment. It explained how the basic idea derives from a mixture of rational and irrational, adult and childish, recent and past images, feelings, desires and thoughts, leading to two different emanations, called *Knuth* and *Alma*. It showed how *Knuth* is able to detect the irregular accents of spoken word, or its internal rhythm. Different possibilities of artistic working and playing with *Knuth* based on these detections were demonstrated, implemented either as a MIDI-based instrument for improvisation, or as a concept for a predefined composition. For the second branch, *Alma's* way of analysing and marking sounding units of different sizes was explained in detail. The implementation of an instrument for improvising with *Alma* was shown, as well as a study for reading a text in partnership with her. The code snippets in this chapter display some ways to write a flexible, always developing live electronics set-up in Csound.

---

[10] Thanks here to Laureline Koenig, Tom Schröpfer, Anna Heintz-Buschart and others who accompanied my journey with Knuth and Alma.

Chapter 20
John ffitch: Se'nnight

Abstract In this chapter, the background and processes are explored that were used in the composition of *Se'nnight*, the seventh work in a sequence *Drums and Different Canons* that explores the onset of chaos in certain sequences.

20.1 Introduction

I have been interested for a long time in sequences of notes that never repeat, but are very similar, thereby combining familiarity and surprise. This led me to the areas of the onset of chaos. Quite separately I get an emotional reaction to small intervals, much less than a semitone. I started a series of pieces with these two interests under the umbrella title *Drums and Different Canons*. The first of this family used a synthetic marimba sound with added bell sounds governed by the Hénon and torus maps (see Section 20.2).

This was followed by five further pieces as shown in Table 20.1. But the seventh piece owes much to the first, which was described in [38].

Table 20.1 Drums and Different Canons Series

#	Date	Title	Style	Main Map	Length
1	1996	Drums & Different Canons#1	Tape	Hénon/Torus	07:00
2	2000	Stalactite	Tape	Lorenz	07:37
3	2001	For Connie	Piano	Lorenz	04:00
4	2002	Unbounded Space	Tape	Hénon	06:50
5	2002/2003	Charles à Nuit	Tape	Hénon	05:02
6	2010	Universal Algebra	Quad	Hénon	05:22
7	2011-2015	Se'nnight	Ambisonic	Hénon	13:40

In this chapter I present the background in near-chaos that is central to the series, describe the initial idea, and how it developed into the final work.

© Springer International Publishing Switzerland 2016

V. Lazzarini et al., *Csound*, DOI 10.1007/978-3-319-45370-5_20

20.2 Hénon Map and Torus Map

The underlying mathematical idea in the whole series of pieces is the use of recurrence relations to generate a sequence of numbers that can be mapped to things such as frequency and duration, these sequences showing near self-similarity.

Central to this is the Hénon map, which is a pair of recurrence equations

$$x_{n+1} = 1 - ax_n^2 + y_n$$
$$y_{n+1} = bx_n$$

with two parameters a and b. For values $a = 1.4$, $b = 0.3$ the behaviour is truly chaotic, and those are the values I used in the first piece in this series. Starting from the point $(0,0)$ the values it delivers look as in Fig. 20.1.

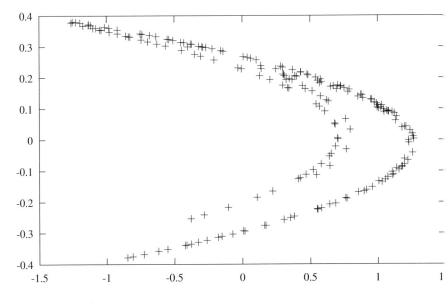

Fig. 20.1 Chaotic Hénon map

This picture shows the values but not the order, which is illustrated in Fig. 20.2.

The paths are similar but never repeat; that is its attraction to me. I will use it as a score generator.

The other map used in the development of *Se'nnight* is the Chirikov standard map, which is on a torus (so I sometimes call it the torus map):

$$I_{n+1} = I_n + K\sin(\Theta_n)$$
$$\Theta_{n+1} = \Theta_n + I_{n+1}$$

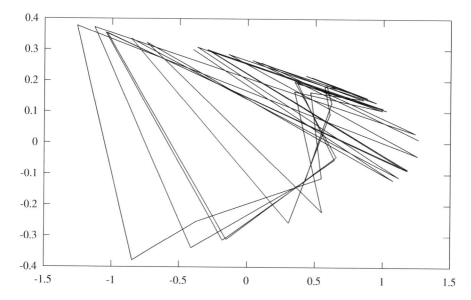

Fig. 20.2 Path of Hénon points

with I and Θ reduced to the range $[0, 2\pi)$. Values from this map have been used as gestures in Drums and Different Canons. In the final version of *Se'nnight* this was not used but remains a potential component and an inspiration.

20.3 Genesis of Se'nnight

I often find titles, and then they generate the ideas or structure of a piece. In the case of *Se'nnight* the title came from a calendar adorned with archaic or little-used words, and this was one of the words that took my attention as one that eventually might become a title. Much later and independently, it occurred to me that a multiple-channel piece might be possible, in the sense that I knew of two locations that had eight speakers, and I might be able to persuade them to play something — I had been somewhat reserved in trying this after hearing the piece *Pre-Composition* by Mark Applebaum which, among other things, mocks the problem of getting an eight-channel piece performed. My personal studio became equipped with quad sound about this time so I decided on a two-dimensional surround sound model.

I did spend a little time wondering about VBAP or ambisonics, or just eight sources; knowing ambisonic pieces and how they affected me I decided on this format, with a HRTF version to assist in the creative process.

The other idea was to revisit the first of the series as it remains one of my favourite works, and seems to have more possibilities in it.

20.4 Instruments

The main instrument in Number 1 was a synthetic marimba built in a modal-modelling way, that originally I found in a set of Csound examples written by Jon Nelson. I have loved the sound of the marimba since hearing one played on Santa Monica Pier. For *Se'nnight* I updated the original instrument a bit, and made provision to position the sound at an angle round the midpoint. Actually two versions were eventually made; one for ambisonics and one for HRTF, but more about that later. Each note is static in space and presumed to be at equal distance.

The other main instrument, reused from the second movement of the first piece, *Gruneberg*, is a synthetic drum, also originating from Jon Nelson, which I tweaked a little.

The third instrument is new, just a use of the `fmbell` with the attack removed. As one can see the instruments are few in number but my interest is largely in investigating the range of sounds and pitches they can produce with parameters driven by the near-chaos sequences. It could be said that this is a minimal method but the whole series does this in different ways. Any complexity is in the score.

20.5 Score Generation

The score is generated by three C programs.

20.5.1 Score1

The first program uses the Hénon map to create a sequence of the drum instrument, using the constants $a = 1.5$, $b = 0.2$. The x values are used to select a pitch (in hundredths of an octave) and location, and y for time intervals between events. The raw values are scaled, `pitch=5.7+x*2.4; duration=fabs(10.0*y)`. There is also code to vary the other parameters of the drum. This mapping was achieved after a great deal of experimentation. The location mapping mainly follows the pitch with some global movement.

The amplitude of the notes is different and, in a strange way, regular. The time between events is quantised to 500 beats per minute, but there is no regular beat. The piece is in 5/4 time with a beat every 7/500 of a minute. The 5/4 time, actually based on the first half of the Hindustani jhaptaal rhythm pattern[1], is manifest as amplitude variations of the first note on or after the beat. This generates a stuttering beat which I like:

```
/* Jhaptaal */
#define beats_per_bar (5)
```

---

[1] dhin na | dhin dhin na | tin na | din din na

```
double vol[beats_per_bar] = { 1.5, 0.5, 1.0, 0.5, 0.0};
int pulse = 7;
int next_beat = 0;

double volume(int tt)
{
    double amp = -6.0;                       /* Base amplitude */

    if (tt >= next_beat) { /* If first event after beat */
                          /* increase amplitude   */
      amp += vol[(tt/pulse) % beats_per_bar];
      next_beat += pulse;   /* and record next beat time */
    }
    return amp;
}
```

The program generate 1,700 events, again chosen after much listening and many variants. The later part of the events has a narrowed location.

```
int main(int argc, char *argv)
{
  double x = 0.0,
         y = 0.0;
  int i;

  initial();
  for (i=0; i<1700; i++) {          /* Some iterations */
    output(x, y);
    henon(&x, &y);
    if (tt>1240) {
      minloc = (tt-1240)*135.0/(1500.0-1240.0);
      maxloc = 360.0-minloc;
    }
  }
  tail();
}
```

For an early version that was all there was, but the wider possibilities beckoned.

20.5.2 Score2 and Score3

The second score section uses much of the structure of the first, but uses a modified bell sound at fixed locations as well as the marimba, using the Henon function in a number of shortish sequences. Again the final form came after many trials and changes, adjusting sounds as well as timings.

I named and archived two further versions of the work before fixing on the final form.

20.6 Start and End

I like my music to have a recognisable ending, and to a lesser extent a start. In the case of the first piece in the series this was a strict cannon of falling tones, as both start and end. For *Se'nnight* I took the same cannon but repeated it from eight different directions around the space. The ending was originally the same, but after consideration and advice the ending was made more reverberant and noticeable. Rather than synthesise this sequence I took the original output, resampled it to 48 kHz, and just used a playback instrument. This was something of a short cut as I usually do everything in Csound.

The other global component that was added to the final assembly was an amplitude envelope for the whole work, mainly for a fade at the end:

```
instr 1
gaenv   transeg  1, p3-15, 0, 1, 15, -1, 0
endin
....
i1  0   1500
```

20.7 Multichannel Delivery

Part of the design was to create a two-dimensional circle around any listener, and ultimately to use ambisonics. But for some of the development in a small studio environment I used the HRTF opcodes; so for the marimba

```
al, ar  hrtfstat a7, p7, 0, "h48-l.dat","h48-r.dat",
        9, 48000
    out  al, ar
```

In the ambisonic rendering these lines read

```
aw,ax,ay,az,ar,as,at,au,av,ak,al,
        am,an,ao,ap,aq
        bformenc1 a7,p7,0
        out      aw, ax,ay,ar,au,av,al,am,ap,aq
```

There are 16 outputs in second-order ambisonics, but six of them relate to height positions which, as I am not using them, are perforce zero. Hence the output is a ten-channel WAV file. Because there is to be further processing this is saved in floating-point format.

There are then simple orchestras to read this high-order audio and output as stereo, quad or octal files:

```
<CsoundSynthesizer>
<CsInstruments>
        sr      =       48000
        kr      =        4800
        ksmps   =          10
        nchnls  =           8

gis init 1.264

instr 1
   aZ      init       0
   aw,ax,ay,ar,au,av,
   al, am, ap, aq diskin2  "ambi.wav", 1
   ao1, ao2, ao3, ao4,
   ao5, ao6, ao7, ao8  bformdec1 4,
           aw, ax, ay, aZ, ar, aZ, aZ, au,
           av, aZ, al, am, aZ, aZ, ap, aq
      out   gis*ao1,gis*ao2,gis*ao3,
         gis*ao4,gis*ao5,gis*ao6,
         gis*ao7,gis*ao8
endin
</CsInstruments>
<CsScore>
i1 0 821
</CsScore>
</CsoundSynthesizer>
```

The variable `gis` is a scaling factor that was determined experimentally, and differs for the various speaker configurations. I have not yet looked into the theory of this, but rendering to floats and using the `scale` utility is a simple pragmatic way to determine suitable scales. As a programmer I wrapped up all the computational details via a `Makefile` mechanism.

20.8 Conclusions

In this chapter I have tried to give some notion of my compositional methods. I rely heavily on small C programs to generate sound sequences, often with a near-chaos component. I also rely on Csound and its commitment to backward compatibility as the software platform.

This piece is rather longer than most of my works but it needed the space, and true to its name it has seven loose sections. To save you looking it up, **Se'nnight** is archaic Middle English for a week.

Chapter 21
Steven Yi: Transit

Abstract This chapter discusses the process of composing *Transit* using Csound and Blue. It will introduce the Blue integrated music environment and discuss how it both augments the features of Csound and fits within a composition workflow.

21.1 Introduction

The act of music composition has always been an interesting experience to me. A large part of my musical process is spending time listening to existing material in a work and making adjustments intuitively. I may be listening and suddenly find myself making a change: a movement of sound to a new place in time, the writing of code to create new material and so on. At other times, I take a very rational approach to analysing the material, looking for relationships and calculating new types of combinations to experiment with. I suppose it is this way for everyone; that writing music should oscillate between listening and writing, between the rational and irrational.

The process of developing each work tends to unfold along its own unique path. In *Transit*, I was inspired by listening to the music of Terry Riley to create a work that involved long feedback delay lines. I began with a mental image of performers in a space working with electronics and worked to develop a virtual system to mimic what I had in mind. Once the setup was developed, I experimented with improvising material live and notating what felt right. I then continued this cycle of improvisation and notation to extend and develop the work.

For *Transit*, I used both Csound and my own graphical computer music environment *Blue* to develop the work. Csound offers so many possibilities but it can become very complex to use without a good approach to workflow and project design. Blue provides numerous features that worked well for my own process of composing this work. The following will discuss the phases of development that occurred while developing this piece and how I employed both Csound and Blue to realise my compositional goals.

© Springer International Publishing Switzerland 2016
V. Lazzarini et al., *Csound*, DOI 10.1007/978-3-319-45370-5_21

21.2 About Blue

Blue is a cross-platform integrated music environment for music composition.[1] It employs Csound as its audio engine and offers users a set of tools for performing and composing music. It provides a graphical user interface that augments Csound with additional concepts and tools, while still exposing Csound programming and all of the features that Csound provides to the user.

Blue is built upon Csound's audio engine and language. All features in Blue, even purely visual ones where users are not exposed to any Csound, generate code using the abstractions available in Csound. For example, a single Blue instrument may generate multiple Csound instrument definitions, user-defined opcodes, score events and function tables. Looking at it from another point of view, Csound provides a small but versatile set of concepts that can support building large and complex systems such as visual applications like Blue.

Blue users can opt to use as many or as few of Blue's features as they like. They can start by working in a traditional Csound way and develop their work using only Csound orchestra and score code. From here, they can choose to use different parts of Blue if it serves their needs. For example, some may use the orchestra manager and convert their text-only instruments into graphical instruments; others may use the score timeline to organise score material visually using its graphic interface; and still others may decide to use the Blue mixer and effects system to assemble and work with their project's signal graph.

Blue plays a major role in my work and is the primary way through which I compose with Csound. Figures, 21.1–21.4 show the primary tools in Blue that I used in writing *Transit*. I will discuss these tools and my approach to composing with Blue below.

21.3 Mixer, Effects and the Signal Graph

For most of my works, I usually begin with the search for sounds. Besides my time spent specifically on composing, I will spend time just working on developing new instruments and effects. In my sound development time, I may write new Csound instruments to try out new designs or I may use previously made designs and explore their parameters to develop new "patches" for the instrument. It is in this time of sonic exploration that a sound may catch my attention and I begin to experiment with it and start to create a foundation for a new piece.

Once composing begins, I develop the signal graph of the piece by routing the output of instruments used to generate the sounds through Blue's mixer system. It is there that I work on signal routing and inserting effects for processing such as reverberation and equalisation.

---

[1] http://blue.kunstmusik.com

Fig. 21.1 Blue's Mixer interface, used to define signal routing and effects processing.

In *Transit*, I used a different approach where I had an image of performers on a stage in mind and began with developing the signal graph using Blue's mixer and effects system. I started by adding Blue Synth Builder instruments (described in Section 21.4) that I had previously developed to my project. Using known instruments allowed me to develop and test the mixer setup first and work on the sounds for the work later.

The primary processing focus for this work was the feedback delay. This is a simple delay line where the output of the delay is multiplied and fed back into the delay. The result is an exponentially decaying echo of any sound fed into the delay effect. Rather than write this effect from scratch, I used the "Tempo-Sync Stereo Delay" by William Light (a.k.a. *filterchild*) that was contributed to BlueShare — a built-in, online instrument and effects exchange platform available within Blue. Blue effects are written using Csound orchestra code and a graphical interface editor available within the program.

Figure 21.1 shows the mixer setup used for the piece. In Blue, each instrument within the project's orchestra has a single mixer channel associated with it. Blue's mixer channels are not bound to any number of audio channels: a Blue instrument may send one or many channels to the mixer and effects will process as many channels as they are designed to take as inputs. Each channel's strip has bins for pre- and post-fader effects and a final output channel can be chosen from the dropdown at the bottom of the channel strip. Channels can either route to sub-channels or the master channel. Sub-channels in turn can route to other sub-channels or the master channel. Blue restricts channel output to prevent feedback within the mixer (i.e.

one cannot route from a channel to a sub-channel and back to the original channel). Users may also insert *sends* as an effect within one of the channel's bins that will send a user-scaled copy of the audio signal to a sub-channel or the master channel.

For *Transit*, there are four main channels shown for each of the four instruments used, a sub-channel named DelayLine that I added and the master channel. Channel 1's output is primarily routed out to the master channel but it also sends an attenuated copy of the signal to the DelayLine sub-channel. Channel 2 and 3's outputs are primarily routed to the DelayLine sub-channel. Channel 4 is only routed to the Master channel and has an additional reverb in its post-fader bin. All of the instrument signals are ultimately routed through the master channel, where a reverb effect (a thin wrapper to the `reverbsc` opcode) is used.

While Blue's mixer and effects system provides the user with a graphical system for creating signal-processing graphs, the underlying generated code uses standard Csound abstractions of instruments and user-defined opcodes. First, each Blue instrument uses the `blueMixerOut` pseudo-opcode to output signals to the mixer. Blue processes instrument code before it gets to Csound and if it finds any lines with `blueMixerOut`, it will auto-generate global audio-rate variables and replace those lines with ones that assign signal values to the variables. These global variables act as an audio signal bus.[2] Next, each Blue effect is generated as a user-defined opcode. Finally, all mixing code is done in a single Blue-generated mixer instrument. Here, the global audio signals are first read into local variables, then processed by calls to effects UDOs. Summing and multiplication of signals is done where appropriate. The final audio signals are written to Csound's output channels to be written to the soundcard or to disk, depending upon whether the project is rendering in real time or non-real time.

Blue's mixer system proved valuable while developing *Transit*. I was able to quickly organise the signal-routing set-up and use pre-developed effects for processing. I could certainly have manually written the Csound code for mixing and effects, but I generally prefer using Blue's mixer interface to help visualise the signal graph as well as provide a faster and more intuitive way to make adjustments to parameters (i.e. fader strengths, reverb times, etc.).

21.4 Instruments

Once the initial signal graph was developed, I turned my attention to the source sounds to the graph. Blue provides its own instrument system that is an extension of Csound's. In Blue, users can write Csound code to write standard Csound instruments, much as one would when using Csound directly. In addition, users can use completely graphical instruments without writing any Csound code, use the Blue Synth Builder (BSB) to develop instruments that use both Csound code and graphi-

---

[2] Blue's mixer system was developed long before arrays were available. Future versions of Blue may instead use a global audio signal array for bussing, to provide easier to read generated code.

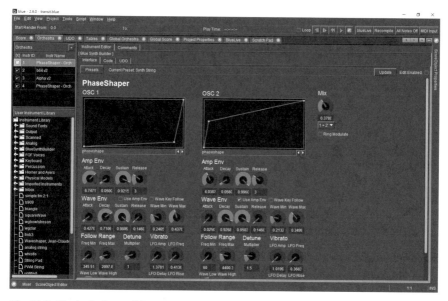

Fig. 21.2 Blue's Orchestra Manager, used to both define and edits instruments used within the work as well as organise the user's personal instrument library

cal widgets (or simply use just the graphical user interface to work with previously developed instruments), or use scripting languages to develop Csound instruments.

For *Transit*, I used three types of instruments I had previously developed: two instances of *PhaseShaper* (a phase-distortion synthesis instrument) and one instance each of *b64* (an instrument design based on the SID synthesis chip, commonly found in Commodore 64 computers) and *Alpha* (a three-oscillator subtractive synthesizer). These instruments are Blue Synth Builder (BSB) instruments where the main code is written in the Csound orchestra language and a graphical interface is developed using BSB's GUI editor. The system lets the user write Csound code using placeholder values inside angle brackets (shown in listing 21.2) within the Csound orchestra code where values from the GUI interface will be used.

While the three instrument types differ in their *internal* design, the instruments share a common approach to their *external* design. By internal design I am referring to the individual synthesis and processing code within each instrument and by external design I am referring to how the instruments are called through their p-fields. For my work, I have developed a basic instrument template using user-defined opcodes that allows instruments to be called with either a five- or eight-p-field note statement or `event` opcode call.

Listing 21.1 Instrument p-field format examples

```
; 5-pfield note format
; p1 - instrument ID
; p2 - start time
```

```
; p3 - duration
; p4 - Pitch in Csound PCH format
; p5 - amplitude (decibel)
i1 0 2 8.00 -12


; 8-pfield note format
; p1 - instrument ID
; p2 - start time
; p3 - duration
; p4 - Start PCH
; p5 - End PCH
; p6 - amplitude (decibel)
; p7 - articulation shape
; p8 - stereo space [-1.0,1.0]
i1 0 2 8.00 9.00 -12 0 0.1
```

Listing 21.1 shows examples of the two different note p-field formats. The five-pfield format is suitable for simple note writing that maps closely to the values one would find with MIDI-based systems: pitch and amplitude map closely to a MIDI Note On event's key and velocity values. I use this format most often when performing instruments using either Blue's virtual MIDI keyboard or a hardware MIDI keyboard.

The eight-p-field format allows for more complex notation possibilities such as defining start and end PCH values for glissandi, articulation shape to use for the note (i.e. ADSR, swell, fade in and out), as well as spatial location in the stereo field. I use this format most often when composing with Blue's score timeline as it provides the flexibility I need to realise the musical ideas I most often find myself exploring.

Listing 21.2 Basic instrument template

```
;; MY STANDARD TEMPLATE
instr x

;; INITIALISE VARIABLES
...

;; STANDARD PROCESSING
kpchline, kamp, kenv, kspace yi_instr_gen \
    i(<ampEnvType>), i(<attack>), i(<decay>), \
    i(<sustain>), i(<release>), <space>

;; INSTRUMENT-SPECIFIC CODE
...

;; OUTPUT PROCESSING
aLeft, aRight    pan2 aout, kspace
```

```
blueMixerOut aLeft, aRight

endin
```

Listing 21.2 shows the basic instrument template I use for my instruments. The template begins with initialising variables such as the number of p-fields given for the event, duration and whether the note is a tied note. Next, the `yi_instr_gen` UDO uses the `pcount` opcode to determine if a five- or eight-pfield note is given and processes according the values of the p-fields together with the values passed in to the UDO from BSB's GUI widgets. `yi_instr_gen` returns k-rate values for frequency (`kpchline`), total amplitude (`kamp`), amplitude envelope (`kenv`) and spatial location (`kspace`). These values are then used by the instrument-specific code and an output signal is generated. The final signal goes through stereo-output processing and the results are sent to Blue's mixer using the `blueMixerOut` pseudo-opcode.

Using a standardised external instrument design has been very beneficial for me over the years. It allows me to easily change instruments for a work as well as quickly get to writing scores as I know the design well. I can also revisit older pieces and reuse score generation code or instruments in new works as I know they use the same format. For *Transit*, I was able to reuse instruments I had previously designed and quickly focus on customising the sounds, rather than have to spend a lot of time re-learning or writing and adjusting instrument code.

In addition, the use of graphical widgets in my instruments allowed me to both quickly see as well as modify current settings for parameters. I enjoy writing signal-processing code using Csound's orchestra language, but I have found that when it comes to configuring the parameters of my instruments, I work much better using a visual interface rather than a textual one. Blue provides me with the necessary tools to design and use instruments quickly and effectively.

21.5 Improvisation and Sketching

After the initial instrument and effects mixer graph was set up, I started experimenting with the various instruments using BlueLive, a feature in Blue for real-time performance. BlueLive offers two primary modes of operation: real-time instrument performance using either Blue's virtual MIDI keyboard or its hardware MIDI system; and score text definition, generation and performance using SoundObjects and the BlueLive graphical user interface.

BlueLive operates by using the current Blue project to generate a Csound CSD file that does not include pre-composed score events from Blue's score timeline. Once the Csound engine is running, Blue will either transform incoming MIDI events into Csound score text or generate score from SoundObjects and send the event text to the running instance of Csound for performance.

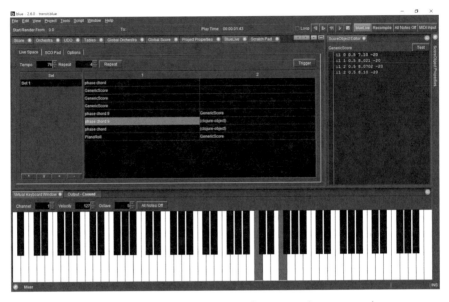

Fig. 21.3 BlueLive, a tool for real-time instrument performance and score generation

When I first start to develop a work, I often find myself using real-time instrument performance to improvise. This serves two purposes: first, it allows me to explore the parameters of the instruments and effects I have added to my project to make adjustments to their settings, and second, it allows me to experiment with score ideas and start developing material for the piece.

Once a gesture or idea takes a hold of my interest, I notate the idea using the Blue-Live interface using SoundObjects. SoundObjects are items that generate Csound code, most typically Csound score. There are many kinds of SoundObjects, each with their own editor interface that may be textual, graphical or both in nature. For *Transit*, I began by using GenericScore and Clojure SoundObjects. The former allows writing standard Csound score text and the latter allows writing code using the Clojure programming language to generate scores.

Figure 21.3 shows the primary interface for BlueLive. A table is used to organise SoundObjects; the user can use it to either trigger them one at a time or toggle to trigger repeatedly. When an object in the table is selected, the user can edit its contents using the ScoreObject editor window.[3]

When composing *Transit*, BlueLive allowed me to rapidly experiment with ideas and explore sonic possibilities of my instruments and effects. It streamlined my workflow by letting me make adjustments and audition changes live without having to restart Csound. It is a key tool I use often when sketching musical material for a work.

---

[3] SoundObjects are a sub-class of a more generalised type of object called ScoreObject. Further information about differences between the two can be found in Blue's user manual.

21.6 Score

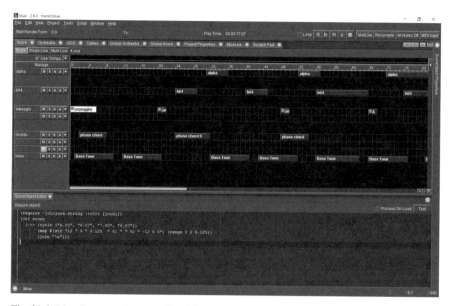

Fig. 21.4 Blue Score, used to organise objects in time for pre-composed works

After notating a number of SoundObjects and experimenting live, I began the process of composing the work by transferring SoundObjects from BlueLive to Blue's score timeline (shown in Figure 21.4). Blue's score timeline is organised into LayerGroups that are are made up of various Layers. Each LayerGroup has its own kinds of layers and unique interfaces for creating and organising musical ideas in time. Each LayerGroup is laid out with space between the layers, much like groups of instruments are organised in orchestral scores.

For *Transit*, I used only SoundObject LayerGroups and used three kinds of SoundObjects: the previously mentioned GenericScore and Clojure objects as well as PianoRolls. PianoRolls offer visual editing of notes in time and are commonly found in digital audio workstation and sequencer software. The choice to use one object rather than another was done purely by intuition: I simply reached for the tool that most made sense at the time while composing.

The ability to choose from a variety of tools for notating scores is an important feature for me in Blue. Each LayerGroup and SoundObject type has its own advantages for expressing different musical ideas. Just as using only text can feel limiting to me, I will also feel confined if I can only use visual tools for notating musical ideas. Sometimes ideas develop more intuitively with one kind of object or another and I am happy to have many options available and to use different kinds of objects together in the same visual timeline system.

I find that visually organising and manipulating objects on a timeline is the optimal way for me to work. I value Csound's score format for its flexibility and expressiveness and often use it for writing single notes or small gestures. However, once a piece grows to a certain level of complexity, I find it difficult to mentally manage all of the parts of a piece using text alone. With Blue, I can directly write Csound score for the parts where it is useful, but I can also leverage all of the other features of Blue's score to aid me in my composing.

One final note about the score in *Transit*: in addition to the organisation of ideas into layers and objects, I also configured the timeline to divide time into equal four beat units at a tempo of 76 beats per minute (BPM). Note that the BPM for the score differs from the BPM for the delay line, which is set to 55 BPM. I found these values by trial and error to create different temporal layers for the work. This provided an interesting polyphonic quality between the notes as written and their processed echoes through time.

21.7 Conclusions

In this chapter, I discussed my own approach to composing using Csound and Blue in the piece *Transit*. Using Blue allowed me to build upon my knowledge of Csound orchestra code to develop and use graphical instruments and effects. These instruments and effects were then used with Blue's mixer system, which provided a way to visualise and develop the signal graph. BlueLive and Blue's score timeline allowed me to perform real-time sketching of material as well as organise and develop my composition in time. Overall, Blue provided an efficient environment for working with Csound that let me focus on my musical tasks and simplified composing *Transit*.

Chapter 22
Victor Lazzarini: Noctilucent Clouds

Abstract This chapter presents a case study of a fixed-media piece composed entirely using Csound. It discusses the main ideas that motivated the work, and its three basic ingredients: a non-standard spectral delay method; a classic algorithm using time-varying delays; and feedback. The source sounds used for the piece are discussed, as well as its overall structure. The case study is completed by looking at how post-production aspects can be seamlessly integrated into the orchestra code.

22.1 Introduction

There are many ways to go about using Csound in fixed-media electronic music composition. An initial approach might be to create orchestras to generate specific sounds, render them to soundfiles and arrange, edit, mix and further process these using a multi-track sequencer. This is often a good starting point for dipping our toes in the water. However, very quickly, we will realise that Csound offers very precise control over all the aspects of producing the final work. Very soon we find ourselves ditching the multi-tracker and start coding the whole process in Csound. One of the additional advantages of this is that the composition work is better documented, and the final program becomes the equivalent to the traditional score: a textual representation of the composition, which can be preserved for further use.

The path leading to this might start with a few separate sketches, which can be collated as the work grows, and it can begin to take shape in various versions. Once the complete draft of the piece is done, we can work on mastering and retouching it to perfect the final sound of the work. It is also important to note that the process of composing can be quite chaotic, sometimes leading to messy code that creates the right result for us, even though it might not use the most orthodox methods. It is good to allow for some lucky coincidences and chance discoveries, which are at times what gives a special touch to the composition.

My approach in composing *Noctilucent Clouds* followed this path, starting with some unconnected sketches, and the exploration of particular processes. Some of

487

the experimentation with these ideas paid off, and the composition of the piece progressed following the most successful results. The overall shape and structure was only arrived at later, once a good portion of the start had already been drafted. This allowed me to grow it from the material, rather than impose it from the outside. Once the piece was completely constructed, I went back to tame some of its excesses, to give a well-rounded shape lacking in the raw instrument output.

22.2 The Basic Ingredients

The processes used in this piece can be grouped into three main methods: spectral delays, variable time-domain delay effects and feedback. Although they are used together, closely linked, it is possible to discuss their operation and function separately. This also allows us a glimpse of the sketching process that goes on in the early stages of composition.

22.2.1 Dynamic Spectral Delays

The original idea for this piece came from discussions on the Csound e-mail list on the topic of delays, and how these could be implemented at sub-band level (i.e. independent frequency bands). I had written a pair of opcodes, `pvsbuffer` and `pvsbufread` (discussed earlier in Chapter 14), that implemented a circular buffer for phase vocoder (PV) data, and noted that these could be used to create spectral signal delays. These are delay lines that are created in the frequency domain, acting on specific sub-bands of the spectrum, applying different delay times to each. The idea is simple: once a buffer is set up, we can have multiple readers at these different bands in the spectrum.

A small number of these can be coded by hand, but if we are talking of a larger number, it could be quite tedious to do. Also, we would like to be able to use a variable number of bands, instead of fixing these. In order to do this, we have to find a way of dynamically spawning the readers as required. This can be done recursively, as shown in various examples in this book, or alternatively we can use a loop to instantiate instruments containing the readers. This turns out to be simpler for this application.

The reader instrument signals will have to be mixed together. This can be done either in the time domain or in the spectral domain. The results are not exactly the same, as the implementation of PV signal mixing accounts for masking effects, and removes the softer parts of the spectrum. However, it is preferable to work in the frequency domain to avoid having to take multiple inverse DFTs, one for each sub-band. This way, we can have one single synthesis operation per channel.

To implement these ideas we can split the different components of the code into four instruments:

1. A source and phase vocoder analysis instrument that will feed the buffer. Only one instance of this is required.
2. A control instrument that will dynamically spawn the readers and exit. Different instances can start groups of readers.
3. The buffer reader instrument, which will be instantiated by 2.
4. A resynthesis instrument, of which only one instance is required.

We will need global variables to act as busses for the PV signals, and to carry the time and buffer references from the writer to the readers. An important implementation detail here is that we will need to be careful when clearing the f-sig bus. We will have to make sure this is done in the correct sequence with the synthesis, otherwise the output will be zero. Normally, this is not required for the streaming PV opcodes, but because we are making the f-sig zero deliberately, it needs to happen exactly after the synthesis.

There are two solutions: one is to fix ksmps to the hopsize, which will align the analysis rate to the k-rate; the other is to make the clearing happen only after an analysis period has elapsed. If this is not done, the clearing step will prevent the streaming mechanism from working. The reason for this is that PV signals and opcodes work using an internal frame counter, and this can get out of sync with the synthesis that happens exactly every hopsize samples. In normal use, we do not clear f-sigs, and the synthesis always happens at the end of a signal graph, so this situation never arises.

Another aspect to note is that we need to find a convenient way to set the individual delay for each band. We can do this by defining a curve in a function table, from which the readers can take their individual delays. The bands themselves can be defined in various ways, but making them have a constant Q is a perceptually relevant method. The reader bands can be defined from a starting point upwards, and we just need to space them evenly until the maximum frequency (e.g. $\frac{sr}{2}$). The bandwidth is then determined by the Q value.

A simplified example extracted from the piece is shown in listing 22.1. In this code, we spawn 120 readers, covering 50 Hz upwards with $Q = 10$. The delay times range from 0.5 to 10 seconds, and are defined in a function table created with GEN 5 (which creates exponential curve segments).

Listing 22.1 Dynamic spectral delays example. Note that the bus clearing is protected to make sure it happens in sync with the synthesis

```
gisiz init 1024
gfmix pvsinit gisiz
gifn1 ftgen 1,0,gisiz,-5,
           10,gisiz/8,
           .5,gisiz/4,
           2,gisiz/8,
           8,gisiz/2,4
instr 1
 asig  diskin2 p4,1
```

```
 fsig pvsanal asig,gisiz,gisiz/4,gisiz,1
 gibuf,gkt  pvsbuffer fsig,10
endin
schedule(1,0,60,"src1x.wav")

instr 2
 kst init p4
 ibands init p5
 kq init p6
 kcnt init 0
 kdel init 0
 even:
  kpow  pow  sr/(2*kst), kcnt/ibands
  kcf = kst*kpow
  kdel tablei (2*kcf)/sr, p7, 1
  event "i",p8,0,p3,kdel,kcf,kcf/kq,p7
  kcnt = kcnt + 1
  if kcnt < ibands kgoto even
 turnoff
endin
schedule(2,0,60,50,120,10,gifn1,3)

instr 3
 icf = p5
 ihbw = p6/2
 idel tablei (2*icf)/sr, p7, 1
 fsig pvsbufread gkt-idel,gibuf,icf-ihbw,icf+ihbw
 gfmix pvsmix  fsig, gfmix
endin

instr 20
 kcnt init 0
 asig pvsynth gfmix
 outs asig
 if kcnt >= gisiz/4 then
  gfmix pvsgain gfmix,0
  kcnt -= gisiz/4
 endif
 kcnt += ksmps
endin
schedule(20,0,60)
```

This example is very close to the early sketches of the dynamic spectral delays that I did for the piece. In their final form, I made four reader variants to be used at different times. These included some extra spectral smoothing and frequency shifting. The reader instruments also feature a choice of two output channels, and interpolation between two function tables to obtain the delay values, controlled by instrument parameters.

22.2.2 Variable Delay Processing

Comb filters are simple, but very interesting devices. With very short delay times and high feedback gain, they resonate at their fundamental frequency and its harmonics; with long delay times, they work as an echo, repeating the sounds going into them. One of the ideas I had been keen to explore in a piece for a long time was the transition between these two states by moving the delay time swiftly between the areas that make these effects come into play. This is reasonably straightforward to achieve with a variable delay processor with an internal feedback (a variable comb filter), which is implemented by the opcode `flanger`.

The result can be quite striking, but it needs to be carefully managed, and tuned to yield the correct results. The main idea is to play with these transitions, so we can encode a series of delay times on a function table, and then read these to feed the time-varying parameter. In order to provide a good scope for manipulation, I use an oscillator to read the table at different speeds and with variable amplitude so that the range of delay times can be expanded or compressed. The output of the oscillator then modulates the flanger.

Listing 22.2 demonstrates the principle. It uses an instrument adapted from the piece, which reads the source file directly, and then places it in the flanger. In the actual work, the signal feeding the variable delay is taken from the two spectral delay mix channels. However, this example shows the raw effect very clearly, and we can hear how it relates to the other materials in the piece.

Listing 22.2 Variable delay line processing, with feedback, moving from echoes to resonant filtering

```
nchnls = 2
gifn1 ftgen 3,0,1024,-5,
                .5,102,
                .5,52,
                .01,802,
                .001,70,.5
instr 10
 asig  diskin2 "src1x.wav",1
 aspeed line p5, p3, p6
 adelnv linseg p4, p3-20,p4,10,p7,10,p4
 a1 oscili adelnv,1/aspeed,gifn1
 asig1 flanger asig*0.2,a1,p8,1
 asig2 flanger asig*0.2,0.501-a1,p8,1
 asi2 dcblock  asig2
 asi1 dcblock  asig1
```

```
  outs asig1,asig2
endin
schedule(10,0,140,1,60,30,0.01,0.99)
```

In the final code, I have four versions with slight variations of this instrument used for different sections of the work. They also incorporate the long feedback path between source input and process output, which is discussed in the next section.

22.2.3 Feedback

The instrument designs discussed in the previous sections are also connected to-gether via a global feedback path, which takes the output of the flanger and mixes it back into the spectral delay input. This places a delay of ksmps samples in the return path. Due to the fact that the spectral delay output is placed at a higher instrument than the flanger, we have one further ksmps block delay. Spectral processing also places a latency equivalent to $N + h$, where N is the DFT size and h is the hopsize, between its input and its output. By making ksmps equal to h and N four times this value, we have a total feedback delay of $7 \times h$.

In the piece, the total feedback delay is equivalent to about 3.73 ms, creating a comb filter with a fundamental at around 268.1 Hz. I also placed an inverse comb in the feedback path, with a varying delay time, which alters some of the high-frequency comb filter peaks. The effect of the feedback line sharpens some of the gestures used in the piece. It needs to be controlled carefully, otherwise it can get very explosive in some situations.

Listing 22.3 shows an input connected to the flanger effect, with a high ksmps to mimic the effect in the piece. It does not include the spectral delays for sake of simplicity, but it demonstrates how the feedback connection can affect the flanger process output. A representation of the total feedback path in the piece is shown in Fig. 22.1.

Listing 22.3 Feedback path from flanger output back to input as used in the piece

```
nchnls=2
ksmps=1792

gifn1 ftgen 3,0,1024,-5,
                .5,102,
                .5,52,
                .01,802,
                .001,70,.5
gafdb init 0
instr 1
 gamix  = diskin2:a("src1x.wav",1) + gafdb
endin
schedule(1,0,140)
```

```
instr 10
 aspeed line p6, p3, p7
 adelnv linseg p5, p3-20,p5,10,p8,10,p5
 a1 oscili adelnv,1/aspeed,gifn1
 asig1 flanger gamix,a1,p11,5
 asig2 flanger gamix,0.501-a1,p11,5
 afdb line  p9,p3,p10
 ak1 expseg 0.001,5,p4,p3-5,p4,1,0.001
 asi2 dcblock  asig2*ak1
 asi1 dcblock  asig1*ak1
 gafdb = asi1*afdb + asi2*afdb
 adel oscili adelnv, -1/aspeed,gifn1
 afdb vdelay  gafdb,adel,1
 gafdb =(afdb + gafdb)*0.5
 asi1 clip asi1,0,0dbfs
 asi2 clip asi2,0,0dbfs
 outs asi2, asi1
endin
schedule(10,0,140,0.1,1,60,30,0.01,0.2,0.2,0.99)
```

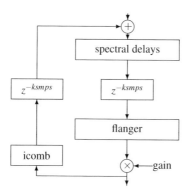

Fig. 22.1 The complete feedback signal path between instruments in the *Noctilucent Clouds* code. The boxes marked z^{-ksmps} denote ksmps-sample delays

22.3 Source Sounds

Conceptually, the source sounds for this piece are not particularly important. Any audio material with a reasonably complex and time-varying spectrum would do.

In practice, however, once I settled for the final sources used, I started to tune the transformations to make the most of their characteristics. Thus, in the process, they became an integral part of the work.

Time-Lines IV

Victor Lazzarini (2006)

Fig. 22.2 The first page of the piano score for *Time-Lines IV*, which is used as the source material for the transformations in *Noctilucent Clouds*

The selection of these was made more or less randomly with various recordings I had at hand. After trying different types of materials, I experimented with the recording of one of my instrumental pieces, *Time-Lines IV* for piano. The actual version used was an early draft containing half of the final work, whose last few bars ended up being edited out of the score. As this seemed to have worked well for the beginning of the piece, I brought in a recording of the second half of the finished piece to complement it. This had the effect of shaping the overall piece in two sections. The start of the piano score is shown in Fig. 22.2.

The piece uses a lot of repeated-note rhythmic figures, which get scrambled once they pass through the spectral delay processing. The thickening of the texture caused by this, the variable delays and the feedback obscures almost completely the original texture, casting a veil over it. This is one of the key ideas I wanted to explore in the piece, how the signal processing can build sound curtains that hide the original sounds from the listener. It also gives the title to the piece: *Noctilucent Clouds* are the visible edge of a much brighter polar cloud layer in the upper atmosphere, a metaphor for the original musical material that lies on the other side of the processing veil.

22.4 Large-Scale Structure

The work is divided evenly into two sections of 241 seconds (4:01 minutes). These two sections form, from the point of view of dynamics, an inverted arch, with a dip and a short silence in the middle. This is also slightly translated to the musical materials, as the big intense sections at the start and end have a certain common textural quality. However, no real repetition of sonic material is audible, even though the underlying piano piece is actually structured as an ABA form. The transformations are worked in different ways, providing the variation between the beginning and the final sections.

The overarching principle of hiding the sources dominates almost the whole of the piece, but not completely, as the piano music is revealed at the end of the work. The idea here is to lift the veil and let the original source show its face. As it coincides with the end of the piano score, it works as a 'composed' end for the piece, and it provides a ready-made solution for the tricky issue of finishing the work. The final few bars of the piano piece are shown in Fig. 22.3, where we can observe that the descending gesture has a definite 'concluding' characteristic.

It is fair to say that, beyond a few general ideas, such as this 'hiding and then revealing' concept, no special effort was made to define the large-scale structure very precisely prior to the actual composition process. This can be attributed to the fact that the underlying piano work was already very well defined in structural terms. Although this music is hidden from the listener, it actually provides the shape for the transformations that were imposed on it, as I tuned and adjusted the instruments to make the most of the source material. This is evident in the fact that the aforementioned dip in the middle is also a feature of the piano score, which is dominated by rhythmically intense and frantic outer sections, and a quieter, steadier middle.

Fig. 22.3 The final bars of the score, where the piano music is finally fully revealed to the listener

22.5 Post-production

Once the piece was finalised, a final step was required to give it a more polished form. It is often the case with some of the more unusual processing techniques (especially with spectral methods), that many of the interesting sound results are somewhat raw and unbalanced, sometimes leaning too heavily on certain frequency bands, or lacking energy in others. The process also led to some sections being too loud in comparison to others, and an adjustment of the overall amplitude curve was needed.

To implement these post-production, mastering requirements, I routed all the audio to a main output instrument, shown in listing 22.4. This allowed me to make all the necessary changes to overall output, without having to modify a very complex code, which relies on very delicately poised and precise parameter settings. In this instrument, I used a graphic equaliser to adjust the overall spectral envelope, and an envelope generator to balance the amplitude of the various sections. This instrument runs from the beginning to the end, and is responsible for switching Csound off when the piece ends (using an 'e' event). It also prints time and rms values each second as it runs.

Listing 22.4 Main output instrument, taking the signals from all sources, and applying equalisation and volume adjustments

```
instr 1000
 al = gaout1 + gaout3 + gaeff1
 ar = gaout2 + gaout4 + gaeff2
 ig1 = ampdb(1)
 ig2 = ampdb(-1.44)
 ig3 = ampdb(-2.88)
 ig4 = ampdb(-3.84)
 ig5 = ampdb(-4.56)
 ig6 = ampdb(-3.36)
 ig7 = ampdb(-1.68)
```

```
ig8 = ampdb(2.88)
ig9 = ampdb(6)
aleq eqfil al,     75,   75,ig1
aleq eqfil aleq, 150,  150,ig2
aleq eqfil aleq, 300,  300,ig3
aleq eqfil aleq, 600,  600,ig4
aleq eqfil aleq,1200,1200,ig5
aleq eqfil aleq,2400,2400,ig6
aleq eqfil aleq,4800,4800,ig7
aleq eqfil aleq,9600,9600,ig8
al   eqfil aleq,15000,10000,ig9
aleq eqfil ar,     75,   75,ig1
aleq eqfil aleq, 150,  150,ig2
aleq eqfil aleq, 300,  300,ig3
aleq eqfil aleq, 600,  600,ig4
aleq eqfil aleq,1200,1200,ig5
aleq eqfil aleq,2400,2400,ig6
aleq eqfil aleq,4800,4800,ig7
aleq eqfil aleq,9600,9600,ig8
ar   eqfil aleq,15000,10000,ig9
again1 expseg 8,30, 2,15,
               1.5,30,1,45, 1,20,
               1,25,  2,55,1,140,
               1,30,   2,30, 1.8,30,
               3,30,3
again2 expseg 8,30,2,15,
                2,30,  2,45,2,20,
                2,25,1,55,1,140,
                1,30,2,30,1.8,30,
                3,30,3
asig_left =     al*again1*0.94
asig_right =    ar*again2*0.8
outs asig_left,asig_right
ktrig = int(times:k())
printf "%ds - L:%.1f R:%.1f \n",
        ktrig,ktrig,
        rms(asig_left),
        rms(asig_right)
gaout1 = 0
gaout2 = 0
gaout3 = 0
gaout4 = 0
gaeff1 = 0
gaeff2 = 0
xtratim 0.1
```

```
if release() == 1 then
 event "e", 0, 0
 endif
endin
schedule(1000,0,482)
```

Feeding this instrument are three sources per channel: the direct spectral envelope output, the flanger signal and an effect bus that is used for reverb in the middle section, and for a string resonator at the end. This approach also allows me to create other versions that can use different routings for the sources, and/or create specific EQ and amplitude adjustments for a given performance situation.

22.5.1 Source Code Packaging

One final detail is how to make the piece available. As a fixed-medium work, a soundfile or a CD recording would appear to be sufficient. However, it is possible to take advantage of the fact that the whole work is defined by its source code and distribute it as a Csound file (CSD). The advantage is that it allows the piece to be studied and even reused in other settings, as well as adjusted for specific performance venues (e.g. equalisation, amplitude control etc.). Most modern computers would have no problem in running the work in real time, so even offline rendering is not at all necessary. All we need is to have Csound installed.

The only issue to be resolved is how to make the two audio source files available. We could package them as an archive, but there is a more flexible way. We can take advantage of the CSD functionality and include the source files together with the code. They are encoded and added as two extra sections with their associated tags at the end of the CSD. When Csound runs it unpacks the files, and performs the piece. Packaging is done with the makecsd utility. From the command line, we can just do

```
makecsd -o noctilucent.csd noctilucent.orc \
                      src1x.wav src2x.wav
```

and the result will be a CSD file *noctilucent.csd* containing the piece and its audio sources. This will be a large file (111 MB) as it contains all the audio data, but any high-quality rendering of the piece would exceed that, so there is no trade-off.

22.6 Conclusions

This chapter explored the process of composing a fixed-media piece using Csound as the only software resource. It employed some non-standard methods such as spectral delays, together with a classic technique of flanging to completely transform its source material. If anything is to be learned at all from this case study, and indeed from this book, it is that experimenting is good, and experimenting with knowledge and skill is even better. Also while we should try to be definite and precise about what we do, it is never a problem to let chance and lucky coincidences play a part in the process.

Another important thing to note is that complicated-looking orchestras are often arrived at in stages, with earlier versions using few components, and more functionality added as the piece gets developed. This allows the work to grow organically from a few basic principles into more complex structures. Simplicity and elegance go together well, and it is from one that we get to the other.

Noctilucent Clouds was premiered in Edinburgh, October 2012. It has also spawned a sister piece, *Timelines + Clouds*, for piano and live electronics, which turns the process inside out, exposing the piano, and creating the 'clouds' around it. It shares some instruments and processing ideas with this piece, and it is designed to be performed with Csound, of course.

References

1. Abe, T., Kobayashi, T., Imai, S.: The IF spectrogram: a new spectral representation. In: Proceedings of ASVA 97, pp. 423–430 (1997)
2. Abelson, H., Sussman, G.J.: Structure and Interpretation of Computer Programs, 2nd edn. MIT Press, Cambridge (1996)
3. Aird, M., Laird, J.: Extending digital waveguides to include material modelling. In: Proceeding of DAFx-01, pp. 138–142. University of Limerick (2001)
4. Aird, M., Laird, J., ffitch, J.: Modelling a drum by interfacing 2-D and 3-D waveguide meshes. In: I. Zannos (ed.) ICMC2000, pp. 82–85. ICMA (2000)
5. Arfib, D.: Digital synthesis of complex spectra by means of multiplication of non-linear distorted sine waves. In: Audio Engineering Society Convention 59 (1978). URL http://www.aes.org/e-lib/browse.cfm?elib=3035
6. Bartlett, B.: A scientific explanation of phasing (flanging). J. Audio Eng. Soc **18**(6), 674–675 (1970). URL http://www.aes.org/e-lib/browse.cfm?elib=1454
7. Beauchamp, J.: Introduction to MUSIC 4C. School of Music, University of Illinois at Urbana-Champaign (1996)
8. Beauchamp, J.: Analysis and synthesis of musical instrument sounds. In: J. Beauchamp (ed.) Analysis, Synthesis, and Perception of Musical Sounds: The Sound of Music, Modern Acoustics and Signal Processing, pp. 1–89. Springer, New York (2007)
9. Beauchamp, J. (ed.): Analysis, Synthesis, and Perception of Musical Sounds: The Sound of Music. Modern Acoustics and Signal Processing. Springer (2007)
10. Begault, D.R.: 3-D Sound for Virtual Reality and Multimedia. Academic Press Professional, Inc., San Diego, CA, USA (1994)
11. Bilbao, S.: Numerical Sound Synthesis: Finite Difference Schemes and Simulation in Musical Acoustics. John Wiley and Sons, Chichester (2009)
12. Bilbao, S., ffitch, J.: Prepared piano sound synthesis. In: Proc. of the Int. Conf. on Digital Audio Effects (DAFx-06), pp. 77–82. Montreal, Quebec, Canada (2006)
13. Blauert, J.: Spatial Hearing : The Psychophysics of Human Sound Localization. MIT Press, Cambridge (1997)
14. Borgonovo, A., Haus, G.: Musical sound synthesis by means of two-variable functions: Experimental criteria and results. In: Proceedings of the International Computer Music Conference, pp. 35–42. ICMA, San Francisco (1984)
15. Boulanger, R. (ed.): The Csound Book. MIT Press, Cambridge (2000)
16. Boulanger, R., Lazzarini, V. (eds.): The Audio Programming Book. MIT Press (2010)
17. Bracewell, R.: The Fourier Transform and Its Applications. Electrical Engineering Series. McGraw-Hill, New York (2000)
18. Bradford, R., Dobson, R., ffitch, J.: Sliding is smoother than jumping. In: ICMC 2005 free sound, pp. 287–290. Escola Superior de Música de Catalunya (2005). URL http://www.cs.bath.ac.uk/~jpff/PAPERS/BradfordDobsonffitch05.pdf

19. Bradford, R., Dobson, R., ffitch, J.: The sliding phase vocoder. In: Proceedings of the 2007 International Computer Music Conference, pp. 449–452 (2007). URL http://cs.bath.ac.uk/jpff/PAPERS/spv-icmc2007.pdf

20. Brandtsegg, Ø.: Feedback piece, live recording (2012). URL https://soundcloud.com/brandtsegg/feed-dokkhuset-2012-03

21. Brandtsegg, Ø.: Feedback piece. in compilation CD "Beyond Boundaries: European Electroacoustic Music", Casa Musicale Eco (2014)

22. Brandtsegg, Ø., Saue, S., Johansen, T.: Particle synthesis, a unified model for granular synthesis. In: Proceedings of the Linux Audio Conference 2011 (2011). URL http://lac.linuxaudio.org/2011/papers/39.pdf

23. Brecht, B.: Leben des Galilei. Suhrkamp Verlag, Berlin (1967)

24. Brun, M.L.: A derivation of the spectrum of FM with a complex modulating wave. Computer Music Journal 1(4), 51–52 (1977). URL http://www.jstor.org/stable/40731301

25. Carty, B.: Movements in Binaural Space. Lambert Academic, Berlin (2012)

26. Chowning, J.: The synthesis of complex spectra by means of frequency modulation. Journal of the AES 21(7), 526–534 (1973)

27. Cook, P.: A toolkit of audio synthesis classes and instruments in C++ (1995). URL https://ccrma.stanford.edu/software/stk/

28. Cook, P.: Physically inspired sonic modelling (PhIsm): Synthesis of percussive sounds. Computer Music Journal 21(3), 38–49 (1997)

29. Cutland, N.: Computability. Cambridge University Press, Cambridge (1980)

30. Dattorro, J.: Effect design, part 1: Reverberator and other filters. J. Audio Eng. Soc 45(9), 660–684 (1997). URL http://www.aes.org/e-lib/browse.cfm?elib=10160

31. Dattorro, J.: Effect design, part 2: Delay line modulation and chorus. J. Audio Eng. Soc 45(10), 764–788 (1997). URL http://www.aes.org/e-lib/browse.cfm?elib=10159

32. Dirichlet, P.G.L.: Sur la convergence des séries trigonometriques qui servent à représenter une fonction arbitraire entre des limites donneés. J. for Math. 4, 157–169 (1829)

33. Djoharian, P.: Shape and material design in physical modeling sound synthesis. In: I. Zannos (ed.) ICMC 2000, pp. 38–45. ICMA (2000)

34. Dodge, C., Jerse, T.A.: Computer Music: Synthesis, Composition and Performance, 2nd edn. Schirmer, New York (1997)

35. Dolson, M.: The phase vocoder: A tutorial. Computer Music Journal 10(4), 14–27 (1986). URL http://www.jstor.org/stable/3680093

36. Dudley, H.: The vocoder. J. Acoust. Soc. Am. 11(2), 169 (1939)

37. Ervik, K., Brandtsegg, Ø.: Creating reverb effects using granular synthesis. In: J. Heintz, A. Hofmann, I. McCurdy (eds.) Ways Ahead: Proceedings of the First International Csound Conference, pp. 181–187. Cambridge Scholars Publishing (2013)

38. ffitch, J.: Composing with chaos. In: R. Boulanger (ed.) The Csound Book: Tutorials in Software Synthesis and Sound Design. MIT Press (2000). On CD-ROM with book

39. ffitch, J.: On the design of Csound 5. In: Proceedings of 4th Linux Audio Developers Conference, pp. 79–85. Karlsruhe, Germany (2006)

40. ffitch, J.: Introduction to program design. In: R. Boulanger, V. Lazzarini (eds.) The Audio Programming Book, pp. 383–430. MIT Press, Cambridge (2010)

41. ffitch, J., Dobson, R., Bradford, R.: Sliding DFT for fun and musical profit. In: F. Barknecht, M. Rumori (eds.) 6th International Linux Audio Conference, pp. 118–124. LAC2008, Kunsthochschule für Medien Köln (2008). URL http://lac.linuxaudio.org/2008/download/papers/10.pdf

42. Flanagan, F., Golden, R.: Phase vocoder. Bell System Technical Journal 45, 1493–1509 (1966)

43. Fourier, J.: Théorie analytique de la chaleur. Chez Firmin Didot, Père et fils, Paris (1822)

44. Friedman, D.H.: Instantaneous-frequency distribution vs. time: an interpretation of the phase structure of speech. In: Proceedings of the ICASSP, pp. 1121–4. IEEE, Los Alamitos (1985)

45. Gardner, W.G.: Efficient convolution without input-output delay. Journal of the Audio Engineering Society 43(3), 127–136 (1995)

46. Gerzon, M.A.: Periphony: with-height sound reproduction. J. Audio Eng. Soc **21**(1), 2–10 (1973). URL http://www.aes.org/e-lib/browse.cfm?elib=2012
47. Globokar, V.: Toucher. Edition Peters, London (1978)
48. Gold, B., Blankenship, P., McAuley, R.: New applications of channel vocoders. IEEE Trans. on ASSP **29**(1), 13–23 (1981)
49. Harris, F.: On the use of windows for harmonic analysis with the discrete Fourier transform. Proceedings of the IEEE **66**(1), 51–83 (1978)
50. Heintz, J.: Läuft es sich in kinderschuhen besser? : Über kompositorisches denken mit und vor dem computer. MusikTexte (141), 43–50 (2014). URL http://joachimheintz.de/lauft-es-sich-in-kinderschuhen-besser.html
51. Heller, A.J., Lee, R., Benjamin, E.M.: Is my decoder ambisonic? In: 125th AES Convention, San Francisco. San Francisco, CA, USA (2008)
52. Hinkle-Turner, E.: Women Composers and Music Technology in the United States: Crossing the Line. Ashgate, Aldershot (2006). URL https://books.google.ie/books?id=FBydHQwZwWkC
53. Howe, H.: A report from princeton. Perspectives of New Music **4**(2), 68–75 (1966)
54. Ingalls, M.: Improving the composer's interface: Recent developments to Csound for the Power Macintosh computer. In: R. Boulanger (ed.) The Csound Book. MIT Press, Cambridge (2000)
55. Jaffe, D.A.: Spectrum analysis tutorial, part 1: The discrete Fourier transform. Computer Music Journal **11**(2), 9–24 (1987). URL http://www.jstor.org/stable/3680316
56. Jaffe, D.A.: Spectrum analysis tutorial, part 2: Properties and applications of the discrete Fourier transform. Computer Music Journal **11**(3), 17–35 (1987). URL http://www.jstor.org/stable/3679734
57. Karplus, K., Strong, A.: Digital synthesis of plucked string and drum timbres. Computer Music Journal **7**(2), 43–55 (1983)
58. Khosravi, P.: Circumspectral sound diffusion with Csound. Csound Journal (15) (2011). URL http://www.csounds.com/journal/issue15/sound_diffusion.html
59. Kleimola, J., Lazzarini, V., Vämäki, V., Timoney, J.: Feedback amplitude modulation synthesis. EURASIP J. Adv. Sig. Proc. **2011** (2011)
60. Kontogeorgakopoulos, A., Cadoz, C.: Cordis Anima physical modeling and simulation system analysis. In: 4th Sound and Music Computing Conference, pp. 275–282. National and Kapodistrian University of Athens (2007)
61. Laird, J., Masri, P., Canagarajah, C.: Efficient and accurate synthesis of circular membranes using digital waveguides. In: IEE Colloquim: Audio and Music Technology: The Challenge of Creative DSP, pp. 12/1–12/6 (1998)
62. Laurens, H.: Electrical musical instrument (1934). URL http://www.google.com/patents/US1956350. US Patent 1,956,350
63. Lazzarini, V.: Introduction to digital audio signals. In: R. Boulanger, V. Lazzarini (eds.) The Audio Programming Book, pp. 431–462. MIT Press, Cambridge (2010)
64. Lazzarini, V.: Programming the phase vocoder. In: R. Boulanger, V. Lazzarini (eds.) The Audio Programming Book, pp. 557–580. MIT Press, Cambridge (2010)
65. Lazzarini, V.: Spectral audio programming basics: The DFT, the FFT, and convolution. In: R. Boulanger, V. Lazzarini (eds.) The Audio Programming Book, pp. 521–538. MIT Press, Cambridge (2010)
66. Lazzarini, V.: The STFT and spectral processing. In: R. Boulanger, V. Lazzarini (eds.) The Audio Programming Book, pp. 539–556. MIT Press, Cambridge (2010)
67. Lazzarini, V.: Time-domain audio programming. In: R. Boulanger, V. Lazzarini (eds.) The Audio Programming Book, pp. 463–520. MIT Press, Cambridge (2010)
68. Lazzarini, V.: The development of computer music programming systems. Journal of New Music Research **42**(1), 97–110 (2013)
69. Lazzarini, V., ffitch, J., Timoney, J., Bradford, R.: Streaming spectral processing with consumer-level graphics processing units. In: Proc. of the Int. Conf. on Digital Audio Effects (DAFx-14), pp. 1–8. Erlangen, Germany (2014)

70. Lazzarini, V., Timoney, J.: New methods of formant analysis-synthesis for musical applications. In: Proc. Intl. Computer Music Conf., pp. 239–242. Montreal, Canada (2009)

71. Lazzarini, V., Timoney, J.: New perspectives on distortion synthesis for virtual analog oscillators. Computer Music Journal **34**(1), 28–40 (2010)

72. Lazzarini, V., Timoney, J.: Theory and practice of modified frequency modulation synthesis. J. Audio Eng. Soc **58**(6), 459–471 (2010). URL http://www.aes.org/e-lib/browse.cfm?elib=15506

73. Lazzarini, V., Timoney, J.: Synthesis of resonance by nonlinear distortion methods. Computer Music Journal **37**(1), 35–43 (2013)

74. Lazzarini, V., Timoney, J., Lysaght, T.: Time-stretching using the instantaneous frequency distribution and partial tracking. In: Proc. of the International Computer Music Conference 2005. Barcelona, Spain (2005). URL http://hdl.handle.net/2027/spo.bbp2372.2005.163

75. Lazzarini, V., Timoney, J., Lysaght, T.: Spectral processing in Csound 5. In: Proceedings of International Computer Music Conference, pp. 102–105. New Orleans, USA (2006)

76. Lazzarini, V., Timoney, J., Lysaght, T.: The generation of natural-synthetic spectra by means of adaptive frequency modulation. Computer Music Journal **32**(2), 9–22 (2008). URL http://www.jstor.org/stable/40072628

77. Le Brun, M.: Digital waveshaping synthesis. J. Audio Eng. Soc **27**(4), 250–266 (1979). URL http://www.aes.org/e-lib/browse.cfm?elib=3212

78. Lefebvre, A., Scavone, G.: Refinements to the model of a single woodwind instrument tonehole. In: Proceedings of the 2010 International Symposium on Musical Acoustics (2010)

79. Lefford, N.: An interview with Barry Vercoe. Computer Music Journal **23**(4), 9–17 (1999)

80. Lorrain, D.: A panoply of stochastic 'cannons'. Computer Music Journal **4**(1), 53–81 (1980). URL http://www.jstor.org/stable/3679442

81. Mathews, M.: An acoustical compiler for music and psychological stimuli. Bell System Technical Journal **40**(3), 553–557 (1961)

82. Mathews, M.: The digital computer as a musical instrument. Science **183**(3592), 553–557 (1963)

83. Mathews, M., Miller, J.E.: MUSIC IV Programmer's Manual. Bell Telephone Labs (1964)

84. Mathews, M., Miller, J.E., Moore, F.R., Pierce, J.R.: The Technology of Computer Music. MIT Press, Cambridge (1969)

85. McAulay, R., Quatieri, T.: Speech analysis/synthesis based on a sinusoidal representation. Acoustics, Speech and Signal Processing, IEEE Transactions on **34**(4), 744–754 (1986)

86. McCartney, J.: Rethinking the computer music language: Supercollider. Computer Music Journal **26**(4), 61–68 (2002)

87. MIDI Manufacturers Association: Midi 1.0 specification (1983). URL http://www.midi.org/techspecs/

88. Mitra, S.K.K.: Digital Signal Processing: A Computer-Based Approach, 2nd edn. McGraw-Hill Higher Education, New York (2000)

89. Moore, F.R.: Table lookup noise for sinusoidal digital oscillators. Computer Music Journal **1**(2), 26–29 (1977). URL http://www.jstor.org/stable/23320138

90. Moore, F.R.: Elements of Computer Music. Prentice-Hall, Inc., Upper Saddle River, NJ, USA (1990)

91. Moorer, J.A.: The synthesis of complex audio spectra by means of discrete summation formulas. J. Audio Eng. Soc **24**(9), 717–727 (1976). URL http://www.aes.org/e-lib/browse.cfm?elib=2590

92. Moorer, J.A.: The use of the phase vocoder in computer music applications. J. Audio Eng. Soc **26**(1/2), 42–45 (1978). URL http://www.aes.org/e-lib/browse.cfm?elib=3293

93. Moorer, J.A.: About this reverberation business. Computer Music Journal **3**, 13–28 (1979)

94. Moorer, J.A.: Audio in the new millennium. J. Audio Eng. Soc. **48**(5), 490–498 (2000)

95. Nyquist, H.: Certain topics in telegraph transmission theory. Transactions of the AIEE **47**, 617–644 (1928)

96. O Cinneide, A.: Introducing pvspitch: A pitch tracking opcode for Csound. Csound Journal **2006**(2). URL http://csoundjournal.com/2006winter/pvspitch.html

97. Oppenheim, A.V., Schafer, R.W., Buck, J.R.: Discrete-time Signal Processing (2nd Ed.). Prentice-Hall, Inc., Upper Saddle River, NJ, USA (1999)
98. Palamin, J.P., Palamin, P., Ronveaux, A.: A method of generating and controlling musical asymmetrical spectra. J. Audio Eng. Soc 36(9), 671–685 (1988). URL http://www.aes.org/e-lib/browse.cfm?elib=5132
99. Pampin, J.: ATS: A system for sound analysis transformation and synthesis based on a sinusoidal plus crtitical-band noise model and psychoacoustics. In: Proceedings of the International Computer Music Conference, p. 402?405. Miami, FL (2004)
100. Park, T.: An interview with Max Mathews. Computer Music Journal 33(3), 9–22 (2009)
101. Pope, S.: Machine Tongues XV: Three packages for software sound synthesis. Computer Music Journal 17(2), 23–54 (1993)
102. Puckette, M.: Formant-based audio synthesis using nonlinear distortion. J. Audio Eng. Soc 43(1/2), 40–47 (1995). URL http://www.aes.org/e-lib/browse.cfm?elib=7961
103. Puckette, M.: The theory and technique of computer music. World Scientific Publ., New York (2007)
104. Pulkki, V.: Virtual sound source positioning using vector base amplitude panning. J. Audio Eng. Soc 45(6), 456–466 (1997). URL http://www.aes.org/e-lib/browse.cfm?elib=7853
105. Randall, J.K.: A report from Princeton. Perspectives of New Music 3(2), 84–92 (1965)
106. Regalia, P., Mitra, S.: Tunable digital frequency response equalization filters. Acoustics, Speech and Signal Processing, IEEE Transactions on 35(1), 118–120 (1987)
107. Ren, Z., Yeh, H., Lin, M.C.: Example-guided physically based modal sound synthesis. ACM Transactions on Graphics 32(1), 1 (2013)
108. Risset, J.C.: An Introductory Catalogue of Computer Synthesized Sounds. Bell Telephone Labs (1969)
109. Roads, C.: Microsound. MIT Press, Cambridge (2001). URL http://books.google.no/books?id=IFW0QgAACAAJ
110. Roads, C., Mathews, M.: Interview with Tongues. Computer Music Journal 4(4), pp. 15–22 (1980)
111. Roberts, A.: MUSIC 4BF, an all-FORTRAN music-generating computer program. In: Proceedings of the 17th Annual Meeting of the AES. Audio Engineering Society, Preprint 397 (1965)
112. Rodet, X.: Time domain formant- wave-function synthesis. Computer Music Journal 8(3), 9–14 (1984)
113. Scavone, G., P.R, C.: Real-time computer modeling of woodwind instruments. In: Proceedings of the 1998 International Symposium on Musical Acoustics, pp. 197–202. Acoustical Society of America (1998)
114. Schottstaedt, B.: The simulation of natural instrument tones using frequency modulation with a complex modulating wave. Computer Music Journal 1(4), 46–50 (1977). URL http://www.jstor.org/stable/40731300
115. Schroeder, M.R., Logan, B.F.: -colorless- artificial reverberation. J. Audio Eng. Soc 9(3), 192–197 (1961). URL http://www.aes.org/e-lib/browse.cfm?elib=465
116. Shannon, C.E.: Communication in the presence of noise. Proc. Institute of Radio Engineers 37(1), 10–21 (1949)
117. Smalley, D.: Space-form and the acousmatic image. Organized Sound 12(1) (2007)
118. Smith, J.O.: A new approach to digital reverberation using closed waveguide networks. No. 31 in Report. CCRMA, Dept. of Music, Stanford University (1985). URL https://books.google.ie/books?id=EpIXAQAAIAAJ
119. Smith, J.O.: Viewpoints on the history of digital synthesis. In: Proc. 1991 Int. Computer Music Conf., Montreal, pp. 1–10. Computer Music Association (1991)
120. Stautner, J., Puckette, M.: Designing multi-channel reverberators. Computer Music Journal 6(1), 52–65 (1982)
121. Steiglitz, K.: A Digital Signal Processing Primer, with Applications to Digital Audio and Computer Music. Addison-Wesley Longman, Redwood City (1996)
122. Tenney, J.: Sound generation by means of a digital computer. Journal of Music Theory 7(1), 24–70 (1963)

123. Tomisawa, N.: Tone production method for an electronic musical instrument (1981). URL http://www.google.com/patents/US4249447. US Patent 4,249,447

124. Vercoe, B.: Reference manual for the MUSIC 360 language for digital sound synthesis. Studio for Experimental Music, MIT (1973)

125. Vercoe, B.: MUSIC 11 Reference Manual. Studio for Experimental Music, MIT (1981)

126. Vercoe, B.: Computer system and languages for audio research. The New World of Digital Audio (Audio Engineering Society Special Edition) pp. 245–250 (1983)

127. Vercoe, B.: Extended Csound. In: Proc. Int. Computer Music Conf. 1996, Hong Kong, pp. 141–142. Computer Music Association (1996)

128. Vercoe, B.: Audio-pro with multiple DSPs and dynamic load distribution. British Telecom Technology Journal 22(4), 180–186 (2004)

129. Vercoe, B., Ellis, D.: Real-time Csound, software synthesis with sensing and control. In: Proc. Int. Computer Music Conf. 1990, Glasgow, pp. 209–211. Computer Music Association (1990)

130. Verplank, B., Mathews, M., Shaw, R.: Scanned synthesis. The Journal of the Acoustical Society of America 109(5), 2400 (2001)

131. Windham, G., Steiglitz, K.: Input generators for digital sound synthesis. Journal of the Acoustic Society of America 47(2), 665–6

132. Wittgenstein, L.: Philosophische Untersuchungen. Suhrkamp Verlag, Frankfurt (1953)

133. Wright, M., Freed, A.: Open sound control: A new protocol for communicating with sound synthesizers. In: Proceedings of the ICMC, pp. 101–104. Thessaloniki, Greece (1997)

134. Wright, M., Freed, A., Momeni, A.: Open sound control, state of the art 2003. In: Proceedings of the 2003 Conference on New Interfaces for Musical Expression (NIME-03), pp. 153–159. Montreal, Canada (2003). URL http://www.music.mcgill.ca/musictech/nime/onlineproceedings/Papers/NIME03_Wright.pdf

135. Yi, S., Lazzarini, V.: Csound for android. In: Proceedings of Linux Audio Developers Conference 2012. Centre for Computer Research in Music and Acoustics, Stanford Univ., USA (2012)

136. Zicarelli, D.: How I learned to love a program that does nothing. Computer Music Journal 26(4), 31–43 (2002)

137. Zoelzer, U. (ed.): DAFx: Digital Audio Effects. John Wiley & Sons, Inc., New York (2002)

Index

© Springer International Publishing Switzerland 2016
V. Lazzarini et al., *Csound*, DOI 10.1007/978-3-319-45370-5

Printed in the United States
By Bookmasters